Springer Series in Chemical Physics

Volume 105

For further volumes:
http://www.springer.com/series/676

The purpose of this series is to provide comprehensive up-to-date monographs in both well established disciplines and emerging research areas within the broad fields of chemical physics and physical chemistry. The books deal with both fundamental science and applications, and may have either a theoretical or an experimental emphasis. They are aimed primarily at researchers and graduate students in chemical physics and related fields.

Boguslaw Buszewski · Ewelina Dziubakiewicz
Michal Szumski

Editors

Electromigration Techniques

Theory and Practice

 Springer

Editors
Boguslaw Buszewski
Ewelina Dziubakiewicz
Michal Szumski
Faculty of Chemistry
Chair of Environmental Chemistry
   and Bioanalytics
Nicolaus Copernicus University
Toruń
Poland

Originally published in Polish under the title "Techniki elektromigracyjne-teoria i praktyka"

ISSN 0172-6218
ISBN 978-3-642-35042-9   ISBN 978-3-642-35043-6   (eBook)
DOI 10.1007/978-3-642-35043-6
Springer Heidelberg New York Dordrecht London

Library of Congress Control Number: 2013932782

Printed on acid-free paper

Springer is part of Springer Science+Business Media (www.springer.com)

Preface

Modern analytical methods allow for the determination of a whole range of compounds existing in various matrices in a precise and thorough way. Particularly important for the correct functioning of natural systems is the determination of pollutants which irretrievably destroy living organisms. Limiting the dissemination of many harmful substances, including xenobiotics, is absolutely essential. The general trend towards a bigger development of the industry, despite the introduction of sustainable development principles or *the green chemistry* programme, does not allow to avoid direct (water, soil, air pollution) or/and indirect hazards (food, plants, animals, a human being). It involves the introduction of a full range of various analytical techniques based on both chemical reactions and physical–chemical phenomena, performed in a preparative, analytical, micro- or/ and nanoscale (Fig. 1).

The development of modern instrumental techniques has led to the lowering of the limit of detection, as well as the increase in the precision and accuracy of measurements. Among all instrumental methods, a primary place is held by separation techniques (chromatography and electromigration techniques). Although they have been introduced into the world literature by a German, Walter Nernst* (extraction), and a Russian botanist, M. S. Tsvet* (chromatography), all this has been achieved in, or is indirectly connected with, Poland. The scope of applicability of separation techniques is connected not only with properties and kinds of determined substances, but mostly with their molecular mass (M_w) (Fig. 2).

Thus, the determination of individual components is connected with a comprehensive approach to individual operations, i.e., taking and preparing of samples, taking into account a selective isolation of individual analytes, a final determination mentioned before, as well as the assessment of analytic procedure corresponding to quality assurance requirements (Quality Control, Quality Assessment). All these actions result from the realization of assumptions of GLP and GMP concepts (*good laboratory practice and good manufacturing practice*), and they refer to the analysis of results, using statistical methods, so-called *validation*. Consequently, one obtains a high reproducibility of data and good precision of

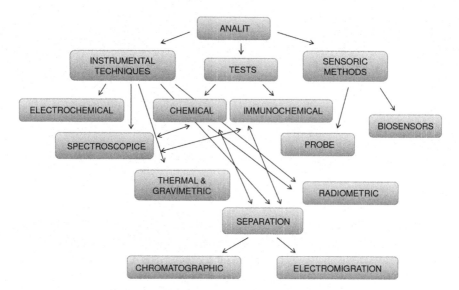

Fig. 1   Types of modern analytical techniques

measurements, with a relatively low individual cost of analysis. The quality assurance of determinations is connected with the automation of analytic process and a combination of chromatographic and electro-migration techniques in *offline* or/and *online* systems, and the construction of modular or hybrid systems, and the recently popular *lab-on-a-chip* systems.

Recently, capillary electromigration techniques have been in the range of interest of both theoreticians and practitioners. Although being developed for 25 years, despite spectacular successes, they are slow in making their way to a routine

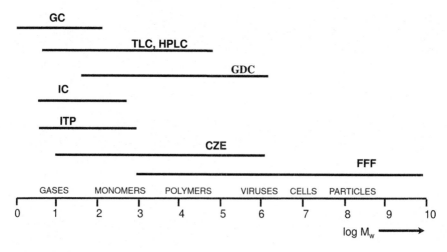

Fig. 2   The scope of applicability of separation techniques

laboratory practice. There are many reasons for this; however, the most important seems to be a strong and established position of the High Performance Liquid Chromatography (HPLC) and two-dimensional gel electrophoresis. Thus, the capillary zone electrophoresis (CZE) may be a technique by choice only in those applications where it is definitely better than HPLC, or when it allows to solve problems that cannot be solved using methods worked out earlier. The strength of capillary electromigration techniques is expressed by the following characteristics: (1) quick separation, (2) good resolution, (3) high sensitivity and selectivity, (4) little requirements as regards the size of a sample, (5) compatibility with detectors used in HPLC, including specific ones, (6) miniaturizability (detection systems and elements where separation processes take place—micro- and/or nanocolumns, chips), and (7) a big flow capacity, both in case of chips and capillaries. These characteristics decide about a privileged position of CZE in the analysis of diversified samples, particularly those containing biologically active substances, taking into account a complicated composition of a matrix. The foregoing particularly refers to the determination of individuals belonging to various types in analysts' range of interests, so-called '-omics' (metabolomics/metabonomics, genomics, transcriptomics, or/and proteomics). This makes the subject of the present study highly interdisciplinary; it is contained in the scope of modern analytics and it also concerns issues from borderlines of pharmacy, biochemistry, environmental chemistry, medical chemistry, and material chemistry. This remains in good harmony with the latest trends in modern analytical chemistry.

We, as the originators, editors, and co-authors of this book, dedicate it to students of all natural sciences, as well as to anyone who wants to broaden his or her knowledge by these fascinating and useful theoretical and practical issues. We shall appreciate any factual comments and suggestions. At the same time, we would like to thank all our colleagues who spared no effort to write the individual chapters. We are extremely thankful to them.

Toruń, Poland,

B. Buszewski
E. Dziubakiewicz
M. Szumski

Contents

1   Outline of the History   1
Bogusław Buszewski and Ewelina Dziubakiewicz
References   3

2   Principles of Electromigration Techniques   5
Ewelina Dziubakiewicz and Bogusław Buszewski
2.1   Electrophoretic Mobility   6
2.2   Electroosmotic Flow   7
2.3   Mechanism of Electrophoretic Separation   10
2.4   Parameters of Electrophoretic Separation   12
          2.4.1   Migration Time   12
          2.4.2   Efficiency   13
          2.4.3   Selectivity   14
          2.4.4   Resolution   14
2.5   Optimization of Electrophoretic Separation Conditions   15
          2.5.1   Polarization of Electrodes   15
          2.5.2   Applied Voltage   16
          2.5.3   Temperature   16
          2.5.4   Capillary   17
          2.5.5   Buffer   19
References   22

3   Equipment   27
Michał Szumski
3.1   HV Power Supply   28
3.2   Capillaries   29
3.3   Thermostating of the Capillary   32
3.4   Vials   32
3.5   Sample Injection   33
3.6   Detection in Electromigration Techniques   34

3.6.1   Spectrophotometric Detection.   36
3.6.2   Fluorescence Detection   39
3.6.3   RI Detection.   49
3.6.4   Electrochemical Detection   50
3.6.5   MS Detection.   57
3.6.6   NMR Detection   60
3.6.7   Summary.   62
3.7   Automation and Robotization.   62
References   66

**4   Qualitative and Quantitative Analysis: Interpretation
of Electropherograms**   69
Michał Szumski and Bogusław Buszewski
4.1   Qualitative Analysis   69
4.1.1   Sources of Errors   70
4.2   Quantitative Analysis   71
4.2.1   External Standard Method (Calibration Curve).   71
4.2.2   Internal Standard Method.   72
4.2.3   Remarks Regarding Peak Shape   74
References   75

5   Micellar Electrokinetic Chromatography.   77
Edward Bald and Paweł Kubalczyk
5.1   Introduction   77
5.2   Principles of Separation.   78
5.2.1   Efficiency   81
5.2.2   Selectivity   82
5.2.3   Retention Coefficient   82
5.3   Factors Influencing the Separation Process.   83
5.3.1   Surfactants.   83
5.3.2   Modifiers.   86
5.3.3   pH   87
5.3.4   Elution Order.   88
5.4   Equipment and Analytical Procedure.   88
5.4.1   Detection.   89
5.4.2   Method Development   89
5.5   Summary.   90
References   91

6   Isotachophoresis.   93
Przemysław Kosobucki and Bogusław Buszewski
6.1   Theoretical Background.   93
6.1.1   Quantitative and Qualitative Analysis   95
6.1.2   Separation Conditions   97

	6.2	Instrumentation.	100
		6.2.1 Separation Column	100
		6.2.2 Detection	102
		6.2.3 Discrete Spacers	104
	6.3	Applications	106
	6.4	Coupling of ITP with Different Analytical Techniques	107
		6.4.1 Isotachophoresis-Capillary Zone Electrophoresis (ITP-CZE)	108
		6.4.2 Isotachophoresis-High Performance Liquid Chromatography (ITP-HPLC)	109
		6.4.3 Isotachophoresis-Mass Spectrometry (ITP-MS)	109
		6.4.4 Isotachophoresis-Nuclear Magnetic Resonance (ITP-NMR)	111
	6.5	Miniaturization	113
	6.6	Summary	115
		References	115
7	**Capillary Isoelectric Focusing**		**119**
	Michał J. Markuszewski, Renata Bujak and Emilia Daghir		
	7.1	Introduction	119
	7.2	CIEF Methodology	120
		7.2.1 Sample Preparation and Injection	120
		7.2.2 Focusing	121
		7.2.3 Two-Step CIEF	122
		7.2.4 Single-Step CIEF	124
		7.2.5 Capillary Choice	124
		7.2.6 Detection Methods in CIEF	125
	7.3	Application of CIEF	126
		7.3.1 CIEF in Hemoglobin Analysis	126
		7.3.2 CIEF in Protein Analysis	127
		7.3.3 CIEF in Glycoprotein and Immunoglobin Analysis	128
		References	130
8	**Two-dimensional Gel Electrophoresis (2DE)**		**133**
	Ewa Kłodzińska and Bogusław Buszewski		
	8.1	Introduction	134
	8.2	Theoretical Basis of 2DE	135
	8.3	Stages of Analytical Procedures Before the 2D Electrophoresis Gel Analysis	137
		8.3.1 Sample Preparation	137
		8.3.2 Protein Solubilization	142
		8.3.3 Chaotropic Substances	142
		8.3.4 Detergents	142

	8.3.5	Reducing Agents	143
	8.3.6	Protein Enrichment	143
	8.3.7	Simplified Fractionation Procedure	144
8.4	First Dimension: Isoelectric Focusing		145
	8.4.1	Forming pH Gradient	146
	8.4.2	Rehydratation and Sample Loading	150
8.5	Second Dimension: Gel Electrophoresis in Denaturizing Conditions		152
	8.5.1	Methods of Protein Dyeing (Staining): Detection	153
8.6	Protein Identification by Mass Spectrometry		154
8.7	Computer in the Visualization of Analysis Results		155
	8.7.1	2D Database	156
8.8	Summary		157
References			157

9   **Electrochromatographic Methods: Capillary Electrochromatograpy** 159
    Michał Szumski

9.1	Theory	160
9.2	CEC Equipment	168
9.3	Problem of Bubble Formation in CEC	171
9.4	Performing CEC Separation	172
9.5	CEC Columns	173
	9.5.1	Packed Columns	175
	9.5.2	Fritless Columns	177
	9.5.3	Monolithic Columns	178
	9.5.4	Polymeric Monoliths	178
	9.5.5	Monolithic Silica Beds	181
	9.5.6	Organic–Inorganic Mixed Mode Monoliths (PSG)	182
	9.5.7	Open-Tubular Columns	184
9.6	CEC Among Contemporary Separation Techniques	186
References		186

10 **Electrochromatography Methods: Planar Electrochromatography** 191
    Adam Chomicki, Tadeusz H. Dzido, Paweł Płocharz
    and Beata Polak

10.1	Introduction	191
10.2	Development of the Method	192
10.3	Chambers for Pressurized Planar Electrochromatography	194
10.4	Advantages and Disadvantages of Pressurized Planar Electrochromatography	196
References		200

11 Non-Aqueous Capillary Electrophoresis............... 203
Michał Szumski and Bogusław Buszewski
11.1 Solvents............................... 204
11.2 Separations in Wide Bore Capillaries 206
11.3 Separation of Large Molecules Insoluble in Water 207
11.4 Separation of Uncharged Compounds 209
11.5 Separation of Optically Active Compounds 210
11.6 Summary.............................. 212
References 212

12 Methods of Analyte Concentration in a Capillary 215
Paweł Kubalczyk and Edward Bald
12.1 Analytes Concentration in Capillary Zone Electrophoresis... 217
12.1.1 Field Amplified Sample Stacking 217
12.1.2 Field Amplified Sample Injection 219
12.1.3 Large Volume Sample Stacking 219
12.1.4 pH-Mediated Sample Stacking 221
12.1.5 Stacking by Transient Isotachophoresis
Mechanism 221
12.1.6 Stacking by Transient Pseudo Isotachophoresis 222
12.1.7 Stacking by Dynamic pH Junction 222
12.1.8 Concentration by Counter-Flow Gradient
Electrofocusing................... 225
12.2 Sample Concentration in Micellar Electrokinetic
Chromatography.......................... 225
12.2.1 Field Amplified Sample Stacking 226
12.2.2 Sweeping....................... 226
12.2.3 Stacking with High Salt Concentration 228
12.2.4 Electrokinetic Sample Injection 228
12.2.5 Field Enhanced Sample Injection 230
12.3 Summary.............................. 230
References 230

13 Stereoisomers Separation 237
Piotr Wieczorek
13.1 Mechanism of Chiral Separation................ 239
13.2 Chiral Selectors 241
13.3 Cyclodextrins.......................... 241
13.4 Chiral Crown Ethers 242
13.5 Chiral Surfactants........................ 243
13.6 Macrocyclic Antibiotics..................... 244
13.7 Chiral Metals Complexes.................... 245
13.8 Natural and Synthetic Polymers 245

13.9   Application of Electromigration Techniques
        on Stereoisomers Analysis 249
13.10   Summary 249
References 250

14 **"Lab-on-a-Chip" Dedicated for Cell Engineering** 253
    Elżbieta Jastrzębska, Aleksandra Rakowska and Zbigniew Brzózka
    14.1   Miniaturization 253
    14.2   Analytical Techniques in Miniaturization. 254
    14.3   Cell Cultures 255
    14.4   Microsystems in Tissue Engineering 256
    14.5   Design and Development of Microsystems. 256
            14.5.1 Construction Materials. 256
            14.5.2 Fabrication of Microsystem 257
            14.5.3 Casting 258
            14.5.4 Hot Embossing. 258
            14.5.5 Injection Molding 258
            14.5.6 Laser Micromachining. 258
            14.5.7 Micromilling 259
            14.5.8 Microsystems Sealing 259
    14.6   Essential Parameters in Microsystem for Cell Culture 259
            14.6.1 Geometry of Microsystem 259
            14.6.2 Sterilization and Culture Condition 260
            14.6.3 Gradient Generation 260
            14.6.4 Other Parameters 262
    14.7   Cell Culture in Microscale 262
    14.8   Cytotoxicity Tests in Microsystems. 264
    References 267

15 **Application of Electromigration Techniques:**
    Metabolomics–Determination of Potential Biomarkers
    Using Electromigration Techniques 271
    Michał J. Markuszewski, Małgorzata Waszczuk-Jankowska,
    Wiktoria Struck and Piotr Kośliński
    15.1   Determination of Nucleosides in Biological Samples
            as Markers of Carcinogenesis. 274
    15.2   Micellar Electrokinetic Chromatography 275
            15.2.1 Selected Methods of Detection. 277
    15.3   Capillary Electrochromatography 278
    15.4   Bioinformatic Method of Analysis
            of the Electrophoretic Data 279
    15.5   Determination of Pteridines Compounds
            in Cancer Diagnosis 282
    References 283

**16 Applications of Electromigration Techniques: Electromigration
Techniques in Detection of Microorganisms**.   287
Ewelina Dziubakiewicz and Bogusław Buszewski
References   296

**17 Applications of Electromigration Techniques: Applications
of Electromigration Techniques in Food Analysis**   299
Piotr Wieczorek, Magdalena Ligor and Bogusław Buszewski
17.1   Polyphenols   300
17.2   Pigments   304
17.3   Vitamins   307
17.4   Food Additives.   307
17.5   Amino Acids, Peptides, Proteins.   308
17.6   Carbohydrates   317
17.7   Nucleic Acids.   317
17.8   Biogenic Amines, Natural Toxins
        and Other Contaminations   318
17.9   Pesticides.   319
17.10   Antibiotics   321
17.11   Summary   323
References   323

**18 Application of Electromigration Techniques
in Environmental Analysis**   335
Edward Bald, Paweł Kubalczyk, Sylwia Studzińska,
Ewelina Dziubakiewicz and Bogusław Buszewski
18.1   Introduction   335
18.2   Detectors   336
18.3   Analytes Concentration in a Sample   337
18.4   Microchips.   338
18.5   Environmental Contaminants Determination.   338
18.6   Pesticides.   339
18.7   Polycyclic Aromatic Hydrocarbons   340
18.8   Phenols   340
18.9   Amines   341
18.10   Carboxylic Acids   342
18.11   Explosives   342
18.12   Pharmaceuticals   344
18.13   Ionic Liquids   345
18.14   Summary   347
References   347

Index   355

Contributors

Edward Bald Faculty of Chemistry, University of Łódź, Lodz, Poland

Zbigniew Brzózka Faculty of Chemistry, Warsaw University of Technology, Warsaw, Poland

Renata Bujak Faculty of Pharmacy, Ludwig Rydygier Collegium Medicum in Bydgoszcz, Nicolaus Copernicus University in Toruń, Toruń, Poland

Bogusław Buszewski Faculty of Chemistry, Nicolaus Copernicus University, Toruń, Poland

Adam Chomicki Faculty of Pharmacy, Medical University of Lublin, Lublin, Poland

Emilia Daghir Faculty of Pharmacy, Ludwig Rydygier Collegium Medicum in Bydgoszcz, Nicolaus Copernicus University in Toruń, Toruń, Poland

Tadeusz H. Dzido Faculty of Pharmacy, Medical University of Lublin, Lublin, Poland

Ewelina Dziubakiewicz Faculty of Chemistry, Nicolaus Copernicus University, Toruń, Poland

Elżbieta Jastrzębska Faculty of Chemistry, Warsaw University of Technology, Warsaw, Poland

Ewa Kłodzińska Faculty of Chemistry, Nicolaus Copernicus University, Toruń, Poland

Piotr Kośliński Faculty of Pharmacy, Ludwig Rydygier Collegium Medicum in Bydgoszcz, Nicolaus Copernicus University in Toruń, Toruń, Poland

Przemysław Kosobucki Faculty of Chemistry, Nicolaus Copernicus University, Toruń, Poland

Paweł Kubalczyk Faculty of Chemistry, University of Łódź, Lodz, Poland

Magdalena Ligor Faculty of Chemistry, Nicolaus Copernicus University, Toruń, Poland

Michał J. Markuszewski Faculty of Pharmacy, Medical University of Gdańsk, Gdańsk, Poland

Paweł Płocharz Faculty of Pharmacy, Medical University of Lublin, Lublin, Poland

Beata Polak Faculty of Pharmacy, Medical University of Lublin, Lublin, Poland

Aleksandra Rakowska Faculty of Chemistry, Warsaw University of Technology, Warsaw, Poland

Wiktoria Struck Faculty of Pharmacy, Medical University of Gdańsk, Gdańsk, Poland

Sylwia Studzińska Faculty of Chemistry, Nicolaus Copernicus University, Toruń, Poland

Michał Szumski Faculty of Chemistry, Nicolaus Copernicus University, Toruń, Poland

Małgorzata Waszczuk-Jankowska Faculty of Pharmacy, Medical University of Gdańsk, Gdańsk, Poland

Piotr Wieczorek Faculty of Chemistry, University of Opole, Opole, Poland

Acronyms

2DE	Two-dimensional electrophoresis
ACN	Acetonitrile
AIDS	Acquired Immuno Deficiency Syndrome
Ala	Alanine
APPI	Atmospheric pressure photoionization
Asn	Asparagine
BGE	Background electrolyte
BMI	Body mass index
CA-15-3	Cancer antigen 15-3
CD	Cyclodextrin
CDT	Carbohydrate-deficient transferrin
CE	Capillary electrophoresis
CEC	Capillary electrochromatography
CE-MS	Capillary electrophoresis mass spectrometry
CHCA	α-cyano-4-hydroxycinnamic acid
CIEF	Capillary isoelectric focusing
COW	Correlation optimized warping
CTAB	Cetyl trimethylammonium bromide
CZE	Capillary zone electrophoresis
DAD	Diode array detector
DHB	2,5-Dihydroxybenzoic acid
DIBR	Detector-to-injection bandwidth ratio
DNA	Deoxyribonucleic acid
DOPA	3,4-Dihydroxyphenylalanine
DTV	Dynamic time warping
DVB	Divinylbenzene
ECD	Electron capture detector
ED	Electrochemical detector
EDTA	Ethylenediaminetetraacetic acid
EFGF	Electric field gradient focusing
EKC	Electrokinetic chromatography

EKS	Electrokinetic supercharging
EOF	Electroosmotic flow
EPO	Erythropoietin
ESI	Electrospray ionization
FASI	Field amplified sample injection
FAASS	Field amplified sample stacking
FEP	Fluorinated ethylene propylene
FITC	Fluorescein isothiocyanate
GABA	Gamma-aminobutyric acid
GC	Gas chromatography
GE	Gel electrophoresis
Hb	Hemoglobin
HEPES	4-(2-Hydroxyethyl) piperazine-1-ethanesulfonic acid
HPLC	High performance liquid chromatography
HPMC	Hydroksypropyl methyl cellulose
IC	Ion chromatography
IEF	Iso electro focusing
ISE	Ion selective electrode
ITP	Isotachophoresis
LE	Leading electrolyte
LED	Light emitting diode
Leu	Leucine
LIF	Laser-induced fluorescence
LLE	Liquid–Liquid extraction
LOD	Limit of detection
LOQ	Limit of quantitation
LVSS	Large volume sample stacking
Lys	Lysine
LysAla	Lysinoalanine
MEEKC	Microemulsion electrokinetic chromatography
MEKC	Micellar electrokinetic chromatography
MES	4-Morpholineethanesulfonic acid
MIP	Molecule imprinted polymers
MS	Mass spectrometry
NACE	Non-aqueous capillary electrophoresis
NMR	Nuclear magnetic resonance
NP	Neopterin
OT-CEC	Open tubular capillary electrochromatography
PAA	Polyacrylamide
PAcA	Poly(acrylamid-co-acrylic acid)
PAH	Polycyclic aromatic hydrocarbons
PC	Polycarbon
PCA	Principal component analysis
pCEC	Pressurized capillary electrochromatography
PCL	Polycaprolactone

PDA	Photodiode array detector
PDMS	Polydimethylsiloxane
PEC	Planar electro chromatography
PEG	Poly ethylene glycol
PEI	Polyethyleneimine
PEO	Poly(ethylene oxide)
PIPES	1,4-Piperazinediethanesulfonic acid
PLGA	Poly(lactic-co-glycolic acid)
PLLA	Poly(l-lactic acid)
PMMA	Poly(metyl methacrylate)
Pro	Proline
PSA	Prostatic specific antigen
PTFE	Polytetrafluoroethylene
PTW	Parametric time warping
PVA	Polyvinyl alcohol
PVP	Polyvinylpirrolidone
rIgG	Recombinant Immunoglobulin
RI	Refractive index
RNA	Ribonucleic acids
RSD	Relative standard deviation
SDS	Sodium dodecyl sulphate
Se-Cys	Selenocysteine
Se-Met	Selenomethionine
SGE	Slab gel electrophoresis
SLM	Supported liquid membrane
SPE	Solid phase extraction
SPME	Solid phase microextraction
TE	Terminating electrolyte
TGF	Temperature gradient focusing
tITP	Transient isotachophoresis
TOF	Time of flight
Tris	Tris(hydroxymethyl)aminomethane
tRNA	TransferRNA
Tyr	Tyrosine
UV–Vis	Ultraviolet visible spectroscopy
Val	Valine
VEGF165	Vascular endothelial growth factor

Chapter 1
Outline of the History

Bogusław Buszewski and Ewelina Dziubakiewicz

Abstract The term 'electrophoresis,' that is, the motion of charged particles under the influence of an applied electric field, was formulated as early as the nineteenth century. The first to use electrophoresis as a technique for mixtures separation was Tiselius, who for his accomplishments was awarded the Nobel Prize in Chemistry 11 years later. Others to contribute to the development of capillary electrophoresis were Hjertén, Jorgenson and Lukacs, as well as Terabe. This chapter presents a brief historical outline of electromigration techniques.

The term 'electrophoresis' is defined as the motion of particles having a charge under the influence of an applied electric field.

Although this term was created and introduced in 1909 by Michaelis [1], the first use of electrophoresis as a technique to separate mixtures is dated for the year 1937 [2] when Tiselius discovered that components of proteins mixture placed in an electric field migrate towards the electrode of the opposite sign, and their mobility depends on their charge. For his achievements, Tiselius was awarded the Nobel Prize in Chemistry 11 years later. Originally, experiments were carried out in quartz U-shaped channels (Fig. 1.1), which were characterized by the use of a sample of big volume, low resolution, and thermal convection caused by the generation of Joule heat. In order to prevent this, stabilizing additions began to be used, i.e., agarose, cellulose nitrate, polyacrylamide, blotting paper, which additionally increased the effect of separation by the fractionation of macromolecules [3].

B. Buszewski · E. Dziubakiewicz (✉)
Faculty of Chemistry, Nicolaus Copernicus University, Toruń, Poland
e-mail: ewelina.dziubakiewicz@gmail.com

B. Buszewski
e-mail: kojo@chem.uni.torun.pl

B. Buszewski et al. (eds.), *Electromigration Techniques*, Springer Series
in Chemical Physics 105, DOI: 10.1007/978-3-642-35043-6_1,
© Springer-Verlag Berlin Heidelberg 2013

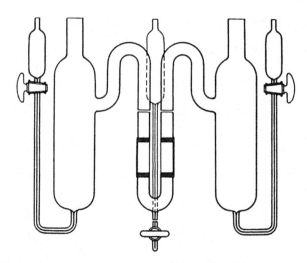

Fig. 1.1 Quartz U-shape tube designed by Tiselius. Reprinted with permission from Tiselius [2]. Copyright 1937 The Royal Society of Chemistry

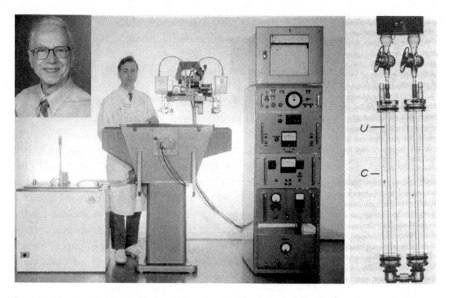

Fig. 1.2 Hjertén in front of one of the first instruments for capillary zone electrophoresis. Reprinted with permission from Hjertén [5]. Copyright 1967 Hjertén

A breakthrough in the development of capillary electrophoresis was the application of glass capillaries, 3 mm in diameter, modified by methylcellulose by Hjertén [4, 5] in 1967, which allowed to eliminate an electroosmotic flow and convection (Fig. 1.2).

Further research confirmed the merits of using capillaries with relatively small internal diameters. In 1983, Jorgenson and Lukacs [6], using silica capillaries of internal diameter below 100 μm, achieved a high efficiency and, at the same time, described the theoretical aspects of the electrophoretic analysis.

The next step in the CE development was the application by Terabe et al. [7], in 1984, of buffered solutions with the addition of ionic micelles, which contributed to the increase in resolution of neutral analytes. Since that time, various modifiers have been used to improve the selectiveness of separation.

Subsequent years brought a dynamic development of this method in the scope of theory and apparatus; owing to its merits it found wide application as a method of analysis recommended for routine determinations. A currently observed progress follows from the miniaturization of the system (capillary, chips) in which the separation takes place.

References

1. L. Michaelis, Electric transport of enzymes, malt distaste and pepsin. Biochemische Zeitschrift **17**, 231–234 (1909)
2. A. Tiselius, A new apparatus for electrophoretic analysis of colloidal mixtures. Trans. Faraday Soc. **33**, 524 (1937)
3. S. Hjertén, High performance electrophoresis. Chromatogr Rev **9**, 122 (1967)
4. S. Hjertén, The history of the development of electrophoresis in Uppsala. Electrophoresis **9**, 3–15 (1988)
5. S. Hjertén, *Free Zone Electrophoresis* (University of Uppsala, Dissertation, 1967)
6. K.D. Lukacs, J.W. Jorgenson, Capillary zone electrophoresis: effect of physical parameters on separation efficiency and quantitation. J High Res Chromatogr **8**, 407–411 (1985)
7. S. Terabe, K. Otsuka, K. Ichikawa, A. Tsuchiya, T. Ando, Electrokinetic separations with micellar solutions and open-tubular capillaries. Anal. Chem. **56**, 111–113 (1984)

Chapter 2
Principles of Electromigration Techniques

Ewelina Dziubakiewicz and Bogusław Buszewski

Abstract Electromigration techniques provide the separation of analyzed sample components owing to external voltage generating electrokinetic phenomena—electrophoresis and electroosmosis. Taking into account the relatively large number of parameters dealt with during electrophoretic analyses, it is essential to know their influence on the achieved separation of analytes. In this chapter the theoretical and practical aspects of a resolution optimization, as well as the effect of different separation parameters on the migration behavior are described. These, among others, include migration time, efficiency, selectivity, and resolution. The influence of electrods polarization, applied voltage, temperature, capillary, background electrolyte, and various additives on the separation is also discussed.

Electromigration techniques, due to their huge analytical potential, are widely applied in the determination of various substances. These techniques provide separation of components of an analyzed sample owing to external voltage generating electrokinetic phenomena—electrophoresis (motion of ions in an electrical field) and electroosmosis (volumetric liquid flow in a capillary caused by an electrical field). The separation takes place in a liquid phase in a solution called (acc. to various literature sources) a buffer, electrolyte or background electrolyte, separation buffer or separation electrolyte. It may be either an aqueous solution or based on pure organic solvents or their mixtures.

Owing to their accuracy and precision of determination, they are widely applied in, e.g., the biochemistry of proteins and nucleic acids, molecular biology, pharmacology, forensic medicine, forensic science, medical diagnostics, and analysis of food

E. Dziubakiewicz (✉) · B. Buszewski
Faculty of Chemistry, Nicolaus Copernicus University, Toruń, Poland
e-mail: ewelina.dziubakiewicz@gmail.com

B. Buszewski
e-mail: bbusz@chem.umk.pl

B. Buszewski et al. (eds.), *Electromigration Techniques*, Springer Series
in Chemical Physics 105, DOI: 10.1007/978-3-642-35043-6_2,
© Springer-Verlag Berlin Heidelberg 2013

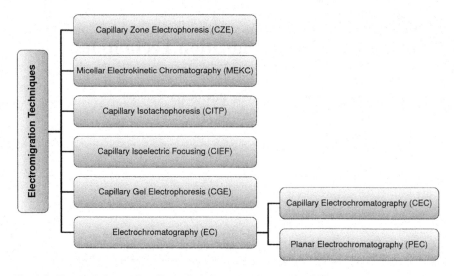

Fig. 2.1 Types of electromigration techniques, according to [11, 20]

[1–10]. All electromigration techniques applied in laboratory practice are derivatives or combinations of three basic types of electrophoreses: Zone Electrophoresis, Isotachophoresis, and Isoelectric Focusing (Fig. 2.1).

2.1 Electrophoretic Mobility

The electrophoretic separation stems from the difference in velocity of charged particles migrating under the influence of the electric field.

In a capillary filled with a separation buffer, the ion is subject to two forces. One is an electrostatic force (F): (Fig. 2.2)

$$F = qE, \qquad (2.1)$$

where q is ions charge and E is strength of electric field [V cm^{-1}], which is a function of the imposed voltage and length of the capillary. The second force is

Fig. 2.2 Forces influencing an ion in separation buffer

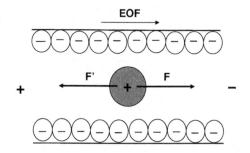

oppositely directed friction force, F', which according to Stokes law for spherical particles is expressed by the equation:

$$F' = -6\pi\eta r v_{ep}, \tag{2.2}$$

where η is solution viscosity [kg m^{-1} s^{-1}], r is ionic radius [m], v_{ep} is electrophoretic velocity [m s^{-1}].

During the electrophoresis (ionic motion), the electrostatic and friction forces are in equilibrium and then the following expression is achieved:

$$qE = -6\pi\eta r v_{ep}. \tag{2.3}$$

Keeping in mind that the electrophoretic mobility $\left(\mu_{ep}\right)$ is proportional to the charge of particle (q), and inversely proportional to its radius (r) and environment viscosity (η):

$$\mu_{ep} = \frac{q}{6\pi r\eta}, \tag{2.4}$$

obtained is the equation for electrophoretic velocity of individual ions:

$$v_{ep} = \mu_{ep}E. \tag{2.5}$$

It can be deduced from Eq. (2.4) that charged species can be resolved according to the differences in their electrophoretic mobility (charge-to-size ratio) which is the basis of the electromigration techniques. This equation proves that the electrophoretic mobility is greater for small particles with a big charge, and smaller for bigger particles with a small charge. And the electrophoretic mobility of neutral particles equal zero, because q = 0. In addition, the ionic mobility is influenced by the changes in temperature which cause changes in the buffer viscosity.

The electrophoretic mobility in a given buffer is a characteristic and constant value for a given ion. A careful selection of the buffer properties makes it possible to control the selectivity of separation by changing the mobility of the determined analytes [11, 12].

2.2 Electroosmotic Flow

An important role during the electrophoretic separation is played by electroosmotic flow (EOF), i.e., the motion of entire liquid by which the capillary is filled, caused by the difference in potentials. It is connected in the presence of electric charges on the internal surface of the capillary and it depends on:

- ionization of functional groups present on their surface and
- adsorption of ions or other buffer components on the inner capillary surface [12].

Depending on the pH of the electrolyte, silanol groups of the internal wall of the quartz capillary are subject to ionization, according to the following equation of reaction:

Fig. 2.3 Diagram of creation of double electric layer

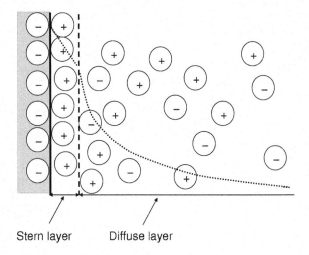

Stern layer   Diffuse layer

$$\equiv Si - OH \rightleftarrows \ \equiv Si - O^- + H^+.$$

As a result of dissociation of silanol groups, the internal wall of the capillary receives a negative charge and this leads to the creation of an electric double layer on the border phases of electrolyte/capillary wall (Fig. 2.3) [13].

The negative charge on the capillary wall (SiO^-) attracts cations from the liquid phase which creates a permanent layer, i.e., the Stern layer. However, not all charges on the surface of a capillary are compensated, which leads to the creation of a second layer, more weakly bounded to the surface, i.e., diffusion layer [14, 15]. Only ions from the diffusion layer take part in electrokinetic phenomena because they are subject to a permanent exchange with the buffer ions from the bulk solution. And the part of the double layer which is directly adjacent to the surface does not participate in the migration.

The electrokinetic potential occurring on the border of two layers (mobile and permanent), called a hydrodynamic slip plane, is called a zeta potential (ζ), which is described by the equation of Smoluchowski [16]:

$$\zeta = \frac{\eta \mu_{EOF}}{\varepsilon_0 \varepsilon_r}, \tag{2.6}$$

where η is solutions viscosity [kg m^{-1} s^{-1}], r is ionic radius [m], μ_{ep} is electrophoretic mobility [cm^2 V^{-1} s^{-1}], $\varepsilon_0, \varepsilon_r$ are dielectric constant in a solution and electric permeability in vacuum.

Zeta potential is proportional to the density of the charge on the internal wall of the capillary, which depends on the pH. Thus, the value of zeta potential differs, depending on the pH value. Above pH 9 silanol groups are completely ionized and the electroosmotic flow is the quickest. Below pH 4 the level of dissociation of silanol groups is low and then EOF is small [17]. The ionic strength of the buffer is also a significant parameter influencing the zeta potential. The increase in the ionic

Fig. 2.4 Profiles of electroosmotic and laminar flow

strength causes the decrease in the double layers thickness, which causes the decrease in zeta potential and slowing of EOF [18].

According to the following equation, the velocity of electroosmotic flow (v_{eo}) depends on the solution viscosity (η), dielectric constant (ε), strength of electric field (E), and the value of zeta potential (ζ) [12]:

$$v_{EOF} = \frac{\varepsilon \zeta E}{4\pi\eta}.$$ (2.7)

Compared to ions velocity (v), the electroosmotic flow can be so strong that solvated ions of the mobile layer move from anode to cathode, regardless of their charge. A value describing the electroosmotic flow is its electroosmotic mobility (μ_{eo}) expressed by the formula:

$$\mu_{EOF} = \frac{v_{EOF}}{E}.$$ (2.8)

After the substitution of Eq. (2.7) into (2.8) we obtain:

$$\mu_{EOF} = \frac{\varepsilon \zeta}{4\pi\eta}.$$ (2.9)

The electroosmotic flow has two important characteristics. First, compared to chromatographic techniques with laminar flow of a parabolic profile, the profile of electroosmotic flow is flat (Fig. 2.4). Second, according to Eq. (2.7), its velocity is not connected with a decrease in pressure (Δp) and it does not depend on the capillaries diameter through which the liquid flows (if double electric layers of opposite capillary walls are not placed on each other). The flat profile of EOF does not directly cause the dispersion of compounds and widening of peaks [12, 19].

The velocity of electroosmotic flow is considerably influenced by conditions of the electrophoresis. Therefore, it should be controlled by the determination of a neutral (without charge) marker velocity. Owing to the fact that this compound does not have an electric charge, its velocity equals the EOF velocity [20]. Examples of markers are: acetone, thiocarbamide, mesityl oxide, and dimethyl sulfoxide.

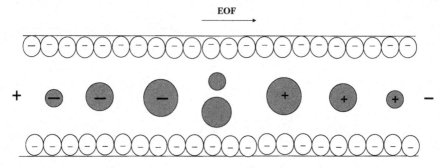

Fig. 2.5 Order of ions migration during the electrophoretic separation

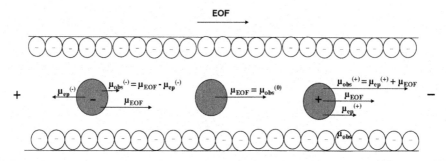

Fig. 2.6 Real versus observed electrophoretic mobility

2.3 Mechanism of Electrophoretic Separation

Under the influence of the electric field strength, particles with electrical charge contained in a sample will move with different velocity, and this is reflected in their separation (Fig. 2.5).

In case of unmodified quartz capillaries, the electroosmotic flow is peformed from anode to cathode (presence of negative charge on the internal surface of the capillary) (Fig. 2.6). Owing to the EOF existence it is possible to determine cations and anions at the same time. If the value of electroosmotic flow is higher $\mu_{obs} > \mu_{ep}$ and in the direction opposite to the electrophoretic mobility of anions, then all ions (both anions and cations) will move in the same direction but with different velocity [12, 21].

A detector placed before the cathode is first reached by cations whose mobility is observed by the sum of the electrophoretic mobility and electroosmotic flow:

$$\mu_{obs}^{(+)} = \mu_{ep}^{(+)} + \mu_{EOF}. \tag{2.10}$$

Second in order from the detector come non-separated neutral individuals, moving with velocity equal to EOF:

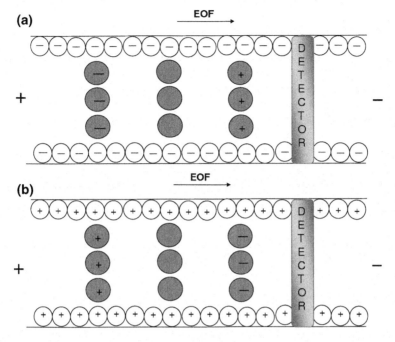

Fig. 2.7 Determination of (**a**) cations with a normal electroosmotic flow, (**b**) anions with a reversed electroosmotic flow

$$\mu_{obs}^{(0)} = \mu_{EOF}, \tag{2.11}$$

and in the end anions migrate, and their mobility is observed in the difference between the electrophoretic mobility and electroosmotic flow

$$\mu_{obs}^{(-)} = \mu_{EOF} - \mu_{ep}^{(-)}. \tag{2.12}$$

The above deliberations are simplified. In reality the following kinds of ionic mobility can be distinguished:

- absolute mobility—being a theoretical constant relating to indefinitely diluted solution in a given temperature,
- effective mobility—related to specific conditions of experiments (e.g., a specific composition and pH of a buffer),
- observed (apparent) mobility—the one described above, constituting the sum of the effective and electroosmotic mobility in specific experiment conditions.

Taking the above deliberations into account, in unmodified quartz capillary the order of components migration is as follows:

- small cations with a big charge,
- bigger cations with a smaller charge,
- non-separated neutral individuals,

- big anions with a small charge,
- small anions with a big charge.

If there is a need to determine anions (whose observed migration times can be very long), a reversed electroosmotic flow is applied to shorten the time of analysis. Owing to this a reversed order of ionic migration is obtained (Fig. 2.7b). For analytes to move from an inlet to an outlet of the capillary, the anode should be located behind the detector which is first reached by anions, then neutral particles, and in the end by cations.

The reversal of EOF is achieved by a specific chemical or dynamic modification of the effective charge on walls of the capillary. The change in direction and velocity of the electroosmotic flow depend on the kind and concentration of a modifier [20, 21].

2.4 Parameters of Electrophoretic Separation

2.4.1 Migration Time

The migration time of analytes (t_m) is a time needed for the transport of the determined components of a sample along the capillary to the detector:

$$t_m = \frac{L_{eff} \cdot L_{tot}}{\left(\mu_{ep} + \mu_{EOF}\right) \cdot U},$$

(2.13)

with the assumption $L_{eff} \sim L_{tot}$ we obtain:

$$t_m = \frac{L^2}{\left(\mu_{ep} + \mu_{EOF}\right) \cdot U},$$

(2.14)

where L_{tot} is the total length of the capillary [cm], L_{eff} is the effective length of the capillary, i.e., a distance from the capillary inlet to a place of detection [cm], μ_{ep}, μ_{EOF} are adequate electrophoretic mobility of the electroosmotic flow [cm^2 V^{-1} s^{-1}], *and U* is voltage [V].

Thus, a shorter capillary, higher voltage, and higher mobility make the migration time of analytes shorter, at the same time shortening the time of the whole analysis [12].

2.4.2 Efficiency

The efficiency in capillary electrophoresis, as in chromatographic techniques, is expressed by the number of theoretical plates determined by the following equation [11]:

$$N = \frac{L^2}{\sigma^2},$$ (2.15)

where L is the length of the capillary [cm], and σ^2 is peak variance.

During the migration of sample components in the capillary, a molecular diffusion can occur, leading to the broadening of band on the electropherogram, defined as the peak variance (σ^2) [12]:

$$\sigma^2 = \frac{2DL^2}{\mu_{obs}U},$$ (2.16)

where D is analyte diffusion coefficient, L is capillary length [cm], and μ_{obs} is observed mobility [cm^2 V^{-1} s^{-1}], U is voltage [V].

The total peak variance is the sum of variances of individual factors influencing its broadening [12]:

$$\sigma^2_{tot} = \sigma^2_{inj} + \sigma^2_{det} + \sigma^2_{tdis} + \sigma^2_{dif} + \sigma^2_{edis} + \sigma^2_{EOF} + \sigma^2_{wall},$$ (2.17)

where σ^2_{inj} is variance resulting from dosing of a sample (e.g., volume, way of dosing, way of preparing of the capillary end) introduced to the capillary, σ^2_{det} is variance resulting from volume and shape of the measuring cell of the detector, σ^2_{tdis} is variance resulting from a thermal dispersion, σ^2_{dif} is variance resulting from a longitudinal diffusion, σ^2_{edis} is variance resulting from electromigration dispersion, σ^2_{EOF} is variance resulting from EOF, σ^2_{wall} is variance resulting from interactions of the analyte with internal surface of the capillary.

After the substitution of Eq. (2.16) into (2.15) we obtain dependence:

$$N = \frac{\mu_{obs}U}{2D},$$ (2.18)

where μ_{obs} is observed mobility [cm^2 V^{-1} s^{-1}], U is voltage [V], and D is analyte diffusion coefficient.

As the above equation proves, it is possible to improve the circuits efficiency by increasing applied voltage, with the assumption of an effective heat removal. On the other hand, a high diffusion coefficient and a small mobility of analytes decrease the circuits efficiency.

2.4.3 Selectivity

The selectivity of the system (α) in case of the capillary electrophoresis can be
expressed as follows [20]:

$$\alpha = \frac{t_{m_2} - t_{EOF}}{t_{m_1} - t_{EOF}}, \tag{2.19}$$

where t_{m1}, t_{m2} are migration times of adjacent peaks [min], t_{EOF} is migration time
of neutral marker [min].
Thus:

$$\alpha = \frac{\mu_1}{\mu_2} \cdot const \tag{2.20}$$

where μ_1, μ_2 are electrophoretic mobility of two analytes.

Thus, the selectivity depends on the ratio of electrophoretic mobilities, and very
often an effective way to change the selectivity is to change the buffer pH [21].

2.4.4 Resolution

A good resolution is achieved when there is a big difference in the electrophoretic
mobility between separated analytes. Thus, the resolution (R_s) can be defined as
follows [11, 12]:

$$R_s = \frac{1}{4} \frac{\Delta\mu_{obs}}{\mu_{s'r}} \sqrt{N}, \tag{2.21}$$

where $\Delta\mu_{obs} = \mu_2 - \mu_1$.
$\mu_{s'r} = \frac{\mu_2 + \mu_1}{2}$ (average mobility of two ions).
Applying Eq. (2.18) we obtain:

$$R_s = 0.177(\mu_2 - \mu_1)\sqrt{\frac{U}{D(\mu_{s'r} \pm \mu_{EOF})}} \tag{2.22}$$

Equation (2.21) allows to assess at the same time two factors which influence
the resolution—selectivity and efficiency. The selectivity is reflected in the electrophoretic mobility of analytes, while the efficiency of the separation process is
determined by the number of theoretical plates [11].

The resolution can be improved by lengthening of the capillary, changing
buffers parameters, or by adding an organic solvent [20].

...AFFECT → / INCREASING... ↓	pH	Temperature	Viscosity	Injection volume	EOF	Diffusion	Current	Ionic strenght	Electrophoretic mobility	Capillary surface	Analyte charge
pH					↑				↕	↕	↕
Temperature	↕		↓	↑	↑	↑	↑	↑	↑		
Viscosity				↓	↓	↓			↓		
Ionic strenght	↑			↕	↓					↕	
Current	↑	↓			↓	↑	↑				
Capillary lenght	↑			↑			↑				
Capillary diameter	↑		↓	↑			↑				
Analyte charge				↕					↕		
Analyte size/shape						↕			↕		↕
EOF						↑					
Electrophoretic mobility				↕							

Fig. 2.8 Relation between parameters in the capillary electrophoresis (*up arrow* parameter's growth, *down arrow* parameter's decrease, *up–down arrow* two-direction influence, according to [11])

2.5 Optimization of Electrophoretic Separation Conditions

Taking into account a relatively large number of parameters dealt with during electrophoretic analyses, it is essential to know their influence on the achieved separation of analytes (Fig. 2.8).

2.5.1 Polarization of Electrodes

The selection of electrodes polarization should be the first operation before the analysis. As it was mentioned before, during the normal polarization solvated ions of a moving layer move from the anode on inlet to the cathode on outlet, in accordance with the electroosmotic flow. If a reversal of the electroosmotic flow direction is required, the electrodes polarization should also be reversed, so the anode should be located behind the detector which is first reached by analytes with a negative charge [11, 21].

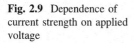

Fig. 2.9 Dependence of
current strength on applied
voltage

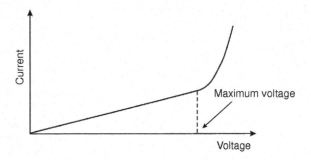

2.5.2 Applied Voltage

The applied voltage largely influences the time and quality of separation. As Eq.
(2.22) proves, the resolution is proportional to the square root of voltage, hence a
fourfold increase in voltage will lead to a twofold increase in resolution [12, 13].
However, benefits resulting from the voltage increase can be lost after exceeding a
maximum voltage which should be utilized for given conditions of the analysis.
This is connected with the influence of a generated Joule heat. The maximum
voltage can be determined from the dependence of the current strength flowing
through the capillary on the applied voltage, according to Ohm's law [13, 20].
According to this law, the strength of flowing current is proportional to the applied
voltage.

As one can see in Fig. 2.9, the applicability of Ohm's law depends on the scope
of applied voltages. In small voltages the proportionality is maintained, and in the
scope of high voltages there is a deviation from linearity which results from a
surplus of generated heat, a decrease in resistance, and increase in current strength.
In a terminal moment the buffer boils in the capillary, which breaks the electric
circuit. Therefore, the point when the proportionality collapses is a value of
maximum voltage [11].

Maximum permissible voltage depends on the composition, concentration, and
pH of the buffer, and on the length and diameter of the capillary (Fig. 2.10).

2.5.3 Temperature

The repeatability of the results of electrophoretic separations depends on maintaining a fixed temperature. The increase in buffer temperature in order to
decrease its viscosity causes acceleration of electroosmotic flow and shortening of
migration time [22, 23]. The lack of a thermostatic system of the capillary may
cause a noticeable, unfavorable influence of generated Joule heat. In a capillary
without a thermostat, depending on its internal diameter, the buffer temperature in
the central part of capillary can be higher than at the walls. This effect can cause

a thermal diffusion, as a result of which analytes in the central, warmer zone, migrate quicker than those in the area of the cooler walls of the capillary [21]. The electropherogram then shows broadened peaks. Besides, the unstable temperature of the sample influences the reproducibility of injection and may lead to the dissolution of sample [13, 17]. In case of processes controlled thermodynamically, the lowering of temperature causes the increase of selectivity. If the selectivity is sufficient, the increase in temperature may be utilized to improve the peak shape, to shorten the time of analysis, or it may cause additional, often favorable, effects, such as conformation changes (separation of chiral compounds) [12].

2.5.4 Capillary

The selection of the internal diameter of capillaries may have a huge influence on the applicability of a given method of analysis. For effective heat removal, a high ratio of surface to volume for capillaries of small diameter is responsible. This allows to apply high voltages, thereby increasing resolution and shortening the time of analysis, or high concentrations of buffers, which may also have a favorable influence on resolution [11, 22].

The increase in the capillary length (Fig. 2.10) extends the migration time and assures a better resolution. This is caused by a longer stay of analytes in the

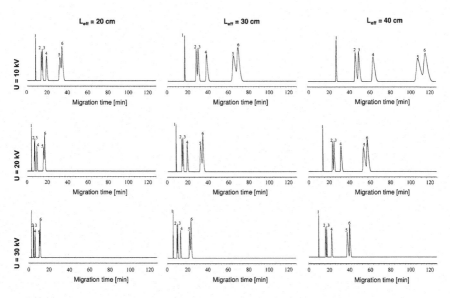

Fig. 2.10 Influence of applied voltage and length of effective capillary tube on the organic acids separation (peaks: *1*—malonic acid, *2*—tartaric acid, *3*—salicylic acid, *4*—citric acid, *5*—apple acid, *6*—formic acid). Conditions of analysis: 5 mmol/L phosphate buffer (pH 2.5)

Table 2.1 Examples of substances added to the buffer to modify EOF

EOF modifier		Ref.
Surfactants	Didodecyldimethylammonium bromide (DDAB)	[26–29]
	Tetradecyltrimethylammonium bromide (TTAB)	[28, 30]
	Cetyl trimethylammonium bromide (CTAB)	[27, 29, 31]
Metal ions	Ca^{2+}, Mg^{2+}, Ba^{2+}, Na^+, K^+	[32–38]
Organic solvents	Methanol, acetonitrile	[26, 27, 39, 40]
Polymers	Pochodne celulozy	[41, 42]
	Polyvinyl alcohol (PVA)	[30, 43–45]
	Poly(ethylene oxide) (PEO)	[46–50]
	Polyethyleneimine (PEI)	[30, 51–53]

Table 2.2 Chemical modification of the internal surface of the capillary

EOF modifier	Influence of modification on EOF	Ref.
Polyacrylamide (PAA)	Reduce/eliminate	[43, 54–60]
Polyvinyl alcohol (PVA)	Reduce/eliminate	[61]
Poly(ethylene oxide) (PEO)	Reduce	[56, 62, 63]
Polyethyleneimine (PEI)	Reverse	[64–66]
Chitozan	Reduce	[67, 68]
Dextran	Reduce	[69–71]
Methylcellulose	Reduce/eliminate	[69, 72]
Cellulose acetate	Reduce	[73]
Hydroxypropylcellulose	Reduce	[56, 74]

capillary. Additionally, the strength of electric field is inversely proportional to the capillary length, hence the increase in the capillary length allows to apply higher voltages [11, 22].

In practice, a chemical character of the internal surface of the capillary is a decisive factor influencing the efficiency and resolution of the system. It depends on a material from which the capillary is made, conditions of conditioning and parameters influencing the direction and size of EOF [12]. The adsorption of analytes on the internal wall of the capillary may cause the broadening of peaks and decrease in resolution. In order to minimize this problem there are a few well-tried ways of running a separation process:

- doing analyses in pH < 2, when silanol groups do not ionize;
- modification of effective analytes charge to prevent their adsorption on the walls;
- adding substances which dynamically modify the internal wall of the capillary (Table 2.1);
- covalent bond of different monomers to the surface of the capillary to block silanol groups (Table 2.2).

Table 2.3 Buffer employed in CE, according to [12]

CE buffers	pKa	pH range
Phosphate	2.12	1.14–3.14
	7.21	6.20–8.20
	12.32	10.80–13.00
Citrate	3.06	3.77–4.77
	4.74	5.40–7.40
	5.40	
Formate	3.75	2.75–4.75
Acetate	4.75	3.76–5.76
MES	6.15	5.15–7.15
PIPES	6.80	5.80–7.80
HEPES	7.55	6.55–8.55
Tris	8.30	7.30–9.30
Borate	9.24	8.14–10.14
CAPS	10.40	9.70–11.10

2.5.5 Buffer

Buffers utilized in the electrophoretic analysis are significant because they assure a correct electrophoretic behavior of individual analytes, a general systems stability, and a satisfactory separation. When choosing a buffer one should take into account an applied detection system, solubility, stability, and the degree of analyte ionization, pH range in which the buffer can be utilized (pKa ±1 units). Also, the influence of various buffer modifiers on the Joule heat dissipation should be considered [12, 24]. Table 2.3 shows examples of the most commonly utilized buffer solutions for CE, along with their pKa values and ranges of pH in which they are used.

2.5.5.1 pH of a Buffer

As it was mentioned before, the degree of silanol groups ionization on the capillary internal wall depends on the buffer pH. Differences in the ionization level may cause differences in electrophoretic mobility and electroosmotic flow [24]. On the other hand, the increase in pH value between 4 and 9 causes the increase of silanol groups ionization, which causes the increase of a negative charge on the capillary walls. Zeta potential is proportional to a charge density on the capillary walls. As a consequence, the increase in potential determines the increase of EOF [17, 24]. One should also remember that the buffer solution pH value may change under the influence of temperature, when buffer ions and organic additions are exhausted [11].

The buffer pH value also influences the ionization of analytes that are weak acids or bases. For weak electrolytes, e.g., acids, a dissociation constant is described by the following equation:

$$K = \frac{C_{H^+} \cdot C_{A^-}}{C_{HA}}, \tag{2.23}$$

where C_{HA} is the undissociated form of a weak acid, and C_{A^-} is the dissociated form of a weak acid.

For a given analyte (weak acid) in a solution of a different pH the dissociation level will be changing (the higher the pH, the greater the ionized form). The electrophoretic mobility for such a substance will be connected with a mobility of a dissociated form and ionization level (α). Terabe et al. [25] found that for two analytes without a big difference in pKa, the optimum pH of separation amounts to:

$$pH_{(optimum)} = pKa - log2 \tag{2.24}$$

At low pH values H^+ ionic concentration increases, and the ions neutralize a negative charge of the analyte (anion). Then a decrease of electrophoretic mobility and shortening of the analysis time is observed. In case of determination of cationic analytes, along with buffer pH increase, the level of dissociation decreases, thus, as a result, their electrophoretic mobility is decreased [24]. It is particularly important to maintain a fixed pH value of the buffer during the electrophoretic separation of proteins and peptides which have amphoteric properties. A resultant electric charge of protein is influenced by the ionization of carboxyl and amine groups. Thus, in the environment with pH > pI the protein has a negative charge which will migrate to anode (+). On the other hand, if pH < pI the protein is charged positively and it will move in the direction of the cathode (−); they will not migrate if pH = pI of a given protein/peptide.

2.5.5.2 Concentration/Ionic Strength of a Buffer

Ionic strength or concentration of buffer solutions used are limited by the length and diameter of the capillary, applied strength of electric field, and by the effectiveness of Joule heat removal. In a situation where the temperature is controlled, the increase in concentration or ionic strength of the buffer reduces EOF [26], which leads to the extension of time of analysis and improvement in resolution (Fig. 2.11).

In case the temperature is not controlled, the increase in buffer concentration or ionic strength increases the buffer conductivity, which leads to an unfavorable generation of heat. Then, the viscosity is decreased and EOF is increased [12]. In order to increase the buffer ionic strength or concentration without increase in heat, the effective length of the capillary should be increased, the capillary internal diameter should be decreased or additions of dual ions should be utilized; these actions do not cause a change in the buffer conductivity. In addition, the increase in buffer ionic strength decreased interactions with analyte/analyte or analyte/internal wall of the capillary [11].

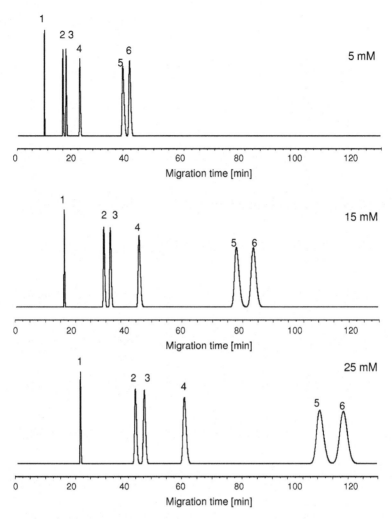

Fig. 2.11 Influence of ionic strength on separation of organic acids (peaks: *1*—malonic acid, *2*—tartaric acid, *3*—salicylic acid, *4*—citric acid, *5*—apple acid, *6*—formic acid). Conditions of analysis: phosphate buffer (pH 2.5); $L_{eff} = 40$ cm; $U = -30$ kV

2.5.5.3 Substances Modifying EOF Added to the Buffer

In order to improve the circuit resolution, substances modifying EOF are added to buffer solutions (Table 2.1). Their function is to prevent the adsorption of analytes on the capillary walls and to rebuild an electric double layer (change of electrokinetic potential and electroosmotic flow). They directly influence the change in buffer viscosity, its solvative properties, and accessibility of silanol groups on the capillary walls [12, 24]. On the other hand, organic solvents, such as methanol or acetonitrile

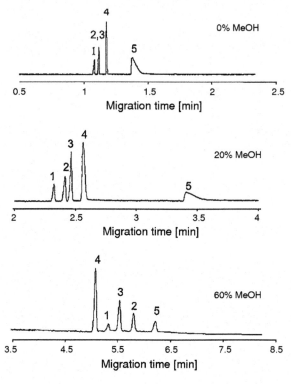

Fig. 2.12 Electropherogram received for a sample containing five non-organic ions. Conditions of analysis: [CTAC] = 0.5 mmol/L, phosphate buffer (pH 8.0), c = 15 mmol/L, L_{eff} = 28.5 cm, U = -15 kV. Peaks on the electropherogram: 1 Br$^-$, 2 NO$_3^-$, 3 NO$_2^-$, 4 I$^-$, 5 SCN$^-$. Reprinted with permission from Diress and Lucy [27] Electroosmotic flow reversal for determination of inorganic anions by capillary electrophoresis with methanol–water buffers. J Chromatogr A 1027:185–191. Copyright 2004 Elsevier

(Fig. 2.12), cause the decrease in buffer conductivity, decrease in thermal diffusion, and a better solubility of analytes. Unfortunately, additions of substances modifying EOF may lead to unwanted modifier/analyte interactions [11, 12].

References

1. M. Blanco, J. Coello, H. Iturriaga, S. Maspochi, M.A. Romero, Analytical control of a pharmaceutical formulation of sodium picosulfate by capillary zone electrophoresis. J. Chromatogr. B **751**, 29–36 (2001)
2. K.D. Altria, Determination of drug-related impurities by capillary electrophoresis. J. Chromatogr. A **735**, 43–56 (1996)
3. W.C. Sung, S.H. Chen, Recent advances in pharmacokinetic applications of capillary electrophoresis. Electrophoresis **22**, 4244–4248 (2001)

4. W. Buchberger, M. Ferdig, R. Sommer, T.D.T. Vo, Trace analysis of rapamycin in human blood by micellar electrokinetic chromatography. Anal. Bioanal. Chem. **380**, 68–71 (2004)
5. H. Nishi, Enantiomer separation of basic drugs by capillary electrophoresis using ionic and neutral polysaccharides as chiral selectors. J. Chromatogr. A **735**, 345–351 (1996)
6. P.G. Righetti, C. Gelfi, M. Conti, Current trends in capillary isoelectric focusing of proteins. J. Chromatogr. B **699**, 91–104 (1997)
7. G. Vanhoenacker, T.D. van den Bosch, G. Rozing, P. Sandra, Recent applications of capillary electrochromatography. Electrophoresis **22**, 1103–4064 (2001)
8. E. Jellum, H. Dollekamp, C. Blessum, Capillary electrophoresis for clinical problem solving: Analysis of urinary diagnostic metabolites and serum proteins. J. Chromatogr. B **683**, 55–65 (1996)
9. F. Kvasnicka, Application of isotachophoresis in food analysis. Electrophoresis **21**, 2780–2787 (2000)
10. G.B. Divall, The application of electrophoretic techniques in the field of criminology. Electrophoresis **6**, 249–258 (1985)
11. J.P. Landers, *Handbook of Capillary and Microchip Electrophoresis and Associated Microtechniques*, 3rd edn. (CRC Press, New York, 2008)
12. M.L. Marina, A. Rios, M. Varcalcel (eds.), *Analysis and Detection by Capillary Electrophoresis* (Elsevier, The Netherlands, 2005)
13. B. Chankvetadze, *Capillary Electrophoresis in Chiral* (Wiley, Chichester, 1997)
14. R.J. Hunter, *Zeta Potential in Colloid Science: Principles and Applications* (Academic, London, 1981)
15. O. Stern, The theory of the electrolytic double-layer. Z. Elektrochem. **30**, 508–516 (1924)
16. M.V. Smoluchowski, *Handbuch der Elektrizitat und des Magnetismus* (Barth, Leipzig, 1921)
17. K.D. Altria, in *Capillary Electrophoresis guidebook. Principles, Operation, and Applications*, ed. by K.D. Altria. Methods in Molecular Biology, vol 52 (Humana Press, Totowa, 1996)
18. B.M. Michov, Ionic mobility parameter. Electrophoresis **6**, 471–475 (1985)
19. R. Wallingford, A. Ewing, Capillary electrophoresis. Adv. Chromatogr. **29**, 1–67 (1989)
20. D.R. Baker, *Capillary Electrophoresis Techniques in Analytical Chemistry* (Wiley, New York, 1995)
21. C. Schwer, E. Kenndler, Capillary electrophoresis. Chromatographia **30**, 546–554 (1990)
22. X. Xuan, D. Sioton, D. Li, Thermal end effects on electroosmotic flow in a capillary. J. Heat Mass Tran. **47**, 3145–3157 (2004)
23. J.H. Knox, K.A. McCormack, Temperature effects in capillary electrophoresis 1 Internal capillary temperature and effect upon performance. Chromatographia **38**, 207–214 (1994)
24. S.F.Y. Li, Capillary electrophoresis—principles, practice and applications. J. Chromatogr. Libr. **52**, 395 (1992)
25. S. Terabe, T. Yashima, N. Tanaka, M. Araki, Separation of oxygen isotopic benzoic acids by capillary zone electrophoresis based on isotope effects on the dissociation of the carboxyl group. Anal. Chem. **60**, 1673–1677 (1988)
26. B.M. Michov, Ionic mobility parameter. Electrophoresis **6**, 471–475 (1985)
27. A. Diress, C. Lucy, Electroosmotic flow reversal for determination of inorganic anions by capillary electrophoresis with methanol-water buffers. J. Chromatogr. A **1027**, 185–191 (2004)
28. J. Melanson, N. Baryla, C. Lucy, Dynamic capillary coatings for electroosmotic flow control in capillary electrophoresis. TRAC-Trends Anal. Chem. **20**, 365–374 (2001)
29. C.A. Lucy, A.M. MacDonald, M.D. Gulcev, Non-covalent capillary coatings for protein separations in capillary electrophoresis. J. Chromatogr. A **1184**, 81–105 (2008)
30. K.K.C. Yeung, C.A. Lucy, Improved resolution of inorganic anions in capillary electrophoresis by modification of the reversed electroosmotic flow and the anion mobility using mixed surfactants. J. Chromatogr. A **804**, 319–325 (1998)

31. A. Cifuentes, M.A. Rodriguez, F.J. Garcia-Montelongo, Separation of basic proteins in free solution capillary electrophoresis: Effect of additive, temperature and voltage. J. Chromatogr. A **742**, 257–266 (1996)

32. H. Kajiwara, Application of high-performance capillary electrophoresis to the analysis of conformation and interaction of metal-binding proteins. J. Chromatogr. A **559**, 345–356 (1991)

33. R. Brechtel, W. Hohmann, H. Rudiger, H. Watzing, Control of the electroosmotic flow by metal-salt-containing buffers. J. Chromatogr. A **716**, 97–105 (1995)

34. M. Mammen, J.D. Carbech, E.E. Simanek, G.M. Whitesides, Treating electrostatic shielding at the surface of silica as discrete siloxide center dot cation interactions. J. Am. Chem. Soc. **119**, 3469–3476 (1997)

35. S. Datta, A.T. Conlisk, H.F. Li, M. Yoda, Effect of divalent ions on electroosmotic flow in microchannels. Mech. Res. Commun. **36**, 65–74 (2009)

36. J.E. Dickens, J. Gorse, J.A. Everhart, M. Ryan, Dependence of electroosmotic flow in capillary electrophoresis on group I and II metal ions. J. Chromatogr. B **657**, 401–407 (1994)

37. B.J. Kirby, E.F. Hasselbrink Jr, Zeta potential of microfluidic substrates: 1 Theory, experimental techniques, and effects on separations. Electrophoresis **25**, 187–202 (2004)

38. T.F. Tadros, J. Lyklema, The electrical double layer on silica in the presence of bivalent counter-ions. Electroanal. Chem. Interfacial Electrochem. **22**, 1–17 (1969)

39. J. Muzikar, T. van de Goor, B. Gas, E. Kenndler, Determination of electroosmotic flow mobility with a pressure-mediated dual-ion technique for capillary electrophoresis with conductivity detection using organic solvents. J. Chromatogr. A **960**, 199–208 (2002)

40. S.R. Bean, G.L. Lookhart, J.A. Bietz, Acetonitrile as a buffer additive for free zone capillary electrophoresis separation and characterization of maize (*Zea mays* L) and sorghum (*Sorghum bicolor* Moench) storage proteins. J. Agric. Food Chem. **48**, 318–327 (2000)

41. Y. Shen, R.D. Smith, High-resolution capillary isoelectric focusing of proteins using highly hydrophilic-substituted cellulose-coated capillaries. J. Microbiol. **12**, 135–141 (2000)

42. B. Verzola, C. Gelfi, P.G. Rightti, Quantitative studies on the adsorption of proteins to the bare silica wall in capillary electrophoresis II Effects of adsorbed, neutral polymers on quenching the interaction. J. Chromatogr. A **874**, 293–303 (2000)

43. H. Engelhardt, M.A. Cunat-Walter, Preparation and stability tests for polyacrylamide-coated capillaries for capillary electrophoresis. J. Chromatogr. **716**, 27–33 (1995)

44. E. Simon-Alfonso, M. Conti, C. Gelfi, P.G. Righetti, Sodium dodecyl sulfate capillary electrophoresis of proteins in entangled solutions of poly(vinyl alcohol). J. Chromatogr. A **689**, 85–96 (1995)

45. W.S. Law, J.H. Zhao, S.F. Li, On-line sample enrichment for the determination of proteins by capillary zone electrophoresis with poly(vinyl alcohol)-coated bubble cell capillaries. Electrophoresis **26**, 3486–3494 (2005)

46. N. Iki, E.S. Yeung, Non-bonded poly(ethylene oxide) polymer-coated column for protein separation by capillary electrophoresis. J. Chromatogr. A **731**, 273–282 (1996)

47. E.N. Fung, E.S. Yeung, High-speed DNA sequencing by using mixed poly(ethylene oxide) solutions in uncoated capillary columns. Anal. Chem. **67**, 1913–1959 (1995)

48. J. Preisler, E.S. Yeung, Characterization of nonbonded poly-(ethylene oxide) coatings for capillary electrophoresis via continuous monitoring of electroosmotic flow. Anal. Chem. **68**, 2885–2889 (1996)

49. M. Girog, D.W. Armstrong, Monitoring the migration behavior of living microorganisms in capillary electrophoresis using laser-induced fluorescence detection with a charge-coupled device imaging system. Electrophoresis **23**, 2048–2056 (2002)

50. E. Kłodzinska, H. Dahm, H. Rożycki, J. Szeliga, M. Jackowski, B. Buszewski, Rapid identification of *Escherichia coli* and *Helicobacter pylori* in biological samples by capillary electrophoresis. J. Sep. Sci. **29**, 1180–1187 (2006)

51. A. Cifuentes, H. Poppe, J.C. Kraak, E.B. Erim, Selectivity change in the separation of proteins and peptides by capillary electrophoresis using high-molecular-mass polyethyleneimin. J. Chromatogr. B **681**, 21–27 (1996)

52. M.S. Nutku, F.B.E. Berker, Polyethyleneimine-coated capillaries for the separation of DNA by capillary electrophoresis. Turk. J. Chem. **27**, 9–14 (2003)
53. M. Spanila, J. Pazourek, J. Havel, Electroosmotic flow changes due to interactions of background electrolyte counter-ions with polyethyleneimine coating in capillary zone electrophoresis of proteins. J. Sep. Sci. **29**, 2234–2240 (2006)
54. S. Hjerten, High-performance electrophoresis: Elimination of electroendosmosis and solute adsorption. J. Chromatogr. **347**, 191–198 (1985)
55. K. Cobb, V. Dolnik, M. Novotny, Electrophoretic separations of proteins in capillaries with hydrolytically-stable surface structures. Anal. Chem. **62**, 2478–2483 (1990)
56. Z. Zhao, A. Malik, M.L. Lee, Adsorption on polymer-coated fused-silica capillary electrophoresis columns using selected protein and peptide standard. Anal. Chem. **65**, 2747–2752 (1993)
57. A. Cifuentes, M. De Frutos, J.C. Santos, J.C. Diez-Masa, Separation of basic proteins by capillary electrophoresis using cross-linked polyacrylamide-coated capillaries and cationic buffer additives. J. Chromatogr. A **655**, 63–72 (1993)
58. R.W. Chiu, J.C. Jimenez, C.A. Monnig, High molecular weight polyarginine as a capillary coating for separation of cationic proteins by capillary electrophoresis. Anal. Chim. Acta **307**, 193–201 (1995)
59. A. Cifuentes, P. Canalejas, J.C. Diez-Masa, Preparation of linear polyacrylamide-coated capillariesStudy of the polymerization process and its effect on capillary electrophoresis performance. J. Chromatogr. A **830**, 423–438 (1999)
60. M. Szumski, E. Kłodzinska, B. Buszewski, Separation of microorganisms using electromigration techniques. J. Chromatogr. A **1084**, 186–193 (2005)
61. M. Gilges, M.H. Kleemiss, G. Schomburg, Capillary zone electrophoresis separations of basic and acidic proteins using poly(vinyl alcohol) coatings in fused silica capillaries. Anal. Chem. **66**, 2038–2046 (1994)
62. N.L. Burns, J.M. van Alstine, J.M. Harris, Poly(ethylene glycol) grafted to quartz: analysis in terms of a site-dissociation model of electroosmotic fluid flow. Langmuir **11**, 2768–2776 (1995)
63. K. Srinivasan, C. Pohl, N. Avdalovic, Cross-linked polymer coatings for capillary electrophoresis and applications to analysis of basic proteins, acidic proteins and inorganic ions. Anal. Chem. **69**, 2798–2805 (1997)
64. J.K. Towns, E.E. Regnier, Polyethyleneimine—bonded phases in the separation of proteins by capillary electrophoresis. J. Chromatogr. **516**, 69–78 (1990)
65. J.T. Smith, El Rassi, Z Capillary zone electrophoresis of biological substances with fused silica capillaries having zero or constant electroosmotic flow. Electrophoresis **14**, 396–406 (1993)
66. M.A. Rodriguez-Delgado, F.J. Garcia-Montelongo, A. Cifuentes, Ultrafast sodium dodecyl sulfate micellar electrokinetic chromatography with very acidic running buffers. Anal. Chem. **74**, 257–260 (2002)
67. Y.J. Yao, S.F.Y. Li, Capillary zone electrophoresis of basic proteins with chitosan as a capillary modifier. J. Chromatogr. A **663**, 97–104 (1994)
68. X. Huang, Q. Wanga, B. Huang, Preparation and evaluation of stable coating for capillary electrophoresis using coupled chitosan as coated modifier. Talanta **69**, 463–468 (2006)
69. S. Hjerten, K. Kubo, A new type of pH- and detergent-stable coating for elimination of electroendosmosis and adsorption in (capillary) electrophoresis. Electrophoresis **14**, 390–395 (1993)
70. Y. Mechref, Z.E. Rassi, Fused-silica capillaries with surface-bound dextran layer crosslinked with diepoxypolyethylene glycol for capillary electrophoresis of biological substances at reduced electroosmotic flow. Electrophoresis **16**, 617–624 (1995)
71. W.C. Yang, M. Macka, P.R. Haddad, Biopolymer-coated fused silica capillaries for high magnitude cathodic or anodic electro-osmotic flows in capillary electrophoresis. Chromatograp **57**, 187–193 (2003)

72. S. Hjertén, K. Elenbring, F. Kilar, J.L. Liao, A.J.C. Chen, C.J. Siebert, M.D. Zhu, Carrier-free zone electrophoresis, displacement electrophoresis and isoelectric focusing in a high-performance electrophoresis apparatus. J. Chromatogr. A **403**, 47–61 (1987)
73. M.H.A. Busch, J.C. Kraak, H. Poppe, Cellulose acetate-coated fused-silica capillaries for the separation of proteins by capillary zone electrophoresis. J. Chromatogr. A **695**, 287–296 (1995)
74. C. Giovannoli, L. Anfossi, C. Tozzi, G. Giraudi, A. Vanni, DNA separation by capillary electrophoresis with hydrophilic substituted celluloses as coating and sieving polymers Application to the analysis of genetically modified meals. J. Sep. Sci. **27**, 1551–1556 (2004)

Chapter 3
Equipment

Michał Szumski

Abstract This chapter describes the most important features of capillary electrophoretic equipment. A presentation of the important developments in high voltage power supplies for chip CE is followed by preparation of fused silica capillaries for use in CE. Detection systems that are used in capillary electrophoresis are widely described. Here, UV-Vis absorbance measurements are discussed including different types of detection cells—also those less popular (u-shaped, Z-shaped, mirror-coated). Fluorescence detection and laser-induced fluorescence detection are the most sensitive detection systems. Several LIF set-ups, such as collinear, orthogonal, confocal, and sheath-flow cuvette, are presented from the point of view of the sensitivity they can provide. Several electrochemical detectors for CE, such as conductivity, amperometric, and potentiometric, are also shown and their constructions discussed. CE-MS and much less known CE (CEC)-NMR systems are also described. The examples of automation and robotized CE systems together with their potential fields of application are also presented.

The capillary electrophoresis apparatus (Fig. 3.1) comprises the following parts:

- high voltage power supply;
- electrodes;
- inlet and outlet vials;
- separation capillary;
- capillary thermostat;
- detector;
- data collection device (e.g., integrator) or a PC with the software for instrument control and data collection.

M. Szumski (✉)
Faculty of Chemistry, Nicolaus Copernicus University, Toruń, Poland
e-mail: michu@chem.umk.pl

B. Buszewski et al. (eds.), *Electromigration Techniques*, Springer Series
in Chemical Physics 105, DOI: 10.1007/978-3-642-35043-6_3,
© Springer-Verlag Berlin Heidelberg 2013

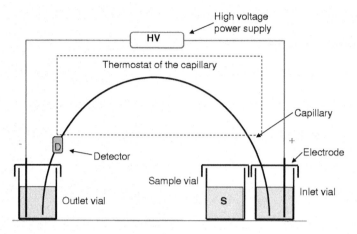

Fig. 3.1 Scheme of capillary electrophoresis system

3.1 HV Power Supply

High voltage power supply (HVPS) provides either electrical voltage, current, or power necessary to perform separation. Most electrophoretic or electrochromato-graphic separations are performed under constant voltage conditions, while isotachophoresis requires constant current.

During the process of capillary electrophoresis , Joule heating is observed, which can significantly influence the conditions inside the capillary. Efficient heat dissipation allows for using voltage up to ±30 kV and currents up to ca. 100 microA (300 microA is a typical system limit in most CE instruments). The power supply units devoted to chip systems usually provide voltage up to ±3 kV and current up to 100 μA [1].

The HVPS should provide the step or ramp gradients of the electrical param-eters mentioned above. Except for rare sinusoidal or pulse changes of voltage during separations, it is believed that voltage ramp (for example from 0 to 25 kV for 300 s) at the beginning of separation may have a positive influence on the resolution. Such phenomenon can be explained by rapid increase in the temper-ature inside the capillary when high voltage is applied. As a consequence, a momentary heat expansion of the buffer can bring the part of the sample back to the inlet vial and disturb the sample plug. It is also very likely that during heat expansion of the buffer it is characterized by a laminar flow (then its profile is parabolic, not flat like in EOF).

High voltage power supplies for chip systems have progressed most signifi-cantly. For example, Li et al. [2] designed an eight-channel HVPS to control the separation processes performed on a chip as shown in Fig. 3.2. Jiang et al. [3] constructed miniaturized HVPSs (4.7 × 5.6 × 2.5 cm) which could be powered from the USB port. The device provided ±1.5 kV while the power consumption did not exceed 1.6 W.

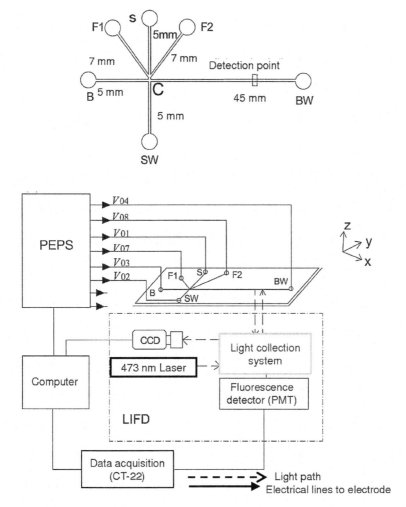

Fig. 3.2 Eight-Path electrode power supply (*PEPS*) for controlling chip separation processes. Reprinted with permission from Li et al. [2]. Copyright 2009 Elsevier

3.2 Capillaries

At the early stage of the development of the miniaturized techniques, including CE, the capillaries were manufactured from materials such as Teflon or Pyrex glass. However, it turned out that the best material, so far, was fused silica. Teflon capillaries are still in use in some ITP devices as no extensive Joule heating is observed in ITP because of lower voltages in comparison to CE as well as conductometric detection which is usually performed after the analytes leave the capillary, so no optical transparency is required. Pyrex glass is a stronger material than fused silica, however, it is no longer in use due to poor optical properties (it is not transparent for the wavelengths below 280 nm).

Because of their brittleness fused silica capillaries are coated with a polyimide protective layer. Such a layer increases the flexibility of the capillaries significantly making them handy in many applications. However, in spectrophotometric detection the polyimide should be removed over a distance of 1–10 mm, which depends on the capillary holder or detection cell used. This place in the capillary is called a detection window. The shorter the detection window the more flexible the capillary remains, and the capillary is less prone to be accidentally broken as it would be if the detection window was longer. Hence, the window should be as short as it is necessary. The detection windows can be made in one of the following manners:

- Thermal degradation of the polyimide in the electric arc using the device designed for removal of polyimide from the GC columns. The method is efficient and very precise, however, some impurities (like dust) present on the capillary or between the capillary and the electrodes can make the arc "creep" along the capillary, which can, in consequence, melt it and bend it (making the capillary useless);
- decomposition of the polyimide in the drop of hot sulfuric acid (using a simple device);
- thermal degradation of the polyimide using a loop of a resistive wire. Temperature of the wire may be adjusted by the applied voltage;
- burning of the polyimide using a butane torch and a mask. A mask is a small metal plate (steel, ca. 5 mm thick) with several holes of different diameters (for example 2, 3, 4, and 5 mm). On one side of the plate the holes are deeply beveled with a drill of large diameter. The bevels make it easier for the butane flame to reach the capillary which is placed only on one of the holes on the flat, non-beveled side of the mask plate. The chosen hole determines the length of the window being created. Because the plate is a relatively thick piece of steel it can be held with bare fingers for the time needed to burn the window (typically up to 5 s);
- the polyimide can be scraped off using a scalpel—this technique requires some manual dexterity and experience, but it allows for production of the windows of precise length without the need of heating the capillary. The method is especially useful when making windows in packed CEC capillary columns, where the sintered internal outlet frit could be damaged under high temperature treatment.

It is not recommended to remove the polyimide using a match or regular lighter flame without using the mask—the obtained windows are relatively long and it is difficult to obtain narrow and reproducible windows.

Several suppliers worldwide provide capillaries of different dimensions. The standard outer diameter which fits most of the commercially available CE devices is 365–375 μm. Internal diameters used in CE and OT-CEC are between 10 and 75 μm and 50–100 μm for packed CEC columns. Capillaries of such dimensions are sold in plastic rolls or sometimes as coils packed into plastic zip bags.

The preparation of the new capillary for CE separations may comprise the following steps:

- preparation of the detection windows and removal of ca. 2 mm of polyimide from both ends;
- secure the capillary in the cassette including proper alignment of the detection window in the detection cell/holder. Place the cassette in the CE apparatus;
- using a program timetable or step-by-step perform the series of rinsing of the capillary with NaOH, HCl solutions, water, and separation buffer;.
- do a series of separations of test solutes to check the stability and repeatability of EOF (use, e.g., acetone as the EOF marker) and voltage–current conditions.

Except for traditional capillaries of circular cross-section, some research has been done on utilization of rectangular- and square-shaped capillaries (Fig. 3.3).

It was found that using them in CE may be advantageous for the following reasons [4, 5]:

- because of high aspect ratio and high surface area to volume ratio in rectangular capillaries much better Joule heat dissipation is observed in comparison to round capillaries;
- because of geometry (flat and ideally parallel opposite walls and flat and ideally perpendicular tangent walls) and longer optical path length they work better in spectrophotometric and fluorescence detection, which can result in increase of sensitivity by ca. one order of magnitude;
- there is less light scattering and deformation of the light beam during the detection.

The literature data report application of capillaries of the following channel dimensions: 50×50, 100×100, 20×200, 30×300, 50×500 and $50 \times 1,000$ µm [4].

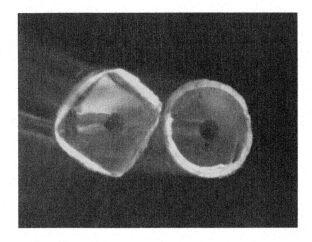

Fig. 3.3 Square and round capillaries. Polyimide coating is visible

3.3 Thermostating of the Capillary

The thermostating of the separation capillary is necessary to dissipate Joule heating. Two solutions were designed for commercially available devices. In the Agilent Technologies CE system the capillary is placed in the special cassette into which the thermostated (Peltier effect) air is blown. Approximately 120 cm long capillary can be placed in the cassette. The capillary is wound alternately onto six pegs. The notches on the pegs keep the capillary in place and prevent the coils to touch each other. The Agilents' solution has two advantages: simple and safe fixing of the detection cells on the capillary and simple exchange of the capillaries. There is no need of having many cassettes to store different capillaries in them. After a short practice the user is able to exchange the capillaries and detection cells within 1–2 min. The disadvantage of the solution is thermostating with air.

Another solution uses cooling with the liquid. Here, the capillary is reeved through a plastic pipe and a detection cell and finally is fixed in the cassette. It is said that the exchange of the capillaries is rather time consuming and the detection window can be easily broken. Some manufacturers (e.g., Beckman–Coulter) suggest to have one cassette for one capillary—once installed the capillary should not be removed. The advantage is liquid cooling which provides fast establishing of the temperature.

3.4 Vials

The inlet and outlet vials contain the separation buffer. In CE it is the same buffer, in ITP they are leading and terminating buffers, respectively. Different CE systems use vials of 0.1 mL to several milliliters of capacity. When filling the vials with the buffer it is good practice to use the pipette to maintain the same level of the liquid in the vials. The difference in the levels may cause siphoning of the buffer from the vial in which the level of the liquid is higher than another. As a consequence the irreproducible migration times may be observed [6]. To maintain the level of the buffer in both vials it is advantageous to have another vial with the buffer, which can be used for rinsing the capillary.

Care should be taken when using vials of very small capacity. Such vials are cone-shaped inside, and if made of hydrophobic plastic (like e.g. polypropylene), it may be difficult to fill them (particularly narrow, conical bottom) with the aqueous buffer due to the surface tension and poor wettability of the inner vial surface. It is not rare to leave small bubbles in the narrow bottom of such low volume plastic vials. If the capillary tip reaches the bubble the sample or buffer does not enter the capillary.

It is crucial to keep the buffer composition constant during separation. Many CE systems provide automatic buffer replenishment in the inlet and outlet vials. The used buffer is sucked out to the special waste bottle, and then the fresh buffer is replenished from another bottle. Such functions allow for full automation and performing serial analyses for long periods.

3.5 Sample Injection

In capillary electromigration techniques the sample can be introduced to the capillary by hydrodynamic or electrokinetic injection. In hydrodynamic methods the sample is sucked or forced into the capillary as a result of a pressure difference between the inlet and outlet of the capillary. The pressures are in the range of several dozens of millibars and are applied for several dozen seconds. The hydrodynamic injection is described by a number being a product of pressure and time, the unit of the number is for example mbar·s (Fig. 3.4).

The approximate volume of the hydrodynamically injected sample can be calculated using Poiseulle equation:

$$V_i = \frac{\Delta p r^4 \pi t}{8\eta L},$$
(3.1)

where Δp is applied injection pressure, r is capillary radius, t is injection time, η is viscosity, and L is total length of the capillary.

The length of the injected plug can be determined from the following equation:

$$L_p = \frac{V_i}{\pi r^2} = \frac{\Delta p r^2 t}{8\eta L}.$$
(3.2)

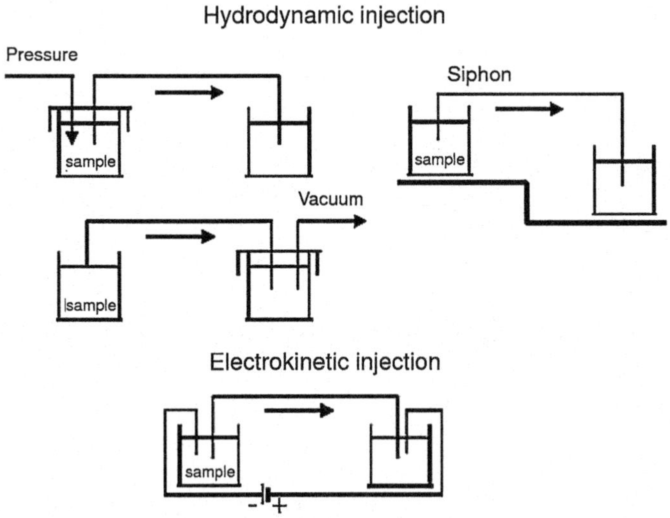

Fig. 3.4 Different sample injection methods used in CE

In electrokinetic injection the electroosmotic flow is predominantly used to introduce the sample into capillary. The method is used particularly in packed or monolithic column capillary electrochromatography, because relatively low pressure used during hydrodynamic mode (typically up to 1–2 bar) would not be enough to push the sample into a column which contains a stationary phase. Electrokinetic injection relies on application of the voltage for a period of time (typically a couple of seconds) to induce EOF which introduces sample into a capillary column. The applied voltage and time are the greater the lower the EOF column/stationary phase can generate. Electrokinetic injection is good and representative for neutral solutes. The charged analytes are usually not injected in the representative way. For example: during electrokinetic injection into a fused silica capillary with cathode at the outlet vial (EOF towards the cathode) the injected sample would contain a representative amount of the neutral analytes and preferential injection of cations. The cations will be introduced because of EOF and their own mobility towards the cathode. Simultaneously, the anions are discriminated because their own mobility will be towards the anode.

Thus, when using open tubular capillaries (CE or OT-CEC) one should rather use hydrodynamic injection and while using columns filled with the stationary phase (packed or monolithic), the injection should be electrokinetic.

3.6 Detection in Electromigration Techniques

Electromigration techniques are liquid phase separation methods, like HPLC, so that in liquid phase the analytes are transported to the detector. Hence, the detection methods are common for these techniques; the most significant differences are in miniaturized design and construction details which are a consequence of much smaller dimension of the separation column/capillary in comparison to HPLC. A wide range of detection methods that offer different sensitivity and possibilities of data interpretation can be used (Table 3.1). However, only some of them became popular and widely used, which translated into commercial availability of ready-to-use instruments devoted to electromigration techniques. The most important detection systems are: UV-Vis absorption (direct and indirect modes), fluorescence, and LIF (laser-induced fluorescence), mass spectrometer, conductivity detector, amperometric detector, radiometric, and refractive index (RI) detector.

The detector applicable to the given analytical task is characterized by the following properties [6]:

- Sensitivity—is a ratio of detector response to the amount of the analyte introduced (slope of the plot response vs. amount; minimum amount on the plot is a limit of detection—LOD); in practice a signal-to-noise ratio, S/N, can be a proper information about LOD—LOD is then defined as the amount of analyte which results in a peak which is two or three times higher than noise.

Table 3.1 Detection methods in electromigration techniques

Detection method	Limit of detection (mol/L)
Absorbance measured directly	$10^{-5} \div 10^{-6}$ (standard optical path—measurement "through the capillary")
	10^{-6} (extended optical path—"Z" and "U"—shaped cells)
Absorbance measured indirectly	$10^{-5} \div 10^{-6}$
Photothermal refraction (PTR)	$10^{-7} \div 10^{-8}$
Direct on-column LIF	10^{-16}
Post-column LIF	10^{-16}
Potentiometric detection	$10^{-7} \div 10^{-8}$
Conductometric detection	$10^{-7} \div 10^{-9}$
Amperometric detection	$10^{-7} \div 10^{-10}$
RI (refractive index) detection	$10^{-6} \div 10^{-8}$ in capillary
	10^{-5} on chip
Raman spectroscopy	$10^{-3} \div 10^{-6}$ (requires preconcentration)
NMR	10^{-3}
Radiometric methods	$10^{-10} \div 10^{-12}$
Mass spectrometry	$10^{-8} \div 10^{-10}$

- Selectivity—shows which compounds (more precisely: which functional groups, atoms or other properties) make the detector respond. The detector responding to some particular group of compounds is selective to these compounds. The detector responding to all compounds is called universal (generic) detector.
- Dynamic range—is a range of amount of the analyte (or its concentration) for which the device provides precise quantitative analysis; if the detector response is linear the dynamic range is called *linearity*. Linearity is determined from the plot of detector response versus the amount (concentration) of the injected analyte and it is confined to the range from LOD to the point where the plot becomes nonlinear. Wide dynamic range of the detector used allows for reliable determination of the analyte in very wide range of concentrations.

Which detectors are better—universal or selective ones? It depends on the analytical task. A universal detector shows the presence and quantity of all compounds present in the sample, which may be useful when unknown sample is analyzed. On the other hand, the detector selective to a particular analyte(s) may be useful in obtaining analytical signal of the analyte of interest in the presence of many other interferents. For example, fluorescence detector is selective only to fluorescing analytes, for any other non-fluorescing compound its response will be zero. As a consequence, the obtained chromatogram or electropherogram will be simpler and easier for interpretation.

3.6.1 Spectrophotometric Detection

Because of the wide use of fused silica capillaries in electromigration techniques (which are transparent to UV and visible light) the most obvious detection technique seems to be spectrophotometric detection. The prerequisite of such detection is that the analyte must possess the system of electrons (a chromophore) able to absorb the radiation in the mentioned region. Although almost all organic compounds can absorb the light of the wavelength between 160 and 180 nm, it may be problematic to carry out measurement at these wavelengths because of strong absorption by the optics and air gases. In practice UV-Vis detectors measure absorbance in the range of 190–900 nm.

In spectrophotometric detection absorbance is a measured parameter, which according to Lambert–Beer law, depends on concentration of the analyte, absorption coefficient, and optical path:

$$A = \varepsilon \cdot c \cdot l, \tag{3.3}$$

where A is absrbance, ε is molar absorption coefficient, c is concentration, and l is length of the optical path (thickness of the absorbing layer).

The measured absorbance and the peak height of the detected compound depends on its absorption coefficient. Under particular analytical conditions, two detected compounds of the same concentration show peaks of different heights only if their ε are different.

In general, two kinds of spectrophotometric detectors are currently in use:

- Variable wavelength detector (VWD)—in such a detector it is possible to collect the data only at one wavelength from the UV-Vis range. The wavelength can be chosen by setting the diffraction grating manually or mechanically (by electric motor). The diffracted light passes the slit which cuts all the unwanted wavelengths. The slit can be adjusted to set a bandwidth (BW) of the beam which enters the detection cell (for eg., if $\lambda = 254$ nm, BW $= 20$ nm are set, it means that the absorbance is measured using a light from the wavelength range of 244–264 nm).
- Diode array detector (DAD, photodiode array detector—PDA)—in such a detector most UV-Vis light provided by lamp(s) (e.g. 190—900 nm) pass the detection cell in which light absorption takes place. The beam is focused using an achromatic lens before the cell. Then the beam, having passed through the cell, goes through a slit and reaches the diffraction grating (unlike VWD, in DAD the grating is fixed); the diffracted light falls onto the diode array. The measurement at the given wavelength is performed by measuring the electrical signal of a diode or group of diodes responsible for a particular wavelengths. Hence, it is possible to collect several signals at different λ simultaneously, which might be useful for qualitative analysis taking the UV-Vis spectra of all the compounds being analyzed.

In electromigration techniques, when measuring absorbance through a detection window of the FS capillary, the optical path equals the internal diameter of the capillary. It is then obvious that using the capillaries of smaller diameters leads to decrease in sensitivity (for eg., the same detector in HPLC measures A over a distance of even $l = 10$ mm, while in CE $l = 0.05$ mm). To increase the sensitivity and keep the cell volume low, one can use several types of extended path cells, which are schematically presented in Fig. 3.5. Figure 3.5a shows typical basic system without a ball lens, which is used in focusing on the beam inside the capillary channel. Sensitivity increases ca. threefold when using a "bubble cell" (Fig. 3.5b) in which the channel is widened (e.g., from 50 to 150 µm; it can be done by controlled blowing using local heating and a micro lathe) to form a "bubble", while the outer diameter remains the same. Because of the flat profiles of EOF and ion migration no band broadening is observed. Detection cells of U and Z shape (Fig. 3.5c, d) can be a part of the separation capillary or they can be a separate part. Usually the beam is focused by a ball lens before it enters the cell. The main part (a double bent capillary to form a cell of the optical path up to even 2 cm, that is 20,000 µm) is usually fixed in a plastic or metal holder for protection. The holder can be designed as an exchangeable detection cell for different types of HPLC detectors. An interesting approach is a *high sensitivity detection cell* (Agilent Technologies) characterized by $l = 2$ mm, which contains a silica cell to which two parts of the separation capillary can be attached using small screws. Another interesting approach is a silver mirror-coated capillary (Fig. 3.5e) where the optical path is increased by multiple reflection of the beam inside the capillary. According to the authors the sensitivity can be increased ca. 40x. The serious technical problem is adjusting the proper incident angle of light beam, manufacturing of such cell, and questionable stability of silver coating. Moreover, the light source is a laser (collimated light) which makes such solution non-versatile [8].

Recently, fiber optics have been extensively used in construction of spectrophotometric detectors. The aim is to increase the versatility of the devices and make them possible to be used in HPLC, micro-HPLC, electromigration techniques in a capillary or chip format. The only exchangeable part is an external detection cell, the parameters of which (optical path, volume, optics) are designed for a given technique. The external cell is connected to the detector's main body using fiber optic. The main body can be placed in a relatively remote place, which allows for free arrangement of other parts like capillaries, valves, wires, or makes it possible to place the detection cell close to another detection system like *on-line* connected mass spectrometer.

3.6.1.1 Direct and Indirect Spectrophotometric Detection

Indirect detection has been designed for detection of those analytes that do not show absorption of UV-Vis light [6]. It was first reported by Hjerten et al. [9]

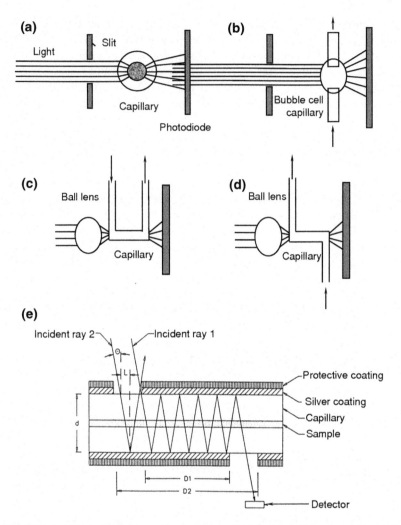

Fig. 3.5 Different types of detection cells used in spectrophotometric detection. **a** Classical on-column (through capillary) detection. **b** "Bubble cell" capillary. **c** U-shaped cell. **d** Z-shaped cell. **e** Multireflection cell. Reprinted with permission from Wang et al. [8]. Copyright 1991 American Chemical Society

and Foret et al. [10]. In indirect detection highly absorbing ions are added to the separation buffer to create high absorbance background signal (Fig. 3.6). Non-absorbing ions being separated displace chromophore ions during the migration. When the zone of non-absorbing analyte reaches the detector the decrease of absorbance is observed (negative peak). To obtain "normal" electropherogram one can change the polarity of the electrical output of the detector, or, which is more common presently— use special functions of the data collection software.

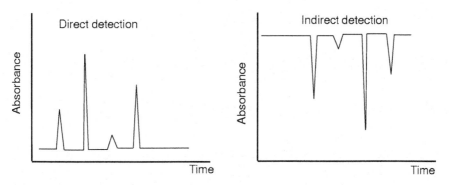

Fig. 3.6 Schematic representation of direct and indirect method of spectrophotometric detection

For detection of non-absorbing anions the following background chromophore ions can be used: chromate, pyromellitic acid, phthalates, and naphthalenesulfonic acid. For the cations imidazole, 1-naphthylamine, quinine, or malachite green are used.

The background ion should be characterized by:

- mobility similar to the analyte's mobility;
- high molar absorption coefficient at wavelength of choice which provides large absorbance drop after the background ions are displaced by the sample ions;
- high absorbance at the wavelengths the analytes do not absorb.

To provide symmetry of the peaks of the analyzed compounds the concentration of the chromophore should be ca 100x higher than analytes. Also, its concentration should provide a compromise between high linearity range (high chromophore concentration is better) and low noise level (low chromophore concentration is better).

3.6.2 Fluorescence Detection

Fluorescence is a phenomenon of emission of light by molecules that have absorbed electromagnetic radiation. The compounds able to fluoresce are called fluorophores. There are two other luminescence phenomena: phosphorescence and chemiluminescence. The major difference between fluorescence and phosphorescence is decay time which is of the order of 10^{-8} s for fluorescence, and is much longer (even several seconds) for phosphorescence.

The energy of the emitted light is lower than the energy of the light absorbed, so the emitted radiation is shifted to longer wavelengths in comparison to absorbed ones (Fig. 3.7).

The analytical signal in fluorimetric detection techniques is fluorescence intensity, I_F, which is usually measured at a particular wavelength (or sometimes a

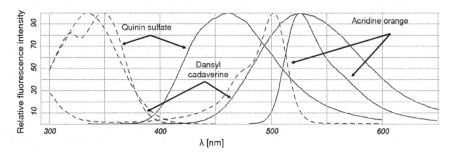

Fig. 3.7 Exemplary absorption (*dashed lines*) and fluorescence (*solid lines*) in the wavelength range from 300 to 650 nm for: quinine sulfate in 0.5 mol/L H_2SO_4, acridine orange bound to DNA and dansylcadaverin in methanol. From: spectra viewer tool from invitrogen: http://www.invitrogen.com/site/us/en/home/support/Research-Tools/Fluorescence-SpectraViewer.html

range of wavelengths). The fluorescence intensity is proportional to the concentration of the analyte in the sample:

$$I_F = KI_0 \varepsilon l c, \tag{3.4}$$

where I_F is fluorescence intensity, K is equipment-dependent constant, I_0 excitation radiation intensity, ε is absorption coefficient, c is concentration of the absorbing/emitting substance, and l is thickness of the absorbing layer.

Quantum yield describes the efficiency of the fluorescence process expressed as a ratio of the photons emitted to the number of photons absorbed:

$$Q = \frac{number\ of\ photons\ emitted}{number\ of\ photons\ absorbend}. \tag{3.5}$$

Quantum yield can be in the range of 0 (no emission) to 1 (100 %—when one photon is emitted after one photon has been absorbed). In practice, for most substances $Q < 1$. Q is not a constant value, and it characterizes a given fluorophore under particular conditions, for eg., in a given solvent or pH.

In fluorescence detector excitation (λ_{ex}) and emission (λ_{em}) wavelengths are set. A fluorescence detector which can work with fused silica capillary is shown in Fig. 3.8. It consists of a light source (deuterium, tungsten, xenon lamp) (1) diffraction grating and slit for excitation light, (2) detection cell (or capillary with detection window), (3) and the second diffraction grating and slit system for emitted light, (4) which is collected at a right angle to the excitation beam. Emitted fluorescence light is directed to a photomultiplier (PMT) in order to amplify the signal. The great advantage of PMTs is the possibility of amplification adjustment in the wide range as well as practically noiseless work.

Recently, much attention has been paid to utilization of light emitting diodes (LEDs) as alternative excitation light sources. LEDs are characterized by several advantages that made them popular light sources in chip systems:

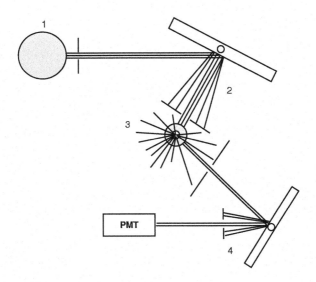

Fig. 3.8 Scheme of fluorescence detector

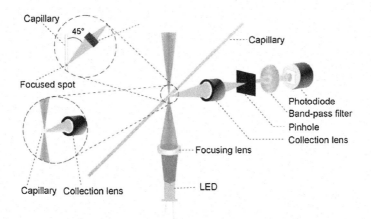

Fig. 3.9 Fluorescence detector for capillary electrophoresis. Key parts are: *blue* LED ($\lambda = 470$ nm) with concave lens, focusing lens, collecting lens, 1 mm pinhole, 535 nm filter, photodiode. Reprinted with permission from Yang et al. [11]. Copyright 2009 Elsevier

- long life span;
- small size;
- low price;
- high stability and light intensity;
- available LEDs for different wavelengths.

Figure 3.9 shows an exemplary simple LED-IF detection system for CE [11].

3.6.2.1 LIF Detectors

According to Eq. 3.4, the fluorescence intensity, I_F, depends on the excitation light intensity I_0, so it is advantageous to use lasers in fluorescence detection (LIF—laser fluorescence detection). Lasers are monochromatic light sources of highly collimated beam and characterized by high intensity. Utilization of lasers as excitation sources allows for decreasing the limit of detection down to the level of $10^{-18} \div 10^{-21}$ mol. Moreover, the laser can be easily focused inside a capillary in a spot smaller than the internal diameter of the capillary. Lasers are also very popular in chip techniques because of easy focusing of the laser beam in any point of the device with or without optical fibers.

The LIF detector for detection in capillaries is schematically presented in Fig. 3.10. The system consists of excitation laser, focusing optics, optical filters, detection cell (capillary), and fluorescence collection system which directs fluorescence beam to the PMT [12–14]. The figure shows two most important LIF arrangements: orthogonal (classical) and collinear, however, other angles, like 45° are also possible in LIF construction. In a collinear detector the excitation laser

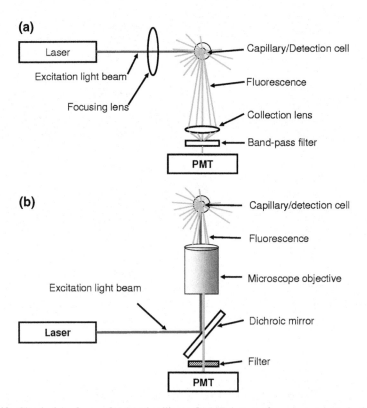

Fig. 3.10 Classical (*orthogonal*) (**a**) and collinear (**b**) LIF detector for measurements in the fused silica capillary

Fig. 3.11 LIF in chip devices. *E* excitation radiation, *F* fluorescence radiation, *L* lens, *O* optical fiber. Reprinted with permission from Fu et al. [15]. Copyright 2006 American Chemical Society

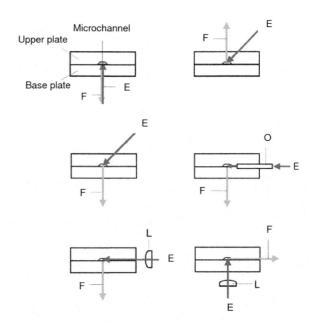

beam and the fluorescence beam use the same focusing/collecting optics which is a microscope objective. An entire microscope can be used or the objective itself after removing it from the microscope. There are also several optical arrangements in chip systems possible (Fig. 3.11) [15].

3.6.2.2 Lasers

A wide range of lasers can be used in LIF detection. When choosing the laser for a given analytical task one should take into account the following:

- the laser wavelength should match or be as close as possible to the absorption maximum of the analyte. However, it should be emphasized that such a criterion is a serious limitation of LIF applicability—in practice, different analytes have different absorption maxima. On the other hand, in many biochemical applications derivatization with fluorescent labels is used (eg., FITC for proteins) and the obtained derivatives have absorption maxima at the same wavelength;
- laser power—the higher the power, the higher the fluorescence intensity. However, extremely high power may induce photochemical decomposition of some analytes;
- Coherence—it should be possible to focus the beam on a very small spot, for eg., in a capillary of chip channel.

Table 3.2 lists several popular lasers used in LIF detection.

Table 3.2 Exemplary lasers used in LIF detection

λ [nm]	Laser source	Power (mW)
266	DPSS	2
325	He/Cd	10
355	DPSS	2
410	Diode	20
473	Diode	30
488	Diode	25
514	Ar	40
532	DPSS CW	15
594	He/Ne	2
633	He/Ne	10
635	CW laser diode	20
650	CW laser diode	15
780	CW laser diode	40

Coupling of CE with UV-LIF suffers from high background level which is connected with the fluorescence of fused silica capillary excited with near UV radiation as well as with fluorescence of contaminants of buffer constituents. Also, the microscope objective can show some fluorescence which results from light scattering. All of these phenomena negatively influence the limits of detection. Several interesting solutions have been worked out to improve sensitivity in LIF detection:

- Confocal system—it was designed to eliminate most light which is of out-of-focus of the fluorescence collection optics (the light which is out of the focal plane is blocked). It is called confocal, because the detection cell (capillary) is placed in one of the objective foci, while in the other there is a pinhole which blocks out-of-focus rays (like in a confocal microscope). Using such a system, Hernandez et al. [16] were able to detect zeptomolar (10^{-21} mol) of fluorescein-derivatized amino acids. Zare reported LOD of $10^{-17} \div 10^{-20}$ mol for polycyclic aromatic hydrocarbons [17]. Confocal system in chip system is shown in Fig. 3.12.
- It is typical in most LIF systems that the fluorescence is collected from "one side" of the capillary/chip channel. In one of the very interesting solutions (by J&M) the fluorescence is collected around the capillary (Fig. 3.13). The detection window is placed in a small unit which contains seven holes radially distributed (45°) around 0.4 mm i.d. central hole (in which detection window is to be placed). To all of the radial holes the optical fibers are connected. One of the fibers is connected to the laser (excitation source), while another six, distributed around the detection window, collects the fluorescence from it. These capillaries are bunched up in a standard SMA connector, which allows to connect it to the PMT. The entire capillary holder is a relatively small part and can be easily connected to the mass spectrometer electrosprayer [18].
- One of the innate source of problems in laser-induced fluorescence detection in fused silica capillaries is the geometry of the capillary itself. The capillary, which is a relatively thick-walled tube of a round cross section, can induce

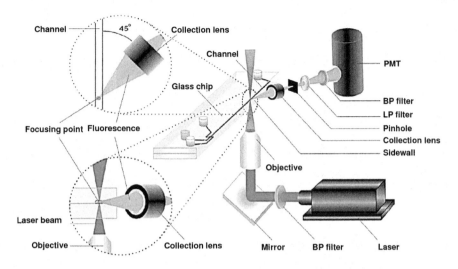

Fig. 3.12 Confocal system in chip device. Reprinted with permission from Fu et al. [15]. Copyright 2006 American Chemical Society

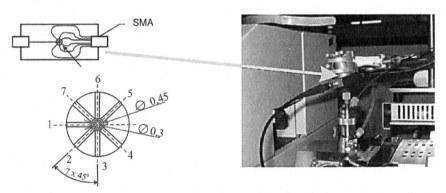

Fig. 3.13 A capillary holder containing optical fibers radially arranged around the capillary. Optical fiber no. 1 is connected to the laser, while fibers 2–7 collect the fluorescence around the capillary. The picture on the *right* presents CE-LIF-ESI-MS coupling. Thanks to courtesy of J&M Analytik AG

problems with focusing the laser inside the capillary channel; moreover, the light scattering on the air-capillary and capillary-analyte (or liquid inside the capillary) borders may have influence on the baseline background level and noise [6]. From this point of view a *sheath flow cuvette* is a much better solution. The cuvette is schematically presented in Fig. 3.14. It is a transparent tube (a flow cytometer cell can be used), usually of square or rectangular cross-section; in such a cuvette there are two perpendicular flat surfaces, through which the excitation beam is delivered and fluorescence is collected. The construction allows the sheath fluid to be pumped through it *slowly* enough to prevent it from mixing with the effluent from the capillary. The sheath fluid and

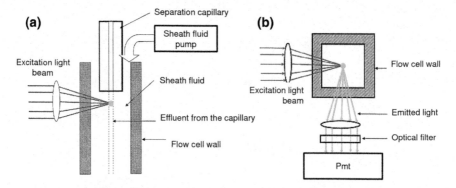

Fig. 3.14 Scheme of the sheath flow detection cell for LIF. **a** longitudinal section; **b** cross-section

the effluent (CE separation buffer) are of the same composition, so any undesirable effects resulting from light refraction and scattering on the border of these two liquids are avoided. Exemplary cell, described by Le et al. [19] was 2 cm long, had 1 mm thick walls, and the internal channel was a square of 200 μm side. The separation capillary of 10 μm internal diameter and 144 μm outer diameter was placed in a cuvette such that the excitation beam was focused 20 μm from the separation capillary outlet. The authors reported LOD of ca. 50 molecules of oligosaccharides detected as their fluorescing tetramethylrodamine derivatives, whose LOD was calculated taking into account injection volume of 5.5 pL and the solution concentration of 10^{-10} mol/L [19].

3.6.2.3 Derivatization in Fluorescence Detection

As mentioned before, the possibility of focusing of the laser beam inside the capillary or chip channel makes the LIF detection perfectly fit the capillary separation techniques like μ-HPLC, CE, or CEC. When using lasers and well-designed optics it is possible to obtain LOD on the level of several hundreds of molecules going through the detection cell. Fluorescence and LIF detection are often connected with derivatization of the analytes, because it is regarded as one of the most sensitive detection methods, compatible with many solvents (the compatibility is often limited in sensitive electrochemical methods).

The derivatization methods can be classified as follows [20]:

- Based on the manner of derivatization:

  - via covalent bind or reagent (fluorophore)—labeling;
  - via non-covalent methods, like ion-pairing, photolysis, redox, or electrochemical reactions or using fluorescent probes.

- Based on the final analytical system:
  - pre-column derivatization—the reaction is performed before separation; the advantage is basically free choice of derivatizing agent, there are no kinetics or number of steps limitations. Generating of derivatives of other constituents of the sample or possibility of contamination during multistep processes can be serious disadvantages of pre-column derivatization;
  - on-column reaction—is performed in column, during separation. One of the advantages is that the sample is non-diluted. Disadvantages are as follows: very fast and quantitative reaction should be chosen, no side-products should be generated, and the derivatives should be stable. The derivatizing agent should not be fluorescent;
  - post-column derivatization—the reaction is conducted after the analytes have been separated but before they pass the detector. The advantage is that the analytes are separated in their native form and generation of side-products is not important. However, special reactors between column and detector are needed, which may influence band broadening.

The choice of the method depends on the separation system (electrophoretic or chromatographic separation mechanism), the analyte, derivatizing reagent, and the aim of the derivatization (Fig. 3.15).

The amine groups (the derivatives are then separated using CE or MEKC) can be derivatized using the following reagents:

- for pre-column derivatization: FITC (fluorescein isothiocyanate), fluorescamine, FMOC (9-fluorenylmethyl chloroformate), OPA (ortophthalic aldehyde), TRITIC (tetramethylrodamine isothiocyanate), AQC (6-aminoquinolyl-N-hydroxysuccinimidyl carbamate), NDA (naphthalene-2,3-dicarboxaldehyde), CBQCA (3-(4carboxybenzoyl)quinoline- 2-carboxaldehyde);
- for on-column derivatization: OPA, NDA;
- for post-column derivatization: OPA, NDA, fluorescamine.

For derivatization of reducing groups (e.g., aldehyde or ketone groups in carbohydrates) one can use: AC (9-aminoacridine), AP (2-aminopyridine), AQ (6-aminoquinoline), ANTS (8-amino-1,3,6-naphthalene trisulfonate), CBQCA, TRSE (5-carboxytetramethylrhodamine succinimidyl ester).

Fluorescent derivatives of carboxylic groups can be obtained using: AAF [5-(aminoacetamido)fluorescein], 7-ANDA (7-aminonaphthale-1,3-disulfonic acid), PDAM (1-pyrenyldiazomethane).

The amine group derivatizing agents are considered the most important because of their wide use in determination of trace amounts of amino acids, peptides, and proteins. The achievable LODs are in the range of pico- and zeptomoles. The great advantage of using such reagents as FITC is their high molar absorption coefficient (for FITC $\varepsilon > 70\ 000$ L/mol cm), high quantum yield and practically no change of λ_{ex} and λ_{em} after reaction with different analytes—it means that one laser and one filters set are needed. The disadvantage of using fluorescein reagents may be strong influence of pH on fluorescence. Nowadays, series of labeling agents covering

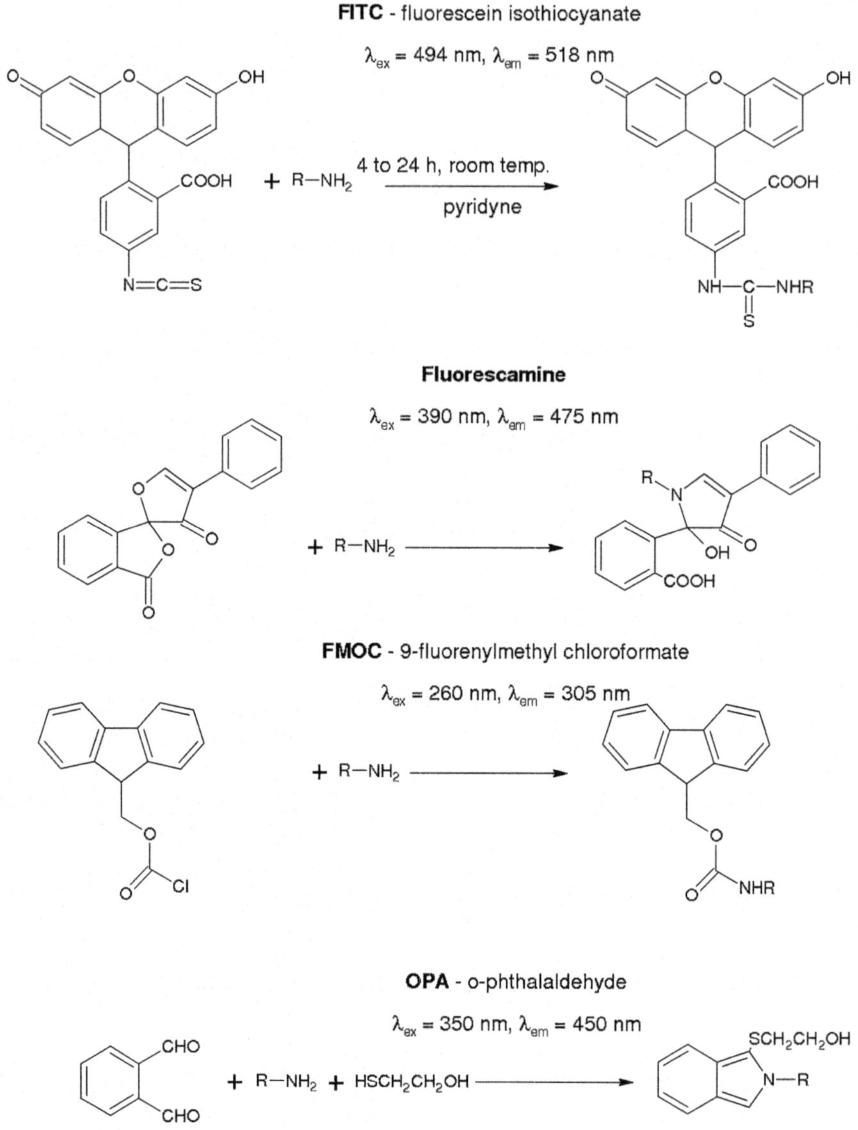

FITC - fluorescein isothiocyanate

$\lambda_{ex} = 494$ nm, $\lambda_{em} = 518$ nm

4 to 24 h, room temp.

+ R—NH$_2$ —————→

pyridyne

Fluorescamine

$\lambda_{ex} = 390$ nm, $\lambda_{em} = 475$ nm

+ R—NH$_2$ —————→

FMOC - 9-fluorenylmethyl chloroformate

$\lambda_{ex} = 260$ nm, $\lambda_{em} = 305$ nm

+ R—NH$_2$ —————→

OPA - o-phthalaldehyde

$\lambda_{ex} = 350$ nm, $\lambda_{em} = 450$ nm

+ R—NH$_2$ + HSCH$_2$CH$_2$OH —————→

Fig. 3.15 Exemplary reagents used for derivatization of amine groups (derivatization of amines, amino acids, peptides, and proteins), according to [23]

excitation wavelengths from UV to red are commercially available. Typically, their fluorescence does not depend on pH and they are synthesized by sulfonation of traditional coumarin, rodamine, or fluorescein dyes. For example, the counterpart of FITS can be Alexa-Fluor 488 (Molecular Probes), which is characterized by almost constant fluorescence intensity in the pH range from 4 to 9.

3.6.2.4 LIF: Summary

Presently, LIF detectors are among the most sensitive detectors used in electro-migration techniques. The most important advantages are extremely low LODs and generally uncomplicated construction. However, many LIF detectors (actually, a straight majority of them) described in the scientific publications are lab-made constructions of the authors. They are based on a variety of ready-to-use modules, fiber optical connections, and different optical and optomechanical parts. Disadvantages of LIF detectors are related to the high price of the laser and PMT; it should be emphasized that some degree of versatility requires having several lasers characterized by different wavelengths, however, changing of the laser may induce changing of other parts of the system like optical filters, dichroic mirrors, etc.

3.6.3 RI Detection

RI (refractive index) detection relies on the measurement of changes of declinations of a light beam caused by the zones of the analytes passing the beam relative to the position of the beam during presence of the buffer/mobile phase. The measurements of RI can be used if the analytes do not absorb radiation in the range of 190–200 nm or do not fluoresce. RI detection can be used, for eg., in detection of simple cations and anions or carbohydrates. RI detector is a concentration sensitive, non-destructive device considered to be universal if only the analyte zone and the buffer refractive indices are different enough to be detected. Despite these advantages RI detectors did not become popular in CE and in HPLC. Under HPLC conditions the disadvantage related to RI detector can be extensive baseline drift and low sensitivity. In capillary techniques the major disadvantage was miniaturization of the RI system without decreasing the sensitivity. It was possible to build relatively stable and sensitive RI systems for CE which allowed for the measurement directly in the capillary [6, 21, 22]. It should be emphasized that RI measurements are strongly temperature dependent—the changes in the range of $10^{-3} \div 10^{-4}$ RI unit/°C may have significant influence on the obtained results. The system should be precisely thermostated because of Joule heating which may induce local RI changes and baseline disturbance.

In RI detectors lasers are used as light sources. In many systems the analytical signal comes from shifts in the interference fringes obtained after diffraction of the beam by the capillary. The fringes and the changes in their position can be detected by a diode array, and the electrical system adapted to such a task is called position sensitive detection (PSD). The exemplary RI detector based on a holographic plate and a laser is presented in Fig. 3.16 [22].

In the recent years a special type of RI detectors was also elaborated, where the capillary is used as a liquid core optical ring resonator (LCORR—Fig. 3.17). The capillary was drawn and etched with HF to decrease its wall thickness to

Fig. 3.16 Exemplary RI
detector for capillary
electrophoresis. Reprinted
with permission from
Krattiger et al. [22].
Copyright 1994 American
Chemical Society

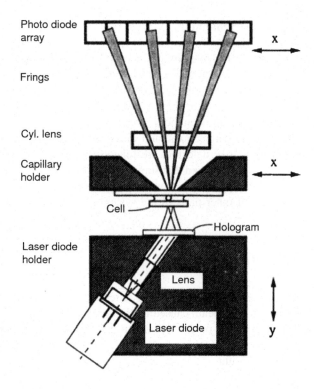

ca. $3 \div 4$ μm. The capillary prepared in such a way was brought in contact with
a perpendicular optical fiber taper (fiber was tapered using a flame to ca. 3 μm).
The light (laser) in the optical fiber induces whispering gallery modes (WGM) in
the thin wall of the capillary. The analytical signal is connected to the shifts of
the resonance wavelengths of the WGM, which in turn depend on changes in RI
of the liquid inside the capillary. The sensitivity of such an RI device is ca. 10^{-6}
RI unit [24].

3.6.4 Electrochemical Detection

Three groups of electrochemical detectors for CE can be distinguished as:
potentiometric, conductometric, and amperometric. Electrochemical detectors are
the most sensitive devices and there are detectors for HPLC systems in the market,
however, in electromigration techniques only conductometric detectors are commercially available (particularly in ITP). The problems that manufacturers are
coping with are connected with the lifetime and preparation of the electrodes,
which should fit the CE capillary internal diameter (75 μm and less) [21].

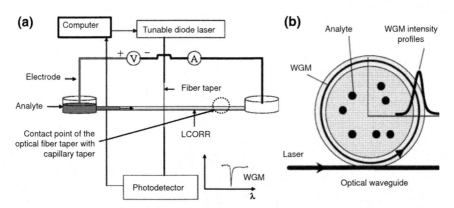

Fig. 3.17 RI detector employing a tapered, thin walled capillary as a liquid core optical ring resonator (*LCORR*). **a** General view of the system. **b** Cross section of the capillary in a place where it touches the optical fiber. Reprinted with permission from Zhu et al. [24]. Copyright 2006 American Chemical Society

3.6.4.1 Potentiometric Detectors

Compared to other electrochemical methods the potentiometric detection concerning electronic instrumentation is the easiest to perform—the measured analytical signal is an electrode potential. Technically, the problem is to construct the electrode whose size fits the internal diameter of the capillary. In potentiometric detection micro ion-selective electrodes (ISE) are used, which should be sensitive to the chosen cation or anion, but in practice they also respond to other species. The ISEs provide relatively low detection limits in low volume samples, hence they can be used in CE as detection systems. In popular liquid membrane electrodes the change in potential (measured against the reference electrode) is observed when the analyte is brought into contact with the hydrophobic membrane containing ionophore. The analyte generates the potential difference between the internal electrolyte and measured solution, i.e., in CE—background buffer. The potential difference depends on the activity (concentration) of the analyte which can be described by the Nikolsky–Eisenmann equation:

$$E = E_0 + \frac{RT}{F} \ln\left(\sum_i K_i^{pot} c_i^{1/z_i}\right). \tag{3.6}$$

K is a selectivity factor, which depends, to some extent, on hydrophobicity of the analyte and thus its affinity to the hydrophobic membrane [25].

Two approaches can be distinguished in potentiometric detection systems for CE. At the beginning the electrodes used were miniaturized versions of the electrodes commonly used in electrochemical measurements. They consisted of glass capillaries with a drawn out tip at which end a membrane containing ionophore was placed. The back of the electrode contained an internal electrolyte and

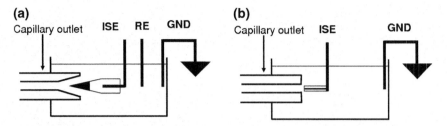

Fig. 3.18 Exemplary systems for potentiometric detection in CE. *ISE* ion-selective electrode, *RE* reference electrode, *GND* ground. Reprinted with permission from Kappes et al. [25]. Copyright 2006 Elsevier

silver wire coated with AgCl as an internal reference electrode. The dimensions of the tip fitted the end of the separation capillary and its position was adjusted with micromanipulator. A micro reference electrode can also be used in such a system (Fig. 3.18a) [25]. In later solutions the construction of the electrodes was simplified and their durability increased. For example, De Backer and Nagels produced the electrodes in the following way: the insulation was removed from the tip of a copper wire (100–250 μm of outer diameter), then it was immersed five times in tetrahydrofurane solution of PVC with ionophore and left to dry after each immersion. Such electrode was used as a detection electrode, while a cathode of a CE system served as an reference electrode (stable reference potential) which was another step to simplify the construction (Fig. 3.18b). The system described has shown linearity in the range of 0–20 pmol injected organic acids (Fig. 3.19) [26].

Because of the short lifetime of electrodes (2–3 days) potentiometric detection s not widely used in capillary electrophoresis and chip systems.

3.6.4.2 Conductometric Detection

In conductometric detection the ability of ions present in the electrolyte to conduct the charge between two electrodes is used. A current is measured between two electrodes after applying a voltage; then according to Ohm's law the resistance of the electrolyte or its conductance can be calculated. It can be expressed as follows:

$$G = \frac{1}{R} = \kappa \frac{A}{l} = \frac{A}{l} \sum \lambda_i c_i, \ \left[\Omega^{-1}\right] = [S] - \text{ simens,} \qquad (3.7)$$

where G is conductance of the electrolyte, R is resistance, κ is specific conductance, A is surface area of the electrodes, l is distance between the electrodes, λ is conductivity of an ion, and c is concentration of an ion.

The conductometric detector measures the differences between conductivity of the separation buffer and migrating zones of the analytes. To avoid electrode oxidation//reduction processes the measurement is conducted using alternating

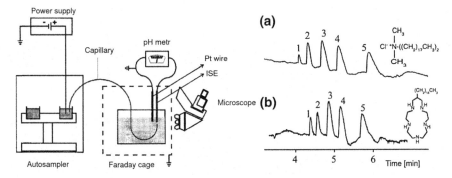

Fig. 3.19 CE with potentiometric detection—separation of organic acid. Resolved analytes: *1* quinic acid, *2* capronic acid, *3* butyric acid, *4* propionic acid, *5* acetic acid. Separation buffer: HEPES pH 7.0, c = 5 mmol/L. Ionophores used: **a** quaternary ammonium salt, **b** macrocyclic pentaamine (15-hexadecyl-1,4,7,10,13,-pentaazacyclohexadecane). Reprinted with permission from De Backer et al. [26]. Copyright 1996 American Chemical Society

current of low (ca. 1 kHz; contact detection—the electrodes are in contact with the solution being measured) or high frequency (ca. 30–40 kHz; contactless detection). Both contact and contactless manners can be effectively used in capillary and chip techniques [21, 25, 27].

In principle three types of constructions for contact conductometric detection have been elaborated (Fig. 3.20). In a classical system the detection is performed between two electrodes arranged perpendicular to separation channel and exactly

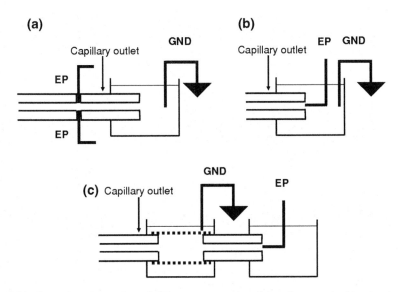

Fig. 3.20 General representation of different constructions in conductometric detection in CE. *WE* working electrode, *GND* ground. Reprinted with permission from Kappes and Hauser [25]. Copyright 2006 Elsevier

opposite (to avoid detection of potential gradient during electrophoretic separation). The group of Zare [28] was first to apply online contact detection in a fused silica capillary. The electrodes (two platinum wires of 25 μm diameter) were placed and fixed with epoxy glue in 40 μm holes drilled in the 50 μm i.d. capillary using a CO_2 laser. In another system the working electrode is placed in some distance from the separation capillary outlet (Fig. 3.20b). Obviously, such systems are not versatile as the exchange of the separation capillary is not a straightforward task.

In conductometric detection the noise depends on the baseline background level which in turn is connected to the composition and concentration of the separation electrolyte. Increasing the concentration of the buffer usually improves the separation but it can decrease the sensitivity of conductometric detection, hence low concentration analytes may not be detected.

Similar to ion chromatography decreasing of the buffer conductivity may have positive influence on LOD. This can be done by conductivity suppression. As long as special suppressing ion-exchange columns or membrane suppressors are used in ion chromatography, in CE a system presented schematically in Fig. 3.20c and in detail in Fig. 3.21 can be used [27]. In such a solution, a membrane (Nafion®) and a negative electrode are placed in the compartment containing regenerant solution. In this compartment the conductivity of the separation buffer is suppressed continuously during CE separation. Another section of the fused silica capillary is used to transport the suppressed solution to the conductometric detection cell. The detection is performed between two wires (arranged to keep $k = A/L$ constant) using pulses of voltage of alternating polarity [27, 29, 30].

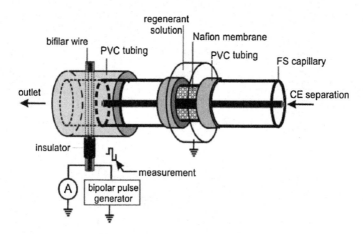

Fig. 3.21 Conductometric detection with suppression in CE. Different arrangements of electrodes with respect to the separation channel of the chip device—a and b—electrodes of different dimensions placed opposite, c—parallel electrodes placed crosswise to the channel. Dimensions of the electrodes: A—23 μm, B—23 μm, C—13 μm, D—23 μm, spaces between electrodes: A–B 7 μm, B–C 12 μm, C–D 22 μm. Reprinted with permission from Guijt et al. [27]. Copyright 2004 Wiley-VCH

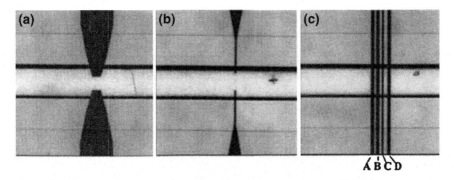

Fig. 3.22 Different variants of arrangements of the electrodes across the separation channel of the chip device; **a** and **b** electrodes of different sizes placed opposite to each other, **c** parallel electrodes placed across the channel, electrode sizes: A—23, B—23, C—13, and D—23 μm, distances between electrodes: A–B 7, B–C 12, and C–D 22 μm. Reprinted with permission from Graß et al. [31]. Copyright 2001 Springer

Contact conductivity detectors can be effectively used in chip systems. Exemplary arrangements of the working electrodes in the separation channel are presented in Fig. 3.22 [31]. As chip systems give more freedom in construction possibilities in comparison to capillary techniques, different electrode materials can be used, for e.g., platinum (as a wire or a layer created by physical vapor deposition, PVD), polymers (polystyrene, nylon) mixed with soot or carbon fibers, carbon. In their chip system Galloway et al. [32] reported LOD 8 nmol/L for alanine which corresponded to 3.4 amol of detected analyte.

Because of their complicated construction, contact conductivity detectors for CE performed in fused silica capillaries are not available on the market presently. In the 1990s, the Thermo Bioanalysis company provided such a detector in their Crystal 1,000 CE setup. Recently, contactless conductivity detectors have become popular.

In contactless conductivity detector (CCCCD—*Capacitively Coupled Contactless Conductivity Detector*) two cylindrical metal electrodes (made of wire or a needle) are placed (with a gap between them) around the fused silica capillary. An electrical signal (alternating current) is supplied through an input electrode, while the output electrode detects the signal after it passes through a section of the capillary between them (this section is a detection cell) (Fig. 3.23). Because of capacitive coupling the high frequency alternating current supplied into input electrode induces current in the output electrode. Distribution of the electromagnetic field generated in the detection cell depends, among other parameters, on the dielectric constant and conductivity of the liquid inside the capillary. The zones of the migrating analytes disturb local conductivity, which can be detected. The possibility of performing the CCCCD measurement does not depend on the length of the electrodes (*l*), but on the distance (*d*) between them [27].

One of the advantages of CCCCD is that it can be easily used in different capillary separation techniques and is ready to use without any special

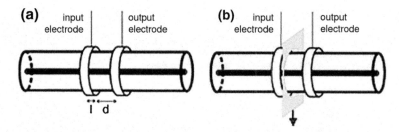

Fig. 3.23 Two systems of cylindrical electrodes for CCCCD **a** without grounded plate, **b** with grounded plate to eliminate the coupling through the air between electrodes. Reprinted with permission from Guijt et al. [27]. Copyright 2004 Wiley-VCH

modification of the capillary. C^4D detector is an example of commercially available CCCCD device, which is designed to ion chromatography, capillary electrophoresis, and chip systems. C^4D is compatible with most popular CE systems.

3.6.4.3 Amperometric Detection

In the amperometric detection electrolysis of a diffusion layer on the electrode is used to obtain the analytical signal. The process of electrolysis is possible when a constant potential is applied between working and reference electrodes. The analytical signal is a current measured between the electrodes, which depends on the concentration of the analyte:

$$I = -AnFD\frac{c}{\delta_N},\tag{3.8}$$

where I is measured current, A is surface area of the electrode, n is number of electrons transferred in the electrode reaction of the analyte, F is Faraday constant, D is diffusion coefficient, c is concentration of the analyte, and δ_N is thickness of the diffusion layer.

The ability of the analyte to undergo the oxidation or reduction reaction at a given potential is a prerequisite of amperometric detection. In other words, a potential high enough to induce the electrode reaction is necessary. Amperometry is then less universal than conductometry, however, much lower LODs are often observed; moreover, changing the potential may have some influence on selectivity. Amperometric detection can be conducted in direct or indirect modes, which is performed in a similar way as in spectrophotometric or fluorescence techniques. Here, the indirect detection relies on addition of an electroactive substance to the separation buffer which results in a high level of background signal. Zones of not electroactive analytes generate negative signals (lower current) when passing through such a detection system. A disadvantage of amperometric detectors is a gradual decrease of their sensitivity caused by deposition of oxidation/reduction reaction products on the electrodes. Thus, the active surface of the electrodes is

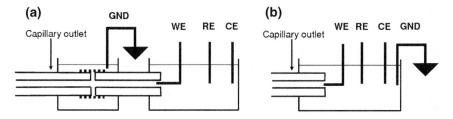

Fig. 3.24 Scheme of the Amperometric detection system for CE. **a** System with porous junction **b** Direct connection. *WE* working electrode, *RE* reference electrode, *CE* counter electrode, *GND* ground. Reprinted with permission from Kappes and Hauser [25]. Copyright 2006 Elsevier

smaller and smaller which leads to decrease of sensitivity or instability of the system (Eq. 3.8). This effect can be minimized by applying pulsed voltage instead of working at a constant potential. The technique is then called pulsed amperometric detection (PAD). The current is measured during a short period of pulse of working potential. In fact, before the working potential is applied the electrode is cleaned by high positive potential which oxidizes the contaminants and the surface of the metal electrode as well, which is followed by application of high negative potential to reduce electrode surface back to pure metal.

In most of the amperometric a system of three electrodes is used: working (WE), reference (RE), and counter (CE) electrode. The electrodes are connected to the potentiostat and they can be arranged as shown in Fig. 3.24 [25]. It is believed that measuring electrode system should be separated from the separation voltage for capillaries larger than ca. 25 µm of internal diameter. In larger capillaries the separation current may have a significant influence on the detection Faraday current. The separation of these two electrical systems (separation capillary with electrodes and detection system) can be done by using porous joint or porous taper etched in the capillary [33]. In the case of the porous joint the analytes reach the detection electrodes being pushed by the EOF from the separation capillary.

In chip systems the separation of electrophoretic and amperometric detection parts can be done using porous decoupler or without it (Fig. 3.25) [34]. Several approaches to place working electrode in a chip can be distinguished: in channel, end channel, and off-channel. Because of possibility of free arrangement of the parts the electrodes in chip systems need not be metal wires. They can be manufactured using different materials (gold, platinum, diamond, palladium, carbon as fibers, paste, film, or dye) and techniques of sputtering, printing, or vapor deposition.

3.6.5 MS Detection

There are numerous advantages of using mass spectrometer as a detector in electromigration techniques. MS is a sensitive and universal detector that can

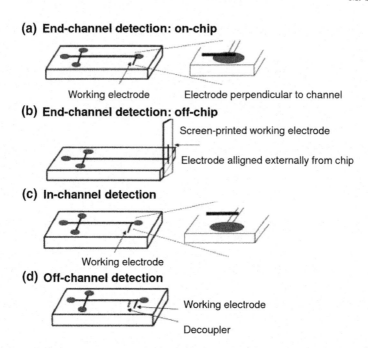

(a) End-channel detection: on-chip

Working electrode   Electrode perpendicular to channel

(b) End-channel detection: off-chip

Screen-printed working electrode

Electrode alligned externally from chip

(c) In-channel detection

Working electrode

(d) Off-channel detection

Working electrode

Decoupler

Fig. 3.25 Different positions of the working electrode in CE chip system. Reprinted with permission from Vandaveer et al. [34]. Copyright 2002 Wiley-VCH

identify all the substances characterized by molecular mass being in the working range of the device. Due to extremely low flow rates in liquid phase capillary techniques there are no significant technical problems in coupling them with MS. Depending on the way of ionization at a first stage of MS analysis the ions of the sample molecules are created as molecular ions, ionic fragments, or complexes. Then the ions are accelerated in a vacuum to high velocities followed by their separation into groups with regard to their mass and finally they are detected and identified. Mass spectrometry is a sample destructive technique [35].

There are several ionization techniques, but only some of them are applied in coupling of capillary techniques with MS: electrospray ionization (ESI), atmospheric pressure chemical ionization (APCI) and continuous flow fast atom bombardment (CF-FAB). The most popular, ESI, perfectly fits CE-MS coupling as it is most effective at very low flow rates (max. 20 µl/min). Electrospray is one of the most gentle ionization methods. The process is performed at room temperature and at atmospheric pressure. The solution flows through a capillary tubing towards the ion source, then at the tubing outlet it is subject to high electric field (several thousand volts, relative to the nearby counter-electrode) which make the fluid sprayed into small droplets of high charge. Depending on the polarity of the electrodes the droplets have positive or negative charge. In ESI only protonated molecules are generated (pseudo-molecular ions) which allows for molecular weight determination with up to 0.01 % accuracy. Capillary electrochromatography is a technique of high significance as for coupling with MS. CEC column can be connected to MS using different types of interfaces (Fig. 3.26) [35, 36].

- Sheathless interface. A sharp and metal coated (e.g. gold) tip of a packed CEC column is directly used to generate a mobile phase spray. No sheath fluid is added, so the eluted analytes are additionally not diluted, which positively influences the sensitivity in comparison to sheath flow and liquid junction interfaces. A sheath gas (for example SF_6) is used in sheathless interface to prevent corona discharge resulting from higher electric field on sharp tips. The lack of sheath fluid influences the choice of buffers compatible with electrospray. In general, the best buffers are those volatile, characterized by low surface tension and low conductivity. Another issue is a lifetime of the metal coating of the tip—it can undergo erosion within several days of working. Additionally, doubly charged adduct ions (like $[M + H + Ag]^+$) can be sometimes observed which makes problems with data interpretation and determination of the molecular mass of the analyte.
- Sheath fluid interface. In this approach the packed capillary column is introduced into the ionization region through a narrow metal tube which is used to

Fig. 3.26 CEC-ESI-MS interfaces. **a** Sheathless interface. **b** Sheath flow interface. **c** Liquid junction interface. Reprinted with permission from Choudhary et al. [36]. Copyright 2000 Elsevier

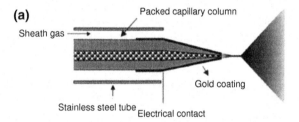

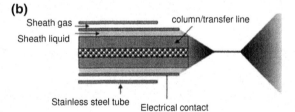

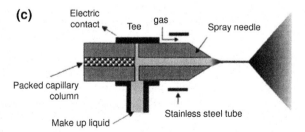

Fig. 3.27 A system for CE-LIF-ESI-MS coupling. Courtesy of J&M Analytik AG

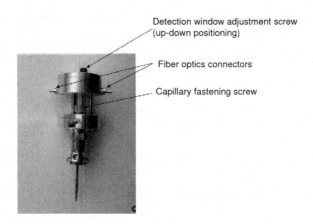

Detection window adjustment screw (up-down positioning)

Fiber optics connectors

Capillary fastening screw

deliver a sheath fluid to the column outlet. As the sheath fluid flows through the tube it is mixed with the effluent from the column and then a stable spray cone is created. Also, as the sheath fluid is in contact with the tube connected to the ground (or a fixed voltage) potential it creates an electrical contact for the CEC column outlet. In such a system an additional third concentric tube can be used to deliver supporting gas which can help in spray formation and scavenging free electrons responsible for corona discharge. Another advantage of using sheath gas is lowering requirements of low surface tension and conductivity of the buffers. One of the problems in using sheath fluid is that usually it is a different liquid than that present in the inlet vial of the CE/CEC system. This makes a discontinuity of the buffer system during CE separation which was found to be a source of changing migration order of the analytes. Moreover, the analytes are diluted in such an interface and ions of sheath fluid compete for charge with the analytes so that the sensitivity may be worse than it is in the sheathless system.

• Liquid junction interface. In this approach a stainless steel T-piece is used to provide electrical contact and to deliver make-up liquid flow. The column outlet and the sprayer needle tip are positioned inside the T-piece which is characterized by a very low dead volume. The make-up liquid, which is delivered through the third T-piece port serves as an outlet buffer vial for CEC system.

CEC-ESI-MS and CE-ESI-MS interfaces are commercially available. Some manufacturers supplies users with ready-to-use ESI-MS sprayer combined with UV-VIS or LIF detection cell (Fig. 3.27) [18].

3.6.6 NMR Detection

Coupling of the capillary electromigration techniques with nuclear magnetic resonance spectroscopy (NMR) has been used by few specialized research groups [37].

NMR is not as sensitive as the techniques mentioned earlier. However, it allows for determination of the structure of the separated analytes and it is non-destructive technique. The most important component of the system is a radio frequency (RF) coil which size should fit the capillary shape and size. The coil is responsible for delivering the radio frequency to the sample to excite the nuclei and then it collects the signals from it. Two types of coils can be distinguished: solenoidal coil and saddle-type (Helmholtz) coil [35, 38]. When using a solenoidal coil the best sensitivity is achieved by keeping the coil axis perpendicular to the static magnetic field (B_0), whereas in saddle coil its axis should be kept parallel to B_0. The solenoidal cell can be lab-made by carefully wrapping a conductive wire around the capillary. The wire then works as an NMR detection cell. The simplicity of such solution allows for easy coupling of NMR with capillary separation techniques (capillary HPLC, CE, CEC). The volume of capillary cells ranges from several nanoliters to several microliters and can be easily changed by using a capillary of different internal diameter (Fig. 3.28).

The CE or CEC coupling with NMR can be characterized as follows [35]:

- Advantages of using CE-NMR or CEC-NMR coupling:

  - high separation efficiency (high number of theoretical plates). For CEC the obtained efficiencies are lower than in CE-NMR;
  - low solvents consumption allows for using fully deuterated solvents that do not give additional signals in ^1H NMR spectrum;
  - separation is performed only in the presence of EOF;
  - the amount of analytes can be greater in CEC in comparison to CE.

- Disadvantages of CE-NMR coupling:

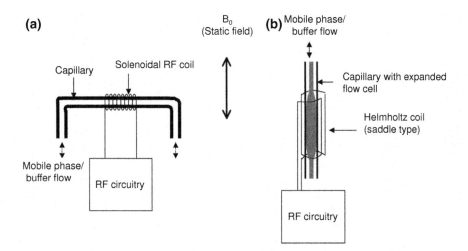

Fig. 3.28 Solenoids used in coupling of capillary separation techniques to NMR. Reprinted with permission from Jayawickrama and Sweedler [38]. Copyright 2003 Elsevier

- in a typical construction the application of solenoidal RF coils (which are more sensitive) is limited to a stopped flow mode because of inducing a local magnetic field;
- long analysis time in a constant flow mode (more than 40 min), it may be extended if stopped flow mode is necessary;
- small detection volume and relatively short time of presence of the analyte in the detection cell make the collection of the NMR spectrum complicated;
- application of too high voltage in order to increase the buffer/mobile phase flow gives the rise to Joule heating which adversely influences the separation efficiency;
- differences in times of passing the detection cell by the analytes have influence on band broadening on the NMR spectrum.

3.6.7 Summary

Detectors used in electromigration techniques allow to obtain numerous analytical data. However, there is no ideal detector among them. Mass spectrometer is close to being ideal—it is sensitive, universal, but allows for selective detection mode to be performed and it allows for obtaining qualitative information about a given analyte. LIF belongs to the most sensitive methods used for detection of fluorescent compounds (direct LIF) or non-fluorescent (indirect LIF or using derivatization); it can be performed in a capillary as well as in microfluidic systems (chip devices). Contactless conductivity detection due to its simplicity (in the terms of combining it with the separation capillary) has become more and more significant during the last years which probably will broaden the application range of conductivity detection in capillary electrophoresis.

3.7 Automation and Robotization

In practice all commercially available capillary electrophoresis systems and their supporting software provide at least basic automation of the separation process. The following can be regarded as basic automation features:

- autosampler for automatic vial management—moving appropriate vial (e.g., with buffer, sample, solutions used for rinsing of the capillary like NaOH, HCL, water, methanol) to the proper position at the inlet and outlet of the capillary. Autosamplers can be in the form of one or two carousels or rectangular tray with depressions holding the vials.
- fraction collection—feature provided by the software relies on such programming of the system (based on such input data like electroosmotic mobility or flow

rate at a given pressure) that allows for collecting each of the resolved analyte into separate vials.

• automatic buffer replenishment.

Several research groups elaborated compact CE devices designed to field analytics [39–41]. Kubáň et al. [39] constructed a simple CE system of $31 \times 22 \times 26$ cm dimensions that was equipped with a capillary ($50 \mu m \times 62$ cm), a high voltage power supply (up to 15 kV and 100 μA) and contactless conductivity detector (C^4D). The 15 kV HV power supply was chosen intentionally for couple of important reasons: under field conditions, often characterized by high humidity, using full-range 30 kV HVPS could be impossible (current leakage), moreover, such device is relatively large and heavy. The device was supplied with 3.2 Ah lead battery. The system was used for determination of some metal cations and selected anions in water (Fig. 3.29). The device did not allow for automated work—for e.g., sample injection was performed by manual moving of the carousel with vials and manual elevating of the sample vial to induce siphon injection [39].

Another example of portable instrument was described by Gerhardt, who constructed relatively small device equipped with automated autosampler, vial lift, and electrochemical (amperometric/voltamperometric) detector (Fig. 3.30) [40].

Chip techniques give more possibilities in terms of automation and robotization. In comparison to capillary techniques, microfluidics are characterized by higher versatility, higher degree of miniaturization, and the possibility of performing complete analytical path from sampling through derivatization to final

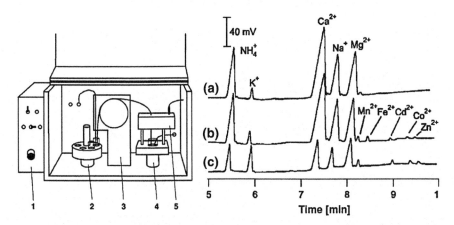

Fig. 3.29 Portable CE instrument and exemplary separation of metal cations. *1* control electronics, *2* sample tray, *3* capillary holder, *4* vial holder, *5* detector cell. Separation conditions: buffer 11 mmol/L His, 50 mmol/L acetic acid, 1.5 mmol/L 18-crown-6, 0.1 mmol/L citric acid, pH 4.1; capillary $d_c = 50 \mu m$, $L_{tot} = 62$ cm ($L_{eff} = 55$ cm); hydrodynamic injection by elevating capillary inlet by 20 cm for 40 s, V = 15 kV. *a* sample of well water, *b* sample spiked with 2.5 mmol/L Mn^{2+}, Cd^{2+}, Co^{2+}, and Zn^{2+} and 5 mmol/L Fe^{2+}, *c* standard solution containing 50 mmol/L NH^{4+}, K^+, Ca^{2+}, Na^+, Mg^{2+}, 5 mmol/L Mn^{2+}, Cd^{2+}, Co^{2+}, and Zn^{2+}. Reprinted with permission from Kubáň et al. [39]. Copyright 2007 Wiley-VCH

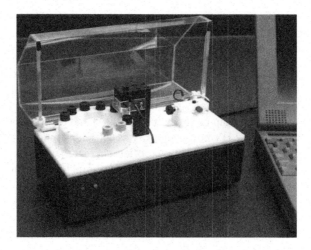

Fig. 3.30 Automated portable CE system. Reprinted with permission from Ryvolová et al. [40].
Copyright 2010 Elsevier

separation and detection. For example, the group of Mathies constructed a portable
instrument which was specially designed for determination of amino-acids in a
Mars-like soil [42]. The device called MOA (Mars Organic Analyzer) which is
presented in Fig. 3.31 consisted of several functional units:

- main multi-layer microfluidic device of 10 cm diameter;
- LIF detector with PMT;
- pneumatic system for sucking and pumping the liquids;
- a sipper used for sampling from the MOD chamber;
- MOD (Mars Organic Detector) chamber in which the soil samples were dried
  under reduced pressure and then the amino acids were sublimed and deposited
  on a fluorescamine coated aluminum disk to generate amino acid fluorescent
  derivatives. The sampling from the disk relied on pumping a small amount of
  the borate buffer out of the sipper and after 60 s the liquid was drawn back to the
  chip.

Limits of detection ranged between 13 nmol/L and 130 pmol/L for different
amino acids.

A fully robotized device (*lab-on-a-robot*) was constructed by Berg et al. [43].
The robot (Fig. 3.32) was built on a mobile platform and was equipped with a
three-channel HVPS, electrochemical detector, modem, compass, and GPS.
FCPGA system was used to control the system. The robot was able to navigate to a
given position location using GPS, then take air samples, and perform full analysis
(injection, separation, and detection) and send wirelessly the obtained chromatogram to the laboratory. Such devices can be used for potentially dangerous sites or
perform fully automated field analyses in different locations [43].

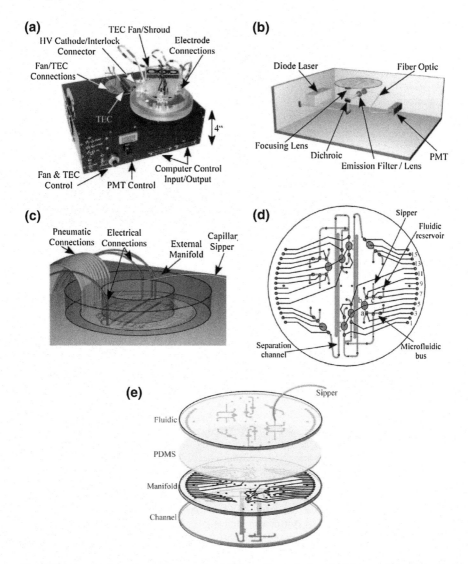

Fig. 3.31 Automatic amino-acids analyzer MOA. **a** General view. **b** Scheme of the detection part of the system (LIF). **c** Electric and pneumatic connections. **d** Arrangement of the channels. **e** Functional layers of the chip. Reprinted with permission from Skelley et al. [42]. Copyright 2005 National Academy of Sciences, USA

Fig. 3.32 Lab-on-a-robot.
Reprinted with permission
from Berg et al. [43].
Copyright 2008 Wiley-VCH

References

1. J.L. Felhofer, L. Blanes, C.D. Garcia, Recent developments in instrumentation for capillary electrophoresis and microchip-capillary electrophoresis. Electrophoresis **31**(15), 2469–2486 (2010)
2. Q. Li, H. Zhang, Y. Wang, B. Tang, X. Liu, X. Gong, Versatile programmable eight-path-electrode power supply for automatic manipulating microfluids of a microfluidic chip. Sens. Actuators B. Chem. **136**(1), 265–274 (2009)
3. L. Jiang, X. Jiang, Y. Lu, Z. Dai, M. Xie, J. Qin, B. Lin, Development of a universal serial bus-powered mini-high-voltage power supply for microchip electrophoresis. Electrophoresis **28**(8), 1259–1264 (2007)
4. T. Tsuda, J.V. Sweedler, R.N. Zare, Rectangular capillaries for capillary zone electrophoresis. Anal. Chem. **62**(19), 2149–2152 (1990)
5. S.X. Lu, E.S. Yeung, Side-entry excitation and detection of square capillary array electrophoresis for DNA sequencing. J. Chromatogr. A **853**(1–2), 359–369 (1999)
6. D.R. Baker, *Capillary Electrophoresis Techniques in Analytical Chemistry* (Wiley, New York, 1995)
7. C.J. Easley, J.M. Karlinsey, J.M. Bienvenue, L.A. Legendre, M.G. Roper, S.H. Feldman, M.A. Hughes, E.L. Hewlett, T.J. Merkel, J.P. Ferrance, J.P. Landers, A fully integrated microfluidic genetic analysis system with sample-in answer-out capability. Proc. Natl. Acad. Sci. U. S. A. **103**(51), 19272–19277 (2006)
8. T. Wang, J.H. Aiken, C.W. Huie, R.A. Hartwick, Nanoliter-scale multireflection cell for absorption detection in capillary electrophoresis. Anal. Chem. **63**(14), 1372–1376 (1991)
9. S. Hjertén, K. Elenbring, F. Kilár, J.L. Liao, A.J.C. Chen, C.J. Seibert, M.D. Zhu, Carrier-free zone electrophoresis, displacement electrophoresis and isoelectric focusing in a high-performance electrophoresis apparatus. J. Chromatogr. **403**, 47–61 (1987)
10. F. Foret, S. Fanali, L. Ossicini, P. Bocek, Indirect photometric detection in capillary zone electrophoresis. J. Chromatogr. **470**, 299–308 (1989)
11. F.-B. Yang, J.-Z. Pan, T. Zhang, Q. Fang, A low-cost light-emitting diode induced fluorescence detector for capillary electrophoresis based on an orthogonal optical arrangement. Talanta **78**(3), 1155–1158 (2009)
12. S. Nagaraj, H. Karnes, Comparison of orthogonal and collinear geometric approaches for design of a laboratory constructed diode laser induced fluorescence detector for capillary electrophoresis. Instrum. Sci. Technol. **28**(2), 119 (2000)

13. A. Malek, M.G. Khaledi, Steroid analysis in single cells by capillary electrophoresis with collinear laser-induced fluorescence detection. Anal. Biochem. **270**(1), 50–58 (1999)
14. L. Hernandez, N. Joshi, E. Murzi, P. Verdeguer, J.C. Mifsud, N. Guzman, Collinear laser-induced fluorescence detector for capillary electrophoresis: analysis of glutamic acid in brain dialysates. J. Chromatogr. A **652**(2), 399–405 (1993)
15. J.-L. Fu, Q. Fang, T. Zhang, X.-H. Jin, Z.-L. Fang, Laser-induced fluorescence detection system for microfluidic chips based on an orthogonal optical arrangement. Anal. Chem. **78**(11), 3827–3834 (2006)
16. L. Hernandez, J. Escalona, N. Joshi, N. Guzman, Laser-induced fluorescence and fluorescence microscopy for capillary electrophoresis zone detection. J. Chromatogr. **559**, 183–196 (1991)
17. C. Yan, R. Dadoo, H. Zhao, R.N. Zare, D.J. Rakestraw, Capillary electrochromatography: analysis of polycyclic aromatic hydrocarbons. Anal. Chem. **67**(13), 2026–2029 (1995)
18. J&M Information materials
19. X. Le, C. Scaman, Y. Zhang, J. Zhang, N.J. Dovichi, O. Hindsgaul, M.M. Palcic, Analysis by capillary electrophoresis-laser-induced fluorescence detection of oligosaccharides produced from enzyme reactions. J. Chromatogr. A **716**(1–2), 215–220 (1995)
20. H.A. Bardelmeijer, H. Lingeman, C. de Ruiter, W.J.M. Underberg, Derivatization in capillary electrophoresis. J. Chromatogr. A **807**, 3–26 (1998)
21. K. Swinney, J. Pennington, D.J. Bornhop, Ion analysis using capillary electrophoresis with refractive index detection. Microchem. J. **62**(1), 154–163 (1999)
22. B. Krattiger, G.J.M. Bruin, A.E. Bruno, Hologram-based refractive index detector for capillary electrophoresis: separation of metal ions. Anal. Chem. **66**(1), 1–8 (1994)
23. M. Albin, R. Weinberger, E. Sapp, S. Moring, Fluorescence detection in capillary electrophoresis: evaluation of derivatizing reagents and techniques. Anal. Chem. **63**, 417–422 (1991)
24. H. Zhu, I.M. White, J.D. Suter, M. Zourob, X. Fan, Integrated refractive index optical ring resonator detector for capillary electrophoresis. Anal. Chem. **79**(3), 930–937 (2006)
25. T. Kappes, P.C. Hauser, Electrochemical detection methods in capillary electrophoresis and applications to inorganic species. J. Chromatogr. A **834**(1–2), 89–101 (1999)
26. B.L. De Backer, L.J. Nagels, Potentiometric detection for capillary electrophoresis: determination of organic acids. Anal. Chem. **68**(24), 4441–4445 (1996)
27. R.M. Guijt, C.J. Evenhuis, M. Macka, P.R. Haddad, Conductivity detection for conventional and miniaturised capillary electrophoresis systems. Electrophoresis **25**(23–24), 4032–4057 (2004)
28. X. Huang, T.K.J. Pang, M.J. Gordon, R.N. Zare, On-column conductivity detector for capillary zone electrophoresis. Anal. Chem. **59**(23), 2747–2749 (1987)
29. S. Kar, P.K. Dasgupta, H. Liu, H. Hwang, Computer-interfaced bipolar pulse conductivity detector for capillary systems. Anal. Chem. **66**(15), 2537–2543 (1994)
30. P.K. Dasgupta, L. Bao, Suppressed conductometric capillary electrophoresis separation systems. Anal. Chem. **65**(8), 1003–1011 (1993)
31. B. Graß, D. Siepe, A. Neyer, R. Hergenröder, Comparison of different conductivity detector geometries on an isotachophoresis PMMA-microchip. Fresenius J. Anal. Chem. **371**(2), 228–233 (2001)
32. M. Galloway, W. Stryjewski, A. Henry, S.M. Ford, S. Llopis, R.L. McCarley, S. Soper, A Contact conductivity detection in poly(methyl methacylate)-based microfluidic devices for analysis of mono- and polyanionic molecules. Anal. Chem. **74**(10), 2407–2415 (2002)
33. S. Hu, Z.-L. Wang, P.-B. Li, J.-K. Cheng, Amperometric detection in capillary electrophoresis with an etched joint. Anal. Chem. **69**(2), 264–267 (1997)
34. W.R. Vandaveer, S.A. Pasas, R.S. Martin, S.M. Lunte, Recent developments in amperometric detection for microchip capillary electrophoresis. Electrophoresis **23**(21), 3667–3677 (2002)
35. M. Nosowicz (2008) Techniki sprzężone w zminiaturyzowanych układach analitycznych. Dissertation, Nicolaus Copernicus University, Toruń

36. G. Choudhary, A. Apffel, H. Yin, W. Hancock, Use of on-line mass spectrometric detection in capillary electrochromatography. J. Chromatogr. A **887**, 85–101 (2000)
37. K. Albert, *On-line LC-NMR and related techniques* (Wiley, Chichester, 2002)
38. D.A. Jayawickrama, J.V. Sweedler, Hyphenation of capillary separations with nuclear magnetic resonance spectroscopy. J. Chromatogr. A **1000**, 819–840 (2003)
39. P. Kubáň, H.T.A. Nguyen, M. Macka, P.R. Haddad, Hauser PC new fully portable instrument for the versatile determination of cations and anions by capillary electrophoresis with contactless conductivity detection. Electroanalysis **19**(19–20), 2059–2065 (2007)
40. M. Ryvolová, M. Macka, M. Ryvolová, J. Preisler, M. Macka, Portable capillary-based (non-chip) capillary electrophoresis. TrAC, Trends Anal. Chem. **29**(4), 339–353 (2010)
41. A. Seiman, J. Martin, M. Vaher, M. Kaljurand, A portable capillary electropherograph equipped with a cross-sampler and a contactless-conductivity detector for the detection of the degradation products of chemical warfare agents in soil extracts. Electrophoresis **30**(3), 507–514 (2009)
42. A.M. Skelley, J.R. Scherer, A.D. Aubrey, W.H. Grover, R.H.C. Ivester, P. Ehrenfreund, F.J. Grunthaner, J.L. Bada, R.A. Mathies, Development and evaluation of a microdevice for amino acid biomarker detection and analysis on Mars. Proc. Natl. Acad. Sci. U. S. A. **102**(4), 1041–1046 (2005)
43. C. Berg, D.C. Valdez, P. Bergeron, M.F. Mora, C.D. Garcia, A. Ayon, Lab-on-a-robot: Integrated microchip CE, power supply, electrochemical detector, wireless unit, and mobile platform. Electrophoresis **29**(24), 4914–4921 (2008)

Chapter 4
Qualitative and Quantitative Analysis: Interpretation of Electropherograms

Michał Szumski and Bogusław Buszewski

Abstract In this chapter the basic information on qualitative and quantitative analysis in CE is provided. Migration time and spectral data are described as the most important parameters used for identification of compounds. The parameters that negatively influence qualitative analysis are briefly mentioned. In the quantitative analysis section the external standard and internal standard calibration methods are described. Variables influencing peak height and peak area in capillary electrophoresis are briefly summarized. Also, a discussion on electrodisperssion and its influence on a observed peak shape is provided.

Capillary electrophoresis, similar to chromatographic techniques, allows for identification of the separated analytes as well as for determination of their quantity in the analyzed sample [1, 2]. Qualitative and quantitative analysis protocols do not differ significantly from those used in chromatography. A migration time and such parameters as peak height or peak surface area are used. CE is also a comparative technique, which means that qualitative and quantitative data can be obtained from the comparison of the analyte peak parameters with the peak of a standard—for identification migration time can be used, and for quantitative analysis—the peak height and peak area.

4.1 Qualitative Analysis

As mentioned above, capillary electrophoresis itself does not provide direct identification of a substance. A basic parameter used here is migration time (t_M) of the analyte. Identification relies on a comparison of the migration time of a peak

M. Szumski (✉) · B. Buszewski
Faculty of Chemistry, Nicolaus Copernicus University, Toruń, Poland
e-mail: michu@chem.umk.pl

B. Buszewski et al. (eds.), *Electromigration Techniques*, Springer Series in Chemical Physics 105, DOI: 10.1007/978-3-642-35043-6_4,
© Springer-Verlag Berlin Heidelberg 2013

obtained for a standard with a migration time of a peak of unknown substance, assuming that both substances underwent electrophoresis under the same conditions. Of course, such comparison only allows the analyst to state with a high probability that unknown substance is identical with the standard. In the next step of identification data obtained from different detectors may be helpful. Diode array detectors (DAD) can take the spectrum in the range of UV–Vis (typically 190–800 nm). The shape of a given spectrum, characteristic minima and maxima, can allow for identification. The spectra of different substances can be gathered in a library (a set of spectra of standards, taken in specific conditions) to be used later by a CE software. The libraries can be purchased from different suppliers, however, they can also be self-made by analysts who possess high standards of purity. There is a serious drawback of the purchased libraries: most of them are dedicated to HPLC, so the spectra are probably taken in organic solvents. The advantage of a self-made library is the full knowledge of the solvent (buffer used, its concentration, contribution of the organic solvent of another additives, like surfactants or polymers) which may have some influence on the spectrum shape. DAD is also helpful in determination of whether the peak represents only one substance or not. In such a case the peak purity is measured, which relies on comparison of spectra taken in different places of a peak (for example at its front, maximum and tail). The identical spectra can prove that the peak represents one substance.

The identification can be provided (again with some high probability) by mass spectrometer provided the measured substance is within its working range.

CE-NMR would identify the substance unequivocally, however, such coupling is rather complicated and has little practical significance in routine analysis.

4.1.1 Sources of Errors

Qualitative analysis can be negatively affected by parameters that influence migration of ions and electroosmosis. These are: voltage, capillary conditioning, modifications of inner capillary wall, buffer exchange, stability of buffer composition during separation, pH and concentration of the buffer, and temperature of separation [3]. If the migration time and/or elctroosmotic flow velocity are irreproducible, an additional marker, not present in the sample, can be added. Then, relative, corrected migration times can be used—they can be calculated by dividing analyte migration time by marker migration time (or opposite). When using the capillaries whose internal surface has been coated with additives like polymers, EOF should be checked from time to time to evaluate coating stability.

4.2 Quantitative Analysis

In capillary electrophoresis quantitative analysis is performed in the same way as in chromatographic methods, i.e., using an external or internal standard method. In both methods the standards of all analytes to be determined are necessary.

4.2.1 External Standard Method (Calibration Curve)

In this method a calibration plot (calibration curve) has to be prepared using the results of measurements of a series of standards of known concentrations. The plot represents the detector response upon changing the concentration of the analyte. The peak parameters used are peak area (S) and height (h). Several rules should be followed when preparing a calibration plot [1–3]:

- Standard solutions have to be prepared for each concentration separately by weighing or pipetting of the substance and dissolving each of the weighed amount in a separate volumetric flask. The solutions should not rather be prepared by dilution of a stock solution of high concentration. For each of the analytes separate series of solutions have to be prepared.
- The concentration of the standard solutions should be expressed in the same units as concentration of the analyte of interest.
- The range of the concentrations of standard solutions should cover the expected concentration of the analyte of interest.
- The calibration plot should be prepared using at least five concentrations and for each concentration the measurement should be done in triplicate.

Ideally, the calibration plot should intersect the origin. If it is otherwise it may be a result of:

- Standard solutions prepared carelessly;
- Wrong peak integration (wrong peak area or height);
- Presence of an interfering substance (e.g. impurity) migrating together with the analyte zone;
- Partial evaporation of a solvent from the standard solution (concentration increased);
- Decomposition of a standard;
- Irreproducible injection volume for different standards;
- Standard solutions concentrations exceed dynamic range of the detector used.

One-point calibration may be used in some cases. Here, the concentration of the standard solution should be close to the expected concentration of the analyte being determined and, importantly, should be within linearity range of the calibration plot (Fig. 4.1). One-point calibration can be used, for e.g., to control some industrial processes, when the expected concentration (or concentration range) is known and predictable.

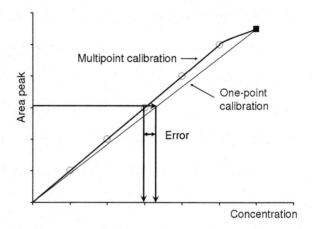

Fig. 4.1 Multipoint and one-point calibration. The plot shows the error that may occur if the one-point calibration outreaches the linearity range. According to [3]

Unknown concentration of the analyte can be read directly from the plot or calculated from the following equation:

$$RF = \frac{c_{std}}{R_{std}} \tag{4.1}$$

$$c_x = RF \cdot R_x \tag{4.2}$$

where RF is response factor, c_{std} is concentration of a standard, R_{std} is response of the detector to the standard concentration c_{std}, c_x is unknown concentration of the analyte, and R_x is response to the unknown concentration of the analyte. A peak area or a peak height can be used as the response, R, for the standard and the analyte.

4.2.2 Internal Standard Method

Internal standard (*IS*) method can be applied when problems with reproducibility of injection or any other problems influencing reproducibility of peak surface area or height occur. In the IS method the compound (called internal standard) of the properties similar to the analytes, but normally not present in the sample, is added to the sample before sample preparation step. Due to this all the possible errors made during sample preparation (extraction, purification, evaporation, etc.) influence all the compounds to the same degree. However, addition of the internal standard does not correct the irreproducibility of S an h during consecutive analyses (for example when temperature is not stable). The following prerequisites must be fulfilled to use the IS method effectively:

- IS must not be present in none of the analyzed samples or used in any step of the sample treatment;
- IS must be well resolved from other analytes and impurities;

- IS should migrate or be eluted close to the analytes to be determined; it should appear in the free space between the peaks of the analytes of interest;
- IS should be available in high purity;
- IS should be stable and should not react with the material of the capillary, buffer constituents or sample constituents;
- the detector response should be similar for the IS and the determined analytes (for example, their UV–Vis spectra or absorption coefficients should be similar);
- chemical properties of the IS should be similar to the analytes, and if, for example, the analytes are derivatized the IS should react in the same way;
- the electrophoretic mobility of the IS should match the mobility of buffer constituents to obtain high peak symmetry;
- the amount of the IS added should allow to obtain ratio of S or h of the IS to the analyte close to the unity.

In the IS method the calibration plot is prepared using the series of standard solutions, but the same amount of the IS is added to each of the standard solutions. Then, on the series of electropherograms IS is represented by the same peak heights or areas, whereas the peaks of the analyte change with their amount in the given standard solution [2, 3]. The calibration plot for the IS method is presented in Fig. 4.2.

The concentration of the analyte can be calculated using the following equation:

$$RF = c_{std} \frac{R_{IS}}{R_{std}} \tag{4.3}$$

$$c_x = RF \frac{R_x}{R_{IS}} \tag{4.4}$$

where R_{IS} is detector response to the IS, R_{std} is detector response to the standard.

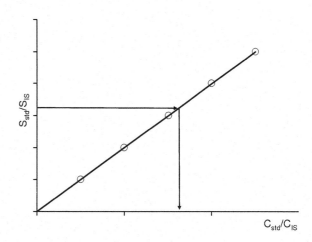

Fig. 4.2 Calibration plot for the internal standard method. According to [3]

4.2.3 Remarks Regarding Peak Shape

In quantitative analysis in CE both peak area and peak height can be used. However, in practice peak surface is rather preferred as it is less sensitive to change of analysis parameters (Table 4.1) and provide broader range of linearity [3].

In chromatographic techniques the analytes eluted from the column move in the detector cell at linear velocity of the mobile phase. It is different in capillary electrophoresis where analytes velocity in the detection cell depend on their electrophoretic mobility. Hence, the "late" peaks (characterized by long migration times) migrate slower, hence their residence time in the detection cell is longer. As a consequence, their peaks are of larger surface area. Each variable influencing the observed migration velocity (for e.g., EOF instability) will influence the reproducibility of peak area of a given peak. As the peak surface is proportional to the migration time one can use "normalized" surface areas which are the quotients of a surface area and migration time of a given peak.

The peak shape in CE can also be influenced by electrodispersion (also referred to in the literature as: electrophoretic dispersion, electrokinetic dispersion, electrodisperssion effects, band broadening caused by conductivity differences, concentration overload) which is a change of a peak shape (broadening) connected with incompatibility of buffer ions mobility (μ_{buf}) and analyte mobility (μ_c) or, in general: connected with differences in conductivity of the buffer and a sample zone. Three cases can be distinguished here (Fig. 4.3) [3, 4]:

- $\mu_c > \mu_{buf}$—the analyte (for e.g., a cation) of the mobility greater than buffer ions generate lower electric field due to higher conductivity. As the cation diffuse it can cross the rear or front boundary of the zone. Whenever it crosses the rear boundary (anodic side of the zone) it becomes exposed to the high electric field (lower conductivity of the electrolyte) and speeds up to the cathode till it gets back to the zone. On the other hand, when the cation crosses the front boundary (cathodic side) it speeds up towards the cathode thus making unstable front boundary and fronted sawtooth peak.
- $\mu_c = \mu_{buf}$—the analyte crossing any boundary will not speed up as there is uniform electric field in the buffer and analyte zone.

Table 4.1 Variables influencing peak height and area

Variable	Make changes in:	
	Height	Surface
Migration velocity (and all parameters influencing it)	No	Yes
Injected amount of analyte	Yes	Yes
Peak shape	Yes	No
Preconcentration in capillary (*stacking*)	Yes	No
Detector response	Yes	Yes
Adsorption of the analyte on the inner wall of the capillary	Yes	Yes

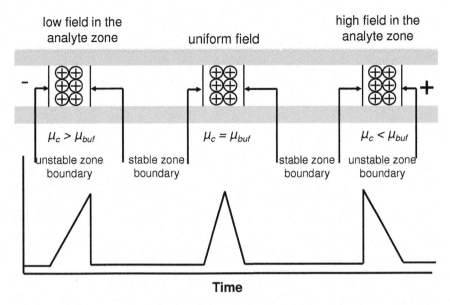

Fig. 4.3 Electrodisperssion and its influence on the peak shape. Reprinted with permission from Heiger [4]. Hewlett-Packard Copyright 1994 Agilent Technologies

- $\mu_c > \mu_{buf}$—in the case that ions are low, when compared to buffer ions, mobility of the rear boundary becomes unstable and tailing. Cations when diffusing across the front boundary are not accelerated because of lower electric field in the buffer hence the front of the peak remains sharp. Diffusion across the rear boundary of the cations decelerate (lower field in the buffer) making the rear boundary unstable and tailing.

The above explanation is rather general, but is in good agreement with most of the obtained results. The problem is described in detail in Ref. [5].

The phenomenon described above is frequently used in some methods of preconcentration (stacking) of the sample in CE. This subject is described in detail in Chap. 11
.

References

1. H.J. Kuss, S. Kromidas (eds.), *Quantification in LC and GC* (Wiley, Weinheim, 2009)
2. W. Wardencki, J. Namieśnik (eds.), *Chromatografia Część VII—Analiza ilościowa, Chemia analityczna* (Politechnika Gdańska, Gdańsk, 2002)
3. D.R. Baker, *Capillary Electrophoresis. Techniques in Analytical Chemistry* (Wiley, New York, 1995)
4. D. Heiger, R. Weinberger (1994) *Determination of small ions by capillary electrophoresis with indirect photometric detection.* Hewlett-Packard Application Note
5. P. Gebauer, P. Bocek, Predicting peak symmetry in capillary zone electrophoresis: the concept of the peak shape diagram. Anal. Chem. **69**(8), 1557–1563 (1997)

Chapter 5
Micellar Electrokinetic Chromatography

Edward Bald and Paweł Kubalczyk

Abstract Since the introduction of micellar electrokinetic chromatography by Terabe, several authors have paid attention to the fundamental characteristics of this separation method. In this chapter the theoretical and practical aspects of resolution optimization, as well as the effect of different separation parameters on the migration behavior are discussed. These among others include fundamentals of separation, retention factor and resolution equation, efficiency, selectivity, and various surfactants and additives. Initial conditions for method development and instrumental approaches such as mass spectrometry detection are also mentioned covering the proposals for overcoming the difficulties arising from the coupling micellar electrokinetic chromatography with mass spectrometry detection.

5.1 Introduction

Capillary Electrophoresis (CE), described and implemented in the late 1970s of the twentieth century as a modern and highly efficient analytical technique, allowed the separation of charged analytes in small amounts quickly and successfully. Although, in comparison to HPLC, it was characterized by higher resolution obtained in a shorter period of time, it could not be applied to separate neutral analytes. The above-mentioned fact imposed a limitation to the applicability of CE until 1984 when Prof. Terabe developed micellar electrokinetic chromatography (MEKC). Professor Terabe admits that the idea of using ionic micelles in CE

E. Bald (✉) · P. Kubalczyk
Faculty of Chemistry, University of Łódź, Łódź, Poland
e-mail: ebald@uni.lodz.pl

B. Buszewski et al. (eds.), *Electromigration Techniques*, Springer Series
in Chemical Physics 105, DOI: 10.1007/978-3-642-35043-6_5,
© Springer-Verlag Berlin Heidelberg 2013

stemmed from Nakagawa's suggestion he had reached 3 years earlier [1]. A neutral analyte obtains apparent electrophoretic mobility when it interacts with an ionic micelle and migrates in electric field with the same velocity as the micelle. Because the analyte distribution equilibrium between the micelle and the surrounding aqueous phase is quickly reached, the migration velocity is determined by the distribution coefficient.

The idea of MEKC is easy to implement. Namely, a surfactant, most often sodium dodecyl sulfate (SDS), is added to a background electrolyte (BGE) used in capillary zone electrophoresis (CZE) in the amount that is sufficient to create micelles, with no need to modify an apparatus. These micelles are called the pseudo-stationary phase as they play the role of the stationary phase in chromatography. Although they migrate inside the capillary tube, they only exist in equilibrium with a solution containing surfactant molecules and cannot be separated. The separation of neutral analytes was initially based on ionic micelles (MEKC). Later, other pseudo-stationary phases (e.g. micro-emulsions) were implemented. All the electrophoretic techniques that involve the use of pseudo-stationary phases [2] belong to (Fig. 5.1) electrokinetic chromatography (EKC).

The fundamentals of MEKC were discussed in numerous papers and handbooks [3–13]. The technique was developed for neutral analytes but soon it was found that it could be practically applied in the selective separation of all substances including the charged ones. For the past 26 years (since MEKC was devised), it has been studied in as many as 3,000 papers.

5.2 Principles of Separation

Micellar EKC (MEKC) is most frequently performed with the use of anionic surfactants, the most popular of which is sodium dodecyl sulphate (SDS). These substances, when added to water in appropriate amounts, create micelles. In the micelles, internally oriented, hydrophobic parts constitute the core while externally oriented hydrophilic parts are in water medium. The MEKC system is composed of two phases: the aqueous phase and the micellar phase also called the pseudo-stationary phase. SDS micelles are of significant negative charge, and thus, exhibit adequate mobility (μ_{ep}) towards the anode in the direction opposite to the electro-osmotic flow (μ_{eo}) in the majority of buffers used in CE. As in neutral and alkaline medium EOF is higher than the electrophoretic migration of micelles, the latter will migrate to the cathode at speed lower than EOF. When migrating, the micelles interact with analytes hydrophobically and electrostatically. The stronger the interaction of the analyte with the micelle, the longer the time of its migration. It results from the fact that the micelle decelerates the analyte migration with EOF. Analytes which do not interact with micelles are carried with EOF. A separation of sample components is shown in Fig. 5.2.

Therefore, MEKC is a way to selectively separate neutral and ionic compounds and retains all the advantages offered by CZE. The migration in MEKC is

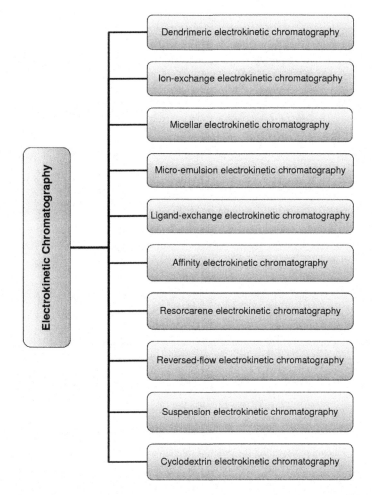

Fig. 5.1 Classification of electrokinetic chromatography based on a kind of pseudo-stationary phase

determined mainly by hydrophobicity but the final result, i.e., analytes separation, is influenced by the charge to mass ratio, hydrophobicity, and the interaction between charges.

The migration time of the analyte (t_R) which interacts with the micelle must be in the range between the migration time of the substance that interacts with the micelle only slightly or no interaction occurs (t_0), and the migration time of the substance that is fully incorporated into the micelle (t_{mc}). Times t_0 and t_{mc} are determined with the use of markers. Methanol can serve as a t_0 marker while Sudan III, Sudan IV, or quinine sulphate can be t_{mc} markers. An electrophorerogram presenting a so-called MEKC time window is shown in Fig. 5.3.

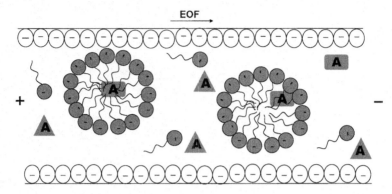

Fig. 5.2 Sample components separation by MEKC using anionic micelles where A = analyte, according to [14]

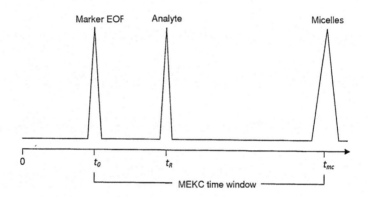

Fig. 5.3 A scheme representing the order of migrations for neutral analytes in MEKC

The necessity to elute the analyte in the time between t_0 and t_{mc} markedly differentiates MEKC from chromatography. In conventional chromatography, the retention coefficient (k) is expressed by the formula:

$$k = \frac{t_R - t_0}{t_0} \tag{5.1}$$

describes the ratio of a number of moles of a component in the stationary phase to its number of moles in the mobile phase. In MEKC, neutral molecules are divided between two moving phases. Hence, the expression of (k) requires modification. Terabe [1] proposed the equation for the retention coefficient in MEKC:

$$k = \frac{t_R - t_0}{t_0(1 - t_R/t_{mc})} \tag{5.2}$$

If t_{mc} approaches infinity, then the expression of (k) becomes identical to the expression of (k) in conventional chromatography; the solid pseudo-phase becomes the solid phase. Consequently, the equation for separation in MEKC takes the form [15]:

$$R_s = \frac{\sqrt{N}}{4}\left(\frac{\alpha-1}{\alpha}\right)\left(\frac{k_2}{1+k_2}\right)\left(\frac{1-(t_0/t_{mc})}{1+(t_0/t_{mc})k_1}\right) \tag{5.3}$$

where N is number of theoretical plates; α is selectivity coefficient equal to k_2/k_1 (from the assumption >1); k_2, k_1 are retention coefficient of 1 and 2 analytes.

The last factor in the above equation is ascribed to the change in effective length of a capillary tube applied in the separation process. Inside the capillary tube, micelles move during the separation process. It results in the fact that the length of the zone in which micelles interact with analytes is shorter than the physical length of the capillary tube. The difference in the lengths is equal to the distance which micelles will pass in time t_R.

5.2.1 Efficiency

In MEKC, capillary tube efficiency is high. When expressed in a number of theoretical plates, it amounts to between 100 and 200 thousands. The diffusion along the capillary tube is the most important determinant of the theoretical plate height, according to formula [10, 16]:

$$H_l = \frac{2\left(D_{aq}+kD_{mc}\right)}{1+(t_0/t_{mc})}\frac{1}{v_{eo}} \tag{5.4}$$

where H_l is the component of the plate height generated by longitudinal diffusion, k- retention coefficient, D_{aq} is diffusion coefficient of solute dissolved in liquid phase, D_{mc} is diffusion coefficient of solute dissolved in micellar phase, t_0 is migration time of the substance that does not interact with the micelle, t_{mc} is migration time of substance completely dissolved in micelle, v_{eo} is the velocity of electro-osmotic flow.

Because D_{mc} of a micelle is one order of magnitude smaller than the diffusion coefficient of a small molecule (analyte), higher N can be expected for the analytes of a higher retention coefficient, which means stronger interacting with micelles. In the case of CZE, the situation is reverse; the number of theoretical plates decreases together with an increase in retention coefficient. Another factor that should be taken into consideration is the length of a sample zone in a capillary tube. High efficiency is obtained if the length of a sample zone does not exceed 1 % of the capillary length unless stacking occurs in the capillary tube. The influence of temperature on efficiency is negligible on the condition that current does not exceed 50 µA. A potential adsorption of sample components to the capillary walls constitutes a threat to high efficiency.

5.2.2 Selectivity

The selectivity coefficient, analogously to HPLC, is an important parameter during separation optimization. As a result, while developing the MEKC analytical method, the choice of a surfactant should be carefully made. The initial experiments are usually carried out with the use of SDS. In the case of a failure, other anionic and cationic surfactants as well as bile acids salts can be applied. Introducing additives such as organic solvents or cyclodextrins supports selectivity optimization. Owing to the high MEKC efficiency, it is assumed that a selectivity coefficient higher than 1.02 assures good separation.

5.2.3 Retention Coefficient

The retention coefficient exists in the third and fourth elements of the above-mentioned equation for resolution. Its optimum value (k_{opt}) giving the biggest product of these components is represented by the formula [17]

$$k_{opt} = \sqrt{\frac{t_{mc}}{t_0}} \qquad (5.5)$$

The t_0 value is easy to determine using methanol, dimethyl sulfoxide, or other compound that does not interact with a micelle. t_{mc} determination can be problematic which results from a difficulty in finding the micelle marker (Sudan III or IV) peak. In such a case, Terabe advises [15] to assume that t_{mc} is four times higher than t_0, and then k_{opt} should be found in the range from 1.7 to 2.0. On the other hand, k is dependent on a surfactant concentration in the form [15]:

$$k = \frac{KV_{mc}}{V_{aq}} \cong K\bar{v}\left(C_{sf} - c_{mc}\right) \qquad (5.6)$$

where K is coefficient of analyte distribution between the micellar phase and the aqueous phase, V_{mc} is volume of micelles, V_{aq} is volume of the aqueous phase, v is specific volume of micelle, C_{sf} is surfactant concentration

The equation suggests an almost proportional dependence between k and the surfactant concentration, which means that the V_{mc}/V_{aq} relation is regulated by the changes in a surfactant concentration. With such an assumption, the optimization of k is an easy task if v (specific volume of micelle) and critical micellar concentration (CMC) are known. If such data are not available, then, assuming a linear dependence between the retention coefficient and surfactant concentration, k can be optimized on the trial-and-error basis by changing the surfactant concentration.

5.3 Factors Influencing the Separation Process

A hybrid nature of micellar electrokinetic chromatography results in the fact that the quality of separation and the final results of quantitative analysis are influenced by more factors than in the case of the simplest and most often applied capillary zone electrophoresis.

5.3.1 Surfactants

Of the large number of commercially available surfactants, only some can be applied practically in MEKC. These compounds, when added to water in amounts slightly exceeding a certain concentration called critical micellar concentration (CMC), create spherical aggregates. In these aggregates, also called micelles, the core is made of hydrophobic fragments of the surfactant structure, whereas the hydrophilic fragments are externally oriented and exist in a water electrolyte. Micelles formation results in lowering the amount of free energy of a system. The micelles are dynamic formations with a lifespan shorter than 10 μs. The process of micelles formation is accompanied by a change in surface tension, viscosity, and the ability to scatter light. Surfactants, and thus, micelles created out of them can be negatively or positively charged. They can migrate according to EOF or in the opposite direction. At the neutral or alkali pH, the EOF flow is usually higher than the migration rate of micelles. Hence, the net migration occurs in the same direction as EOF, i.e., towards the cathode. Due to a number of reasons, SDS is the most popular surfactant in MEKC. Among these reasons are high stability, low absorption in the ultraviolet region, high ability to dissolve, and the commercial availability of the high quality reagent. The reagent CMC amounts to 8 mmol/L in water; for buffer solutions, it is even 3 mmol/L. In most cases, 10–50 mmol/L SDS solutions are applied. Still, higher concentrations, even up to 100 mmol/L, give good results provided current intensity does not exceed 50 μA.

Micelles exhibit the ability to order analytes by hydrophobic and electrophoretic interactions. According to Terabe [10, 18], three kinds of analyte interaction mechanisms with micelles can be differentiated, as presented in Fig. 5.4. These are introducing the analyte to the hydrophobic core, the analyte adsorption on the surface or palisade layer, and implementing the analyte as the co-surfactant. Highly hydrophobic nonpolar compounds such as aromatic hydrocarbons are introduced to the core. In such a case, a slightly positive correlation can be observed between the separation coefficient and the length of an alkyl surfactant chain. Higher selectivity in the separation of those kinds of analytes is obtained using bile acid salts to create the pseudo-stationary phase. It is supposed that the majority of analytes interact with micelles through the surface, palisade layer, and polar groups. Separation selectivity can be significantly improved by applying mixed micelles composed of ionic and non-ionic surfactants (Fig. 5.4b).

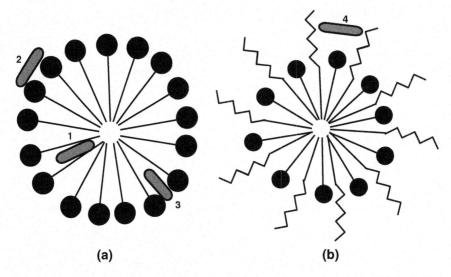

Fig. 5.4 Analytes interactions with micelles: **a** ionic; **b** mixed ionic–non-ionic. *1* Interaction with a hydrophobic core, *2* interaction with the surface, *3* interaction as a co-surfactant, and *4* interaction with the non-ionic surface, according to [16]

In addition to anionic and cationic ones, non-ionic, zwitterion, and mixed micelles are used in MEKC. Some of them are presented in Table 5.1. In practice, the applicability of a surfactant in MEKC is determined by its solubility and CMC. Those of high CMC are inappropriate owing to big amounts of generated current deteriorating separation results. Mixed micelles and additives applied broaden a variety of pseudo-stationary phases that can be applied in MEKC. Surfactants interact with analytes, but they can also adsorb on capillary walls modifying EOF. Depending on a charge, EOF can be increased, reduced, or reversed, which is depicted in Fig. 5.5. The phenomenon occurs at surfactant concentrations below CMC.

Naturally occurring bile acids salts (Fig. 5.6) are also helpful during hydrophobic analytes separation. They function as chiral selectors. A molecular structure of their micellar aggregates markedly differs from the structures of surfactants with long alkyl chains. Hydroxyl groups remaining on one plane cause the micelles to have hydrophobic and hydrophilic regions. Their internal part is less hydrophobic in comparison to the SDS micelle core. They are characterized by a smaller number of aggregation and higher tolerance towards organic modifiers. The critical micellar sodium cholate concentration changes markedly when addition of methanol exceeds 30 % while in SDS such changes occur at the concentration above 10 %.

An increase in the surfactant concentration usually prolongs the migration time of most compounds as the probability of getting in contact with micelles rises. Such a dependence is observed for hydrophobic and electrostatic interactions. Lowering the pH can enable ionic pair formation with negatively charged micelles,

Table 5.1 Some surfactants use in MEKC

Surfactant	
Anionic	
Sodium decyl sulfate	$CH_3(CH_2)_9OSO_3^- Na^+$
Sodium dodecyl sulfate (SDS)	$CH_3(CH_2)_{11}OSO_3^- Na^+$
Sodium tetradecyl sulfate (STS)	$CH_3(CH_2)_{13}OSO_3^- Na^+$
Sodium dodecyl sulfonate	$CH_3(CH_2)_{11}SO_3^- Na^+$
Cationic	
Dodecyltrimethylammonium chloride (DTAC)	$CH_3(CH_2)_{11}Na^+(CH_3)_3Cl^-$
Dodecyltrimethylammonium bromide (DTAB)	$CH_3(CH_2)_{11}Na^+(CH_3)_3Br^-$
Cetyltrimethylammonium chloride (CTAC)	$CH_3(CH_2)_{15}Na^+(CH_3)_3Cl^-$
Cetyltrimethylammonium bromide (CTAB)	$CH_3(CH_2)_{15}Na^+(CH_3)_3Br^-$
Cationic fluorosurfactant (Fluorad FC 134)	$CF_3(CF_2)_7SO_2NH(CH_2)_3N(CH_3)_3I$
Non-ionic and zwitterionic	
Octyl glucoside	
3-[3-(chloroamidopropyl)dimethylammonio]-1-propanesulfonate (CHAPS)	
Chiral	
Sodium N-dodekanoyl-L-valinate (SDVal)	
Bile salts	
Sodium cholate	
Sodium taurocholate	

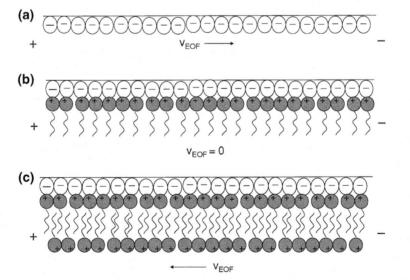

Fig. 5.5 Influence of a cationic surfactant on electroosmotic flow: **a** normal flow with no surfactant added; **b** adsorption of cationic surfactant on internal walls of a capillary which results in stopping EOF; **c** creating a double layer as a result of hydrophobic activity of aliphatic chains which brings about EOF reverse

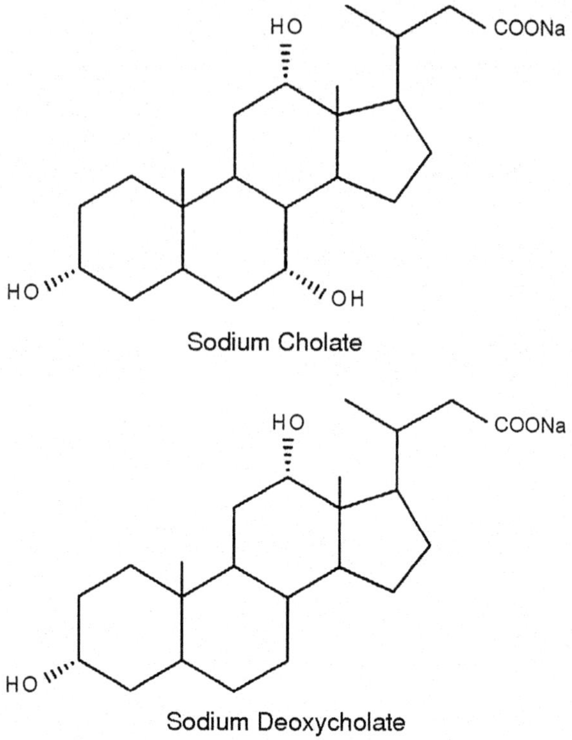

Sodium Cholate

Sodium Deoxycholate

Fig. 5.6 Bile acid salts

which is meaningful during the separation of mixtures of compounds with amino groups. A retention coefficient of a neutral compound rises together with an increase in a surfactant concentration.

5.3.2 Modifiers

Cyclodextrins (CD) are macro-cyclic oligosaccharides obtained by enzymatic starch digestion. They are composed of 6, 7, or 8 glucopyranose units and are called α-, β-, and γ- CD. They are of toroidal shape with a relatively hydrophobic internal part that allows the formation of inclusive complexes with analytes that match the internal CD sizes. The external surface is hydrophilic. The complexes result from hydrogen bonds formation, van der Waals forces activity, or hydrophobic interactions. They, however, influence CD separation only when an analyte size accurately corresponds to their internal sizes. Too big analytes do not form complexes, and in the case of small analytes, the molecular contact with CDs is insufficient for influencing the separation. The phenomenon of inclusion has been applied in a range of separation methods [19, 20].

Cyclodextrins are added to the BGE in MEKC in order to weaken the interactions with micelles. Consequently, the separation of strong hydrophobic analytes is improved. Adding CDs can change selectivity mainly in relation to aromatic isomers. CDs possess an optically active hydrocarbon structure and, when added, they enable enantiomers (particularly those neutral ones) separation. In practice, the most commonly used ones are β-CD and γ-CD at concentrations ranging between 10 and 40 mmol/L. A surfactant molecule can get inside CD causing a change in an inclusive complexes formation constant between an analyte and a CD.

Organic modifiers that mix with water are useful in MEKC, similar to HPLC. Adding an organic solvent reduces EOF flow, and therefore, elution window is extended. A more important role the organic solvent plays is based on a change in an analyte distribution coefficient between the micelle and the BGE. A modifier makes the aqueous phase more "friendly" to hydrophobic analytes, and so the migration time increases.

In MEKC, a number of solvents are used. Most often methanol and acetonitrile are added in amounts between 5 and 25 %. Aprotic solvents such as tetrahydrofurane, dimethyl sulfoxane, dimethylformamide are less commonly used. The amount of a modifier added is limited by its influence on the number of aggregation as well as micelles dissociation. When a modifier concentration exceeds certain value, the separation which proceeds according to the MEKC mechanism changes into the CZE separation mechanism.

Reagents used to form ionic pairs in MEKC act in a way similar to that in reversed-phase HPLC [21]. The addition of tetraalkylammonium salts (cationic ion pair reagents) improves the ability to separate many ionic analytes by shortening cations migration times and prolonging anions migration. It suggests the possibility of forming ion pairs with anions which leads to their more efficient incorporation to the micellar phase.

Urea influences the micelles formation [22, 23] and, when added, improves separation selectivity in MEKC [24], reduces the EOF and migration time of the micelle. In 8 mmol/L urea solution, viscosity increases by 66 % and current drops by three times. Retention coefficient is reduced for most analytes.

5.3.3 pH

The pH value influences electrokinetic velocities and ionizable compounds charges. SDS charge and electrophoretic micelles flow (v_{ep}) do not change together with the change in the pH. EOF changes with the change in the pH, and consequently, micelles migration velocity (v_{mc}) is changed [25]. At pH 5, the net velocity of micelles approximates zero. At pH below 5, the direction of the micelles flow is reversed because electrophoretic micelles flow is higher than that of EOF. Analytes separation in solutions with pH $\approx$ 5 is not advised due to low reproducibility of retention times. At pH $\approx$ 3 hydrophobic analytes and ionic pairs formed reach the detector before other components of an analyzed mixture do.

The pH influence on an analyte charge is essential and should be considered when designing method selectivity. At a low pH value, amino compounds will undergo protonating, bringing about strong affinity towards micelles. Analytes of weak acids character will then undergo ionic suppression.

5.3.4 Elution Order

In MEKC analysis with the use of SDS micelles, a highly probable elution order is as follows: anions–neutral compounds–cations. Anions spend more time in the electrolyte phase mainly as a result of their repulsion by anionic micelles. The higher the negative charge of an ion, the shorter the retention time. Neutral analytes incorporated into micelles as a result of hydrophobic interactions with their cores elute in the secondary order. Cations keep their tendency to elute as the last ones mainly due to electrostatic attraction. Ionic pairs with micelles are formed which leads to a generalization: the higher the positive charge, the higher the retention.

The above rules may not apply to particular analytical tasks. Strong hydrophobic interactions may overcome electrostatic attraction or repulsion. As far as non-homologous structure analyte mixtures (e.g., drugs mixtures) are concerned, the elution order in MEKC can be significantly changed with no obvious correlation between migration rate and a charge to mass relation or hydrophobicity.

5.4 Equipment and Analytical Procedure

MEKC does not require any apparatus modification usually applied in CE. A capillary tube of dimensions typical to CE (internal diameter 50 or 75 μm), made of fused silica is filled with a buffer containing a dissolved ionic surfactant. Phosphate, borate, and Tris buffers at the concentration of 20–50 mmol/L are most commonly used. Regarding the pH, it should be remembered that electro-osmotic flow (EOF) almost completely ceases at the pH value below 2; above 7, it is fast and stable. Together with a rise in the buffer concentration, its buffer capacity also grows. Moreover, there is a risk of generating significant amounts of current and Joule heat with all the negative consequences for quality of separation. For MEKC, the risk is particularly big owing to the presence of a surfactant in the ionic buffer. It is advised to avoid using buffers with potassium because it can lead to the surfactant precipitation in the form of a potassium salt after the ionic exchange with the buffer. Current obtained during the MEKC separation can be higher than that obtained during CZE. It results in the need to refill an electrolyte in order to prevent changes in the pH. If cationic surfactants such as CTAB, which cause a change in EOF are used, the capillary regeneration has to be done with much care.

In an anodic vial, CTAB undergoes electrolysis, and bromine which contaminates the electrolyte is evolved. It should be remembered when refilling the electrolyte.

5.4.1 Detection

Detection in MEKC does not differ much from that applied in CZE. UV–Vis, fluorescence, and electrochemical detectors are used. The problems that emerge while connecting a mass detector (MS) to CZE are thus more real in the case of MEKC in which problems result from the presence of naturally non-volatile surfactants in the BGE. Surfactants and strong electrolytes pollute and lower the ionizing ability of interfaces. To omit the problem, a separation with micelles migrating in the opposite direction or volatile surfactants was carried out.

In practice, it occurred that the MEKC-MS procedure can be successfully applied if a non-volatile surfactant and the BGE concentrations do not exceed 20 and 10 mmol/L, respectively [26, 27]. Another good solution is to implement fotoionization at the atmospheric pressure, resistant to the surfactant concentration even up to 50 mmol/L [28].

5.4.2 Method Development

Terabe [3] gives initial operating conditions (Table 5.2) to start with when elaborating a separation method for neutral or slightly ionizable analytes by means of MEKC. As he claims, for the BGE, a slightly alkali borate buffer should be chosen as it enables keeping small current due to low electrophoretic borate ion mobility. The sample can be dissolved in any solvent that mixes with water. In the case of sample preconcentration in a capillary tube, the sample solution must be prepared according to the sample preconcentration technique requirements (stacking,

Table 5.2 Initial conditions of the analysis by MEKC, according to [3]

Capillary	50–70 µm × 20–50 cm, standard, uncoated
BGE	50 mmol/L SDS in 50 mmol/L borate buffer (pH 8.5–9.0) or 50 mmol/L SDS in 50 mmol/L phosphate buffer (pH 7.0) (sodium salt)
Voltage	10–25 kV (current below 50 µA)
Temperature	25 °C or ambient
Solvent of sample	Water, methanol or other solvent miscible with water
Sample concentration	0.1–1.0 mg/mL (or lower detectable concentration)
Sample introduction	Hydrodynamically on the anodic capillary side
Volume of sample introduced	Less than 1 % of the capillary length
Detection	UV–Vis

sweeping). In the first experiments, relatively high concentrations of standard analytes solutions should be used in order to make their detection easier. Migration times depend on EOF and the effective capillary tube length. When the separation is not satisfactory after the first attempt, selectivity should be manipulated by making changes in a series of parameters. The first step is to determine retention coefficients using time window parameters. If k is higher than 10, micelles concentration should be lowered, methanol or cyclodextrin added, or other surfactant applied. When k is too low, a surfactant concentration should be increased for a neutral analyte, an ionic pair reagent added, or a surfactant exchanged with a cationic one for anionic analytes.

Detailed directions concerning selectivity manipulation on the basis of migration behaviors are described in literature [8]. For the resolution optimization process in MEKC, computer modeling based on physicochemical models describing migration behaviors [29–31] can be applied; in the chemometric approach, polynomial equations can be useful [32–36].

5.5 Summary

As a method, MEKC turned out to be more universal. It also showed its higher capability of selective separation when compared to CZE. Nevertheless, it tends to exhibit low concentration sensitivity of detection, which is a common feature of all the capillary electromigration techniques. These inconveniences are minimized, however, by the development of various techniques of analyte concentration in an electrophoretic system and outside it, as well as sensitive detection methods. Preconcentration techniques and detectors used in electrophoresis are discussed in other chapters of the book. Also, comments on some papers selected from more than 2,000 research works on MEKC applications can be found. The method is employed in, e.g., pharmaceutical analysis, bio-analysis, environmental, or food analysis.

Intensive studies on the improvement of reproducibility and repeatability of migration times as well as peak heights and areas are being carried out. Providing stable EOF, hence significant in almost all CE modes, requires even more attention in the case of MEKC as its magnitude is determined by a greater number of parameters.

MEKC is comparable with HPLC in the reversed phase system, but in MEKC, there is a broader range of factors influencing separation selectivity. Green chemistry enthusiasts emphasize smaller amounts of reagents as well as organic solvents used. MEKC requires a small sample volume (a few nanolitres), however, to introduce such a volume into a capillary tube, at least a few microliters of a sample solution in a conical-shaped vial are required.

References

1. S. Terabe, K. Otsuka, K. Ichikawa, A. Tsuchiya, T. Ando, Electrokinetic separations with micellar solutions and open-tubular capillaries. Anal. Chem. **56**, 111–113 (1984)
2. S. Terabe, Electrokinetic chromatography: An interface between electrophoresis and chromatography. Trends Anal. Chem. **8**, 129–134 (1989)
3. S. Terabe, *Micellar Electrokinetic Chromatography* (Beckman, Fullerton, 1992)
4. J. Vindevogel, P. Sandra, *Introduction to Micellar Electrokinetic Chromatography* (A. Hüthig, Heidelberg, 1992)
5. U. Pyell (ed) (2006) Electrokinetic chromatography. Theory, instrumentation and applications. Wiley, Chichester
6. S. Terabe, N. Chen, K. Otsuka, Micellar electrokinetic chromatography. Adv. Electrophoresis **7**, 87–153 (1994)
7. M.G. Khaledi, Micellar electrokinetic chromatography, in *High Performance Capillary Electrophoresis*, ed. by M.G. Khaledi (Wiley-Interscience, New York, 1998)
8. S. Terabe, Selectivity manipulation in micellar electrokinetic chromatography. J. Pharm. Biomed. Anal. **10**, 705–715 (1992)
9. K. Otsuka, S. Terabe, Micellar electrokinetic chromatography. Bull. Chem. Soc. Jpn. **71**, 2465–2481 (1998)
10. S. Terabe (2008) Micellar electrokinetic chromatography, ed. by J.P. Landers handbook of capillary and microchip electrophoresis and associated microtechniques, 3rd ed., CRC Press, Taylor and Francis Group, Boca Raton, London, New York
11. J.R. Mazzeo, Micellar electrokinetic chromatography, in *Handbook of Capillary Electrophoresis*, 2nd edn., ed. by J.P. Landers (CRC Press, Boca Raton, 1996)
12. M. Silva, MEKC: An update focusing on practical aspects. Electrophoresis **28**, 174–192 (2007)
13. S.F.Y. Li (ed) (1993) Capillary electrophoresis. In: J. Chromatography Library. vol. 52, Elsevier, Amsterdam
14. Technical Bulletin in Polish/Hewlett-Packard (1997) 12-5965-5984E
15. S. Terabe, K. Otsuka, T. Ando, Electrokinetic chromatography with micellar solution and open-tubular capillary. Anal. Chem. **57**, 834–841 (1985)
16. S. Terabe, K. Otsuka, T. Ando, Band broadening in electrokinetic chromatography with micellar solutions and open-tubular capillaries. Anal. Chem. **61**, 251–260 (1989)
17. J.P. Foley, Optimization of micellar electrokinetic chromatography. Anal. Chem. **62**, 1302–1308 (1990)
18. S. Terabe, Micellar electrokinetic chromatography. Anal. Chem. **76**, 240A–246A (2004)
19. C.P. Ong, C.L. Ng, N.C. Chong, H.K. Lee, S.F.Y. Li, Analysis of priority substituted phenols by micellar electrokinetic chromatography. Environ. Monitor Assess **19**, 93–103 (1991)
20. S.K. Yeo, C.P. Ong, S.F.Y. Li, Optimization of high-performance capillary electrophoresis of plant growth regulators using the overlapping resolution mapping scheme. Anal. Chem. **63**, 2222–2225 (1991)
21. N. Nishi, N. Tsumagari, S. Terabe, Effect of tetraalkylammonium salts on micellar electrokinetic chromatography of ionic substances. Anal. Chem. **61**, 2434–2439 (1989)
22. P. Mukerjee, A. Ray, The effect of urea on micelle formation and hydrophobic bonding. J. Phys. Chem. **67**, 190–192 (1963)
23. M.J. Schick, Effect of electrolyte and urea on micelle formation. J. Phys. Chem. **68**, 3585–3592 (1964)
24. S. Terabe, Y. Ishihama, H. Nishi, T. Fukuyama, K. Otsuka, Effect of urea addition in micellar electrokinetic chromatography. J. Chromatogr. **545**, 359–368 (1991)
25. K. Otsuka, S. Terabe, Effects of pH on electrokinetic velocities in micellar electrokinetic chromatography. J. Microcol. Sep. **1**, 150–154 (1989)

26. G.W. Somsen, R. Mol, G.J. de Jong, On-line micellar electrokinetic chromatography–mass spectrometry: Feasibility of direct introduction of non-volatile buffer and surfactant into the electrospray interface. J. Chromatogr. A **1000**, 953–961 (2003)
27. R. Mol, E. Kragt, I. Jimidar, G.J. de Jong, G.W. Somsen, Micellar electrokinetic chromatography–electrospray ionization mass spectrometry for the identification of drug impurities. J. Chromatogr. B **843**, 283–288 (2006)
28. R. Mol, G.J. de Jong, G.W. Somsen, Atmospheric pressure photoionization for enhanced compatibility in on-line micellar electrokinetic chromatography–mass spectrometry. Anal. Chem. **77**, 5277–5282 (2005)
29. M.G. Khaledi, S.C. Smith, J.K. Strasters, Micellar electrokinetic capillary chromatography of acidic solutes: Migration behavior and optimization strategies. Anal. Chem. **63**, 1820–1830 (1991)
30. J.K. Strasters, M.G. Khaledi, Migration behavior of cationic solutes in micellar electrokinetic capillary chromatography. Anal. Chem. **63**, 2503–2508 (1991)
31. C. Quang, J.K. Strasters, M.G. Khaledi, Computer-assisted modeling, prediction, and multifactor optimization in micellar electrokinetic chromatography of ionizable compounds. Anal. Chem. **66**, 1646–1653 (1994)
32. J. Vindevogel, P. Sandra, Resolution optimization in micellar electrokinetic chromatography: Use of Plackett-Burman statistical design for the analysis of testosterone esters. Anal. Chem. **63**, 1530–1536 (1991)
33. Y. He, H.K. Lee, Orthogonal array design experiments for optimizing the separation of various pesticides by cyclodextrin-modified micellar electrokinetic chromatography. J. Chromatogr. A **793**, 331–340 (1998)
34. K. Persson-Stubberud, O. Åström, Separation of ibuprofen, codeine phosphate, their degradation products and impurities by capillary electrophoresis: I method development and optimization with fractional factorial design. J. Chromatogr. A **798**, 307–314 (1998)
35. S. Mikaeli, G. Thorsén, B. Karlberg, Optimisation of resolution in micellar electrokinetic chromatography by multivariate evaluation of electrolytes. J. Chromatogr. A **907**, 267–277 (2001)
36. H. Wan, M. Öhman, R.G. Blomberg, Chemometric modeling of neurotransmitter amino acid separation in normal and reversed migration micellar electrokinetic chromatography. J. Chromatogr. A **916**, 255–263 (2001)

Chapter 6
Isotachophoresis

Przemysław Kosobucki and Bogusław Buszewski

Abstract Isotachophoresis (ITP) (*iso* = equal, *tachos* = speed, *phoresis* = migration) is a technique in analytical chemistry used to separate charged particles. It is a further development of electrophoresisand a powerful separation technique using a discontinuous electrical field to create sharp boundaries between the sample constituents. In this chapter the basic background (including theory, instruments, analysis) is presented. Applications and coupling of ITP with other analytical techniques (NMR, HPLC, MS, CZE) are important components of the mentioned chapter. Authors do not forget the miniaturization of isotachophoresis. Isotachophoresis shows its superiority to conventional separation techniques when the maximum resolution is achieved with the latter. The choice of the experimental parameters remains complex but is very rewarding when successful. In water or wastewater samples no sample pretreatment or only filtration is required before analysis. In analysis of solid samples (soils, sediments) before analysis, extraction techniques are necessary. Thanks to its low cost and high rate isotachophoresis could be useful in routine analysis.

6.1 Theoretical Background

The process of the separation of ions (cations or anions) that use suitable electrolytes from the zones in an external electric field according to decreasing the electrophoretic mobility and move with equal speed is called isotachophoresis (ITP). The word "isotachophoresis" can be divided into components (from Greek: *iso*—constant, *tachos*—speed, *phoresis*—migration) [1].

P. Kosobucki (✉) · B. Buszewski
Faculty of Chemistry, Nicolaus Copernicus University, Toruń, Poland
e-mail: pkosob@chem.umk.pl

B. Buszewski et al. (eds.), *Electromigration Techniques*, Springer Series
in Chemical Physics 105, DOI: 10.1007/978-3-642-35043-6_6,
© Springer-Verlag Berlin Heidelberg 2013

The speed of the movement of the above-mentioned compounds in the electric field is directly proportional to the intensity of this field which can be expressed by Eq. (6.1).

$$v = \mu \cdot E \qquad (6.1)$$

where v is the linear speed of the movement of the chemical compound along the capillary [m s^{-1}], μ is the proportionality factor called the electrophoretic mobility, characteristic for a species in a particular solvent [m^2 V^{-1} s^{-1}], and E is the intensity of the electric field expressed as a gradient of this field along the capillary per length unit [V m^{-1}].

The isotachophoresis uses two different buffer systems. The leading electrolyte (LE) contains ions with mobility greater than the ions of the sample and the counter-ion has a large buffering capacity. The terminating electrolyte (TE) contains ions with mobility lower than the ions of sample [2]. Hence, it is necessary to meet the following condition in order that the separation by the isotachophoresis may occur:

$$\mu_{LE} > \mu_{anality} > \mu_{TE} \qquad (6.2)$$

The sample placed between the LE and TE forms a local balance system. The leading electrolyte ions must have the highest electrophoretic mobility in the whole isotachophoretic system, because they are usually light ions of small size. Hydronium cations (H_3O^+) from the dissociation of a strong acid (H_2SO_4) are most commonly used in the analysis of cations and chloride anions (Cl^-) used in the analysis of anions. The terminating electrolyte (TE) in the analysis by ITP method must have the lowest electrophoretic mobility in the entire isotachophoretic system to reach the detector finallly. Heavy, organic ions of large size characterized by low mobility, for e.g., citrates ($C_6H_5O_7^{3-}$), are generallly used in the analysis of anions such as tetrabutylammonium cation (TBA+) in the analysis of cations. The purity of electrolyte solutions that must be prepared from deionized water is very important from a practical point of view. The presence of even small ionic impurities in electrolyte solutions cause the appearance of additional zones that greatly hamper both qualitative and quantitative analysis.

The schematic course of the separation process is shown in Fig. 6.1. The upper part of Fig. 6.1 shows the LE and TE among which a mixture of ions A and B was placed. The next parts of the diagram show the state of the separation in the consecutive intervals. The lower part of the diagram shows the balance state. One kind of ions forms each zone and concentrations of the substance in these zones are in equilibrium with the concentration of the leading electrolyte. The length of the particular zones do not change any more, when a diameter of the capillary and the composition of the leading electrolyte are constant [1].

The concentration of a particular component (for example A) is determined according to Kohlrausch's law Eq. (6.3) and is equal to the concentration of the leading electrolyte. It can be calculated if the concentration of the leading

Fig. 6.1 Scheme of separation of **a** and **b** anions by isotachophoresis

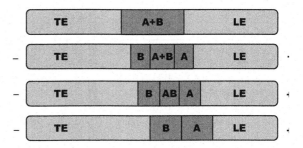

electrolyte zone, the value of the mobility of LE ions, and pK zones, both the leading zone and the following zone, are known. In conclusion, the ion concentrations in the zones are in e balance with the preceding zone and the zone of the leading electrolyte. The concentration in the zone of the terminating electrolyte should also be constant.

$$\frac{c_A}{c_B} = \frac{\mu_A}{\mu_A + \mu_B} \cdot \frac{\mu_B + \mu_Q}{\mu_B} \cdot \frac{Z_A}{Z_B} \tag{6.3}$$

where $\mu_{A,B,Q}$ is the mobility of ions A^-, B^- and the counter-ion $Q+$, $c_{A,B}$ is concentration of ions A^- and B^-, *and* $Z_{A,B}$ are charges of ions A^- and B^-.

Therefore, all zones move with equal speed and at the same time individual types of ions have different effective mobility each zone has its own potential gradient, which gives it the same speed. The potential gradient is constant in each zone, but increases from one zone to another. All zones are ranked according to their effective mobility. The increase in the potential gradient is closely related to the reduction in the concentration of the right kind of ions in the appropriate zone. Thus, the observed boundaries of the zones between separated substances are very sharp [3].

Table 6.1 shows the composition of the buffers most commonly used in isotachophoresis [4]. Obviously, these buffers do not limit the potential uses of ITP in any way. It is possible to modify their composition (both qualitative and quantitative) depending on the analytical problems which can be encountered in practice. It is necessary to bear in mind that the mixture of separated components must be soluble in water or methanol and must have ions where mobilities are contained between the mobilities of LE and TE—Eq. (6.2).

6.1.1 Quantitative and Qualitative Analysis

The result of the isotachophoretic analysis is a graph showing changes in the value of the detector signal within the time isotachophoregram or generally electropherogram. Figure 6.2 shows an example of isotachophoregram of a mixture of anions obtained by the conductometric detector.

Table 6.1 The most often used buffers in isotachophoresis

Symbol	Composition	pH	Applications
LE − 1	10 mM HisHCl + His	6.0	Anions analysis
LE − 2	10 mM HCl + BALA	3.2	Anions analysis
LE − 3	10 mM HCl + BALA	3.5	Anions analysis
LE − 4	10 mM HCl + BALA	3.9	Anions analysis
LE − 5	10 mM HCl + BALA	4.5	Anions analysis
LE − 6	10 mM HCl + BTP	6.0	Anions analysis in wine
LE − 7	10 mM HCl + BTP	9.0	Anions analysis in high pH
LE + 1	10 mM NH_4^+ + CH_3COOH	5.2	Cations "screnning"
LE + 2	10 mM NH_4^+ + CH_3COOH + 30 % PEG	5.2	Alkali cations analysis
TE − 1	5 mM MES + His		
TE − 2	5 mM caproic acid		Universal TE for anions analysis
TE − 3	5 mM glycin		
TE + 1	5 mM acetic acid		Universal TE for cations analysis
TE + 2	5 mM $TBA^+ClO_4^-$		

Where HisHCl is histidine hydrochloride, His is histydine, BALA is β-alanine, BTP is bis–tris propane, PEG is polyethylene glycol, MES is morpholinoethanosulphonic acid, and TBA^+ - ClO_4^- is tetrabutyloammonium perchlorate. Reprinted with permission from Krivankova et al. [4]. Copyright 1987 Elsevier

The qualitative analysis is possible to be conducted on the basis of the height (h) of zones in relation to the baseline (RSH called *Relative Step Height*, RSH = h/H) and by comparing to the patterns of analytes (Fig. 6.2). The value of RSH amounts from 0.000 to 1.000.

The quantitative analysis is possible to be conducted on the basis of the length (L) of the identified zones and based on previously prepared calibration curves. The quantitative analysis of the most commonly used method is the calibration curve. It requires the preparation of series of standard solutions, their analysis, the

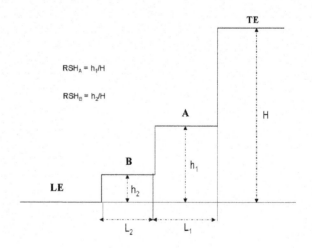

Fig. 6.2 Isotachophoregram with parameters to qualitative and quantitative analysis

$RSH_A = h_1/H$

$RSH_B = h_2/H$

Fig. 6.3 Calibration curve

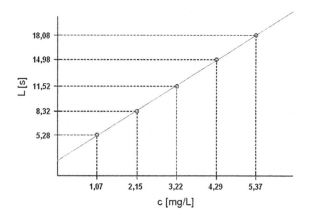

identification of zones, and the indication of the appropriate calibration curve $L = f(c)$ (Fig. 6.3).

In practice, in the case of determining the concentration of one ion in the sample, the method of addition of the internal standard may be used. It involves the analysis of two samples, one of them is the tested sample, the other is the tested sample with the addition of the determined ion (of known volume and concentration). The increase of the length (L) of one of the zones (tested ion) is observed at the isotachophoregram, simple calculations allow to determine the concentration of the tested ion.

It may be taken into consideration why the migration time t_m is not used in the qualitative analysis. It is connected with the properties of the tested sample. If the tested sample contains a small number of determined ions and their concentrations are small, then the total analysis time is, for e.g., 10 min (the typical time of the determination of anions in water samples). If the analyzed mixture contains a larger number of ions (more steps) and their concentrations are higher (longer L), respectively, the total time of the analysis increases. The observed effect shows that the same ions migrate at different times and the qualitative analysis based on t_m is impossible to be performed. RSH is constant and invariable in the same conditions of the course of the analysis.

In summary, the value of the detector signal for a given zone (zone height, h) is the qualitative information and the zone length (L) is the quantitative information.

6.1.2 Separation Conditions

The separation of ions in the isotachophoresis is possible only in the case of sufficient differences in the effective electrophoretic mobility. The effective electrophoretic mobility of ions depends on several parameters:

- degree of dissociation;
- pK value;
- changes of the temperature (Joule's heat);
- changes in pH during the process;
- solvation of ions;
- the effect of relaxation;
- ion radius;
- ion charge;
- dielectric constant of solvents;
- viscosity of solvents [3].

These parameters should be taken into consideration during selecting proper, working conditions for the separation by ITP. In practice the separation can be performed in four ways:

- Using the differences in absolute mobilities of ions, the pH value is chosen in such a way that all ions that are present in the tested sample are completely dissociated;
- On the basis of differences in the pK values of the components of the tested sample, the pH of electrolytes is chosen in such a way that most of the ions are not fully dissociated, resulting in increasing the differences of the values of the effective mobility of ions. This method is used most commonly when the ions have almost the same ionic mobilities, but their pK values are different;
- When the ions have almost the same ionic mobilities and they differ a little with the value of the pK, the solvent can be changed, usually from water into methanol;
- Adding the complexing substances (for example the α-hydroxyisobutyric acid, HIBA) to the leading electrolyte causes that determined ions form complexes with different effective mobilities.

The selection of an appropriate pH is not a simple matter. The pH of the zones largely depends on the pK value of buffer ions and ions of the sample. During the separation of strong acids, the pH is nearly equal to pK_1, but in the case of the weak acids, there are large shifts caused by values that are important for next dissociation constants. When the pH of the zone is lower than the pH of the previous zone, the effective mobilities of ions that are left behind will increase and within the time the zones will extend and mixed zones will be formed. The pH of the system can be chosen in such a way that the ions in the given zone will have a higher mobility than the effective mobility of the leading ion. It is difficult to maintain the conditions of the separation at the low pH values in the case of separation of cations and at high pH values in the case of separation of anions. It is caused by the fact that the pH increases during the separation of anions while conducting the process and reduces during the separation of cations. These effects relate mainly to weakly dissociated substances. Maintaining the constant pH is particularly important during the separation of weak acids and alkalis [1].

The isotachophoretic separation of ions in unbuffered systems often leads to unexpected effects that disrupt the course of analysis. The influence of the background electrolytes (leading and terminating) is so huge that after some time the phenomenon of the expulsion (elimination) may appear. Zone limits become more and more blurred, zones become longer, and they have changed heights of signals. These effects are caused by electrode reactions, particularly when the electrolytes in the electrode compartments are not renewed on time [5].

The process of the isotachophoresis is sometimes disrupted. Zones can be moved as a result of changes in the pH (counter-ion with too small capacity), electrode reactions or membrane reactions (due to the current flow), changes in the temperature (insufficient thermostating), CO_2 permeation through Teflon walls of the capillary (pH changes), electroosmosis or the formation of hydrodynamic flow (no savers of EOF, such as hydroxyethylocellulose).

The isotachophoresis in contrast to capillary zone electrophoresis (CZE) uses a constant current flowing through the capillary that is connected with changes in the potential of the electric field during the analysis.

In practice, the observed separation in the isotachophoresis depends on three main factors [1]:

- The length of the column—a substance injected into the column occupies a certain volume and thus knowing the diameter of the column it is possible to determine the length of the column occupied by the sample (l_{MIX}). There is a relationship binding this length with the column length required for the complete separation of the analyzed mixture (l_{SEP}):

$$l_{SEP} = l_{MIX} \frac{\bar{\mu}_B}{\bar{\mu}_A - \bar{\mu}_B}, \text{ when } |\bar{\mu}_A| > |\bar{\mu}_B| \qquad (6.4)$$

where: $\bar{\mu}_A \bar{\mu}_B$ are effective electrophoretic mobilities of A and B ions.

- The buffer pH—the effective mobility of ions depends on the degree of the dissociation and pK values, as well as the temperature and pH value:

$$\bar{\mu}_i = \frac{\mu_i}{1 + 10^{pKi - pHmix}} \qquad (6.5)$$

where u_i is electrophoretic mobility of i ions, pKi is dissociation coefficient i ion.

- Working current—working current value (I) is directly related to the speed of movement of ions (v) in the electric field:

$$v_i = \frac{I \cdot \kappa}{S} \cdot \mu_i \qquad (6.6)$$

where I is applied current (A), κ is conductivity of the medium (S m^{-1}), S is capillary area (m^2), and μ_i is electrophoretic mobility of i ions.

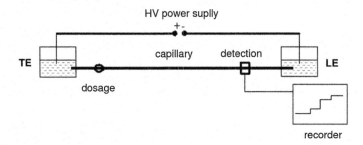

Fig. 6.4 Instrumentation for isotachophoresis

6.2 Instrumentation

The block diagram of the equipment for the isotachophoresis is shown in Fig. 6.4. The electrolytes compartments of a leading electrolyte (LE) and a terminating electrolyte (TE) are connected to a high voltage power supply (up to 20,000 V).

The sample is injected between the LE and TE by the dispenser that can be the sample loop with volumes from 30 μl to 200 nl. If it is necessary, the standard chromatographic syringe can also be used.

6.2.1 Separation Column

The analytical column is a capillary, empty inside, usually made of fluoropolymers (for example FEP Fluorinated Ethylene Propylene, PTFE Polytetrafluoroethylene, Teflon) [6]. Thanks to the fact that the capillary is made of inert materials, the capillary tube is resistant to high temperatures, cleaning is not a problem, and there is no adsorption of analytes on the internal walls.

Two variants of the isotachophoresis work can be found in the analytical practice that result from the construction of the capillary:

- a single column isotachophoresis (sometimes called one-dimensional)—ITP. Then one column at a specified internal diameter filled with the LE (Fig. 6.4);
- a double column isotachophoresis (also called two-dimensional)—ITP–ITP. Then two columns are used. The upper column is called the preseparation column and usually has an internal diameter several times larger (for example 800 μm) than the lower column. The lower column is called the analytical column and usually has an internal diameter of 300 μm, shown schematically in Fig. 6.5.

Both columns (preseparation and analytical) have universal conductivity detectors, and more, the analytical column has the ability to connect the specific spectrophotometric detector (UV–VIS).

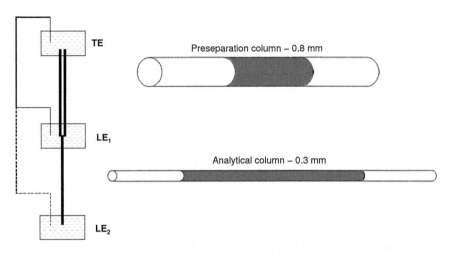

Fig. 6.5 Column coupling system

Such a structure of the system (a combination of capillaries of different diameters) gives the user several unique advantages that can be used in practice:

- The limit of detection (LOD) of analytes reduces. It results from the fact that the ions in the capillary occupy a well-defined, constant volume. In a capillary with larger internal diameter, the volume of this "hypothetical cylinder" is less than in the capillary with smaller diameter (Fig. 6.5). Therefore, in the analytical column, these zones can be quantified wherein the detector placed in the pre-separation column is not able to record. At the same time the relative error of the determination is smaller;

- The application of combined capillaries allows to perform the quantitative determination of components of mixtures in which the ion concentration ratio is 1:10,000. Properly controlling of the analysis (changing columns), macrocomponents can be determined in the preseparation column. Microcomponents are detected in the analytical column. This considerably shortens the total analysis time (typically about 30 min), without switching the columns it can be up to 2 h (depending on the complexity of the analyzed mixture).

- Two columns allow us to determine a wider range of analytes in a single analytical run. The two different leading electrolytes are applied, various for the preseparation column (LE_1) and the analytical column (LE_2) that do not mix together.

- The equipment can operate online in the coupling mode ITP-CZE (isotachophoresis—capillary zone electrophoresis), it will be discussed in a wider way later in this chapter. Thanks to that it is possible to combine the advantages of the ITP method that allows the concentration of the samples in the first column with the more sensitive CZE method in which the concentrated ions are analyzed in the second column. In this way, we reduce the detection limits and eliminate the negative influence of the matrix.

6.2.2 Detection

The contactless detection and contact detection can be used in the isotachophoretic technique. If we apply a thermoelement with thin wires to the capillary, their changes of signals will characterize the individual zones. Differentiating of these signals can determine the zone limits. This is called the contactless detection method. Such a detection is universal, but less selective. The detection sensitivity can be increased by lengthening the zones by narrowing the capillary. Thermodetectors successfully detect amounts of compounds up to a few µg in the sample. Another type of non-contact detection that is used in the isotachophoresis is the optical detection based on the measurement of light absorption in the individual zones. The optical detectors have much better ability to detect short zones and they also have satisfactory dynamic parameters. UV detectors are not universal, because they can only be used if the given substance has a light absorption in the UV range (it has chromophores).

In the analytical isotachophoresis, the detection system should detect as short zones as possible of separated ions, i.e., it should have high sensitivity and separation ability. Furthermore, it should detect additional components on the basis of some universal property of the zones. The detection system should be capable of differentiating the zones, even when there is only a slight difference in the given property. It should also produce a rapid and proper response to the quality as well as to the number of individual ions. The contact detectors satisfy these conditions.

The contact detectors that are available nowadays have electrodes immersed inside the capillary and they also exist in two types. The first type of the contact detector measures the conductivity of the zones, the second type of the contact detector measures directly the electrical gradients in the individual zones. The conductivity detector can detect 50-fold smaller amount of the substance compared with the thermometric detector. What is more, the range of concentrations of compounds in the sample may also be much greater and the obtained isotachophoresis is characterized by very sharp and clear levels. Another type of contact detection uses a direct measurement of the electrical gradient in the zones. The detection is carried out directly in the capillary by measuring the voltage drop between two low-resistance electrodes that are placed in a small distance from each other in the direction which is the same as the movement of the zones. Thanks to the fact that the measured current value is very small, the polarization of the electrodes and depolarization are completely eliminated. The dynamic parameters of such a detector are satisfactory even at the fast high-voltage isotachophoresis analysis. Figure 6.6 shows diagrams of the detectors used in the isotachophoresis [6].

Detectors used in isotachophoresis can be divided into two main groups: universal and specific detectors (Fig. 6.6). Universal detectors provide signals from all the analytes present in the sample that reach its cell. A characteristic feature of specific detectors is the ability to response to only one (or several) selected compounds on the basis of certain characteristics of the analytes without the response to others.

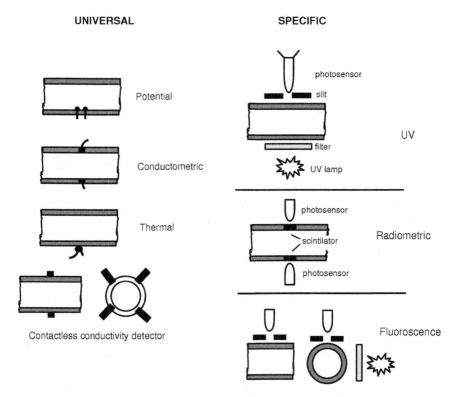

Fig. 6.6 Detectors used in isotachophoresis. Reprinted with permission from Villa Labeco [6]. Copyright 2012 Villa Labeco

The universal detectors used in ITP include: the conductivity detector, thermal detector, contactless conductivity detector (high-frequency conductivity), and potential gradient. The specific detectors include: the UV detector, radiometric detector and fluorescent detector. The characteristics of selected detectors used in the isotachophoresis are summarized in Table 6.2 [7–17].

The data in Table 6.2 show that the limits of detection (LOD) and the limits of quantification (LOQ) are comparable for universal detectors and specific detectors. A conductivity detector is the most commonly used universal detector in isotachophoresis and in other analytical separation techniques used for the separation of ions (Ion Chromatography, IC) (Fig. 6.7) [13].

Operation of conductivity detector based on the fact that in the range of small concentrations, the proper conductivity is a linear function of the electrolyte concentration of the analyzed ions. It can be used for the analysis of all substances that reach the detector cell in the ion form and the conductivity of the leading electrolyte (LE) determines the baseline. When ions of the sample reach the measuring cell, the conductivity increases in proportion to the concentration of flowing ions and remains constant during the movement of these ions through the

Table 6.2 Limit of detection (LOD) and limit of quantification (LOQ) for selected detectors used in isotachophoresis, according to [7–17]

Detector	LOD [µg/L]	LOQ [µg/L]
Conductometric	10–50	30–150
Thermal	20–50	50–130
Contactless conductivity	10–20	30–50
Potential	10–50	30–60
UV–VIS	10–100	30–300
Radiometric	10–100	30–250
Fluoroscence	20–50	60–150

cell. When the ions of the terminating electrolyte reach the detector, the observed conductivity is the highest which indicates the analysis is complete [13].

The dependence of the conductivity solution on the concentration of ions with a specific electric charge for the dilute solutions is described by Kohlrausch's law, given by Eq. (6.7):

$$\kappa = \frac{\sum \lambda_i^0 \cdot Z_i \cdot c_i}{1,000} \tag{6.7}$$

where κ is specific conductivity of electrolytes [µS/cm]; λ_i^0 is molar conductivity [S cm^2 mol^{-1}], z_i is ion charge; c_i is ion concentration [mol/L].

The choice of the detector depends primarily on the properties of the determined substance. The photometric detector allows us to analyze only the ions that absorb the radiation in the visible range or the ultraviolet range. The amperometric detection is used for the determination of ions that cannot be detected by the conductivity detector due to the low degree of dissociation and they undergo the redox reactions.

6.2.3 Discrete Spacers

In the analysis of samples (for e.g., urine, serum, plant extracts) that are characterized by a rich matrix of substances absorbing in the ultraviolet range, we meet some difficulties in the identification of substances; large systematic error indications and the recorded isotachophoregram is difficult to interpret. Substances are separated if they form successively migrating zones with a minimum length or they move in the spaces limited by the components that form the proper zones [6]. Despite this fact the photometric detector cannot distinguish them properly (Fig. 6.8a). This problem can be solved in two ways:

- The mixture of ampholytes is added to the sample (amphoteric electrolytes, for example Ampholine, Servalyt) that contain many chemical substances migrating

Fig. 6.7 Diagram of
operations of conductivity
detector. Reprinted with
permission from Rocklin
[13]. Copyright 1991 Elsevier

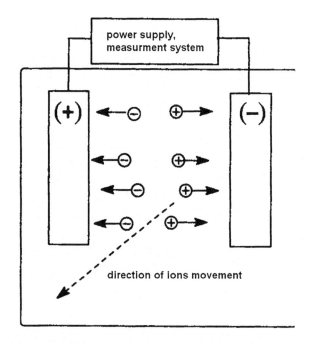

in the isotachophoretic conditions in the range of the mobility between the leading ion and terminating ion Eq. (6.2). These substances that do not absorb the radiation of the given wavelength, form a range of zones and the sample components absorbing the radiation take in such a separating mixture the position in accordance with their electrophoretic mobilities. In this way the number of components of the analyzed sample that can be detected by the UV detector increases considerably. This method is often used in the analysis of proteins (Fig. 6.8b, c);

• The mixture of ion compounds prepared from migrating components in the range of the mobility between the leading ion and terminating ion (Eq. 6.2) that do not absorb the radiation of the given wavelength and mix in concentrations that are not much higher than these necessary to form their own zones. During analyzing the sample with the mixture of the properties mentioned above, the parameters of the determination can be improved thanks to using the photometric detector. The mixtures of ion substances used for this purpose are called "discrete spacers"—Fig. 6.8d, e [6].

After adding the substances mentioned above, we observe the significant improvement of the signal from the UV detector. At the same time, its interpretation is much simpler; the reverse situation is in the case of the conductivity detector (considerable complication of the isotachophoregram).

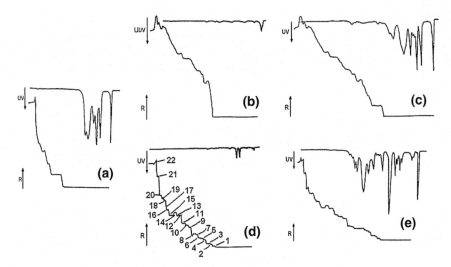

Fig. 6.8 Application of discrete spacers in isotachophoresis, where: **a** urine analysis (5 ml, dilution 1:10); **b** Servalyt (1 ml, dilution 1:10); **c** urine with Servalyt (1 ml, dilution 1:10); better separation in comparison with **a**; **d** mixture of spacers (30 ml, c = 5·10^{-3} mol/L): 1-sulphate, 2-chlorite, 3-sulfonamide, 4-nitriltriacetate, 5-dichloroacetate, 6-acetate, 7-malonian, 8-cysteinian, 9-tartrate, 10-formic, 11-citric, 12-maleinian, 13-isocitric, 14-lactate, 15-N-acetyloserinian, 16-N-acetyloglutamic, 17-N-acetyloglicyic, 18-N-acetyloleucic, 19-β-bromopropionate, 20-succinate, 21-glutaric, 22-adypic, **e** urine (5 ml) with mixture of spacers (30 ml, c = 5·10^{-3} mol/L; dilution 1:10), better separation in comparison with **a**. Reprinted with permission from Villa Labeco [6]. Copyright 2012 Villa Labeco

6.3 Applications

The isotachophoresis can be applied in the analysis of water, sewage, fodder, food, medicines, body fluids, soil, sediment, etc. [18–22]. ITP is an alternative analytical technique used in the analysis of ion substances in the case of the ion chromatography (IC) or to a smaller extent in the case of the capillary zone electrophoresis (CZE) [23].

The isotachophoresis is not a popular technique in research laboratories. There are several research groups worldwide and in Poland that have been continually improving the equipment and looking for new applications. Several review publications on the isotachophoresis have been written [24–28]. In Poland, one of the first publications was the work of Medziak and Waksmundzki in 1979 [3].

Figure 6.9 shows the example of the analysis of water for the presence of cations (Na^{+}, K^{+}, Ca^{2+}, Mg^{2+}) and anions (Cl^{-}, NO$_3^{-}$, SO$_4^{2-}$, NO$_2^{-}$, PO$_4^{3-}$, F^{-}) [21].

A good separation of tested ions was obtained in a short time. The preparation of samples for the analysis is not a problem and it is very simple (filtration, degassing, and dilution). In practice, several other electrolyte systems are used for the analysis of inorganic ions tested in water samples. Some of them were subjected to the full validation and they were presented in the form of standards [29, 30].

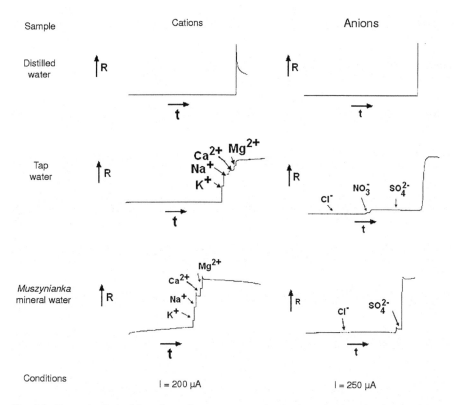

Fig. 6.9 Determination of ions in various water samples by isatochophoresis, cations: LE—10 mmol/L CH_3COOH + 30 % PEG, pH 5.40; TE—10 mmol/L $(CH_3COO)_2Zn$; anions: LE—8 mmol/L Cl^- + 3 mM BTP + BALA, pH 3.55; TE—2 mmol/L CITR. Reprinted with permission from Kosobucki and Buszewski [21]. Copyright 2006 Taylor & Francis Group, LLC

As mentioned above, the isotachophoresis can be conducted in non-aqueous conditions. The separation of organic acids (homologs from C_1 to C_{10}) using methanol leading electrolyte (Fig. 6.10) may be example [31].

In conclusion, isotachophoresis is a useful method for the determination of cations and anions both organic and inorganic. However, it requires the use of different pairs of electrolytes (the leading and terminating) depending on the aims of the analysis.

6.4   Coupling of ITP with Different Analytical Techniques

The coupled techniques give the opportunity to use highly specific and selective detectors and at the same time allow to solve many technical problems (ITP-CZE) or analytical (ITP-NMR, ITP-MS).

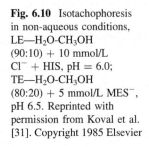

Fig. 6.10 Isotachophoresis
in non-aqueous conditions,
LE—H$_2$O-CH$_3$OH
(90:10) + 10 mmol/L
Cl$^-$ + HIS, pH = 6.0;
TE—H$_2$O-CH$_3$OH
(80:20) + 5 mmol/L MES$^-$,
pH 6.5. Reprinted with
permission from Koval et al.
[31]. Copyright 1985 Elsevier

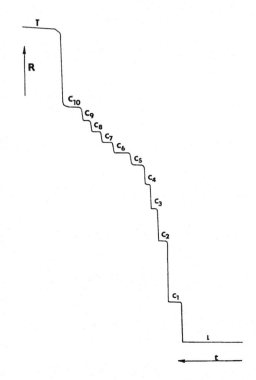

6.4.1 Isotachophoresis-Capillary Zone Electrophoresis (ITP-CZE)

The capillary electrophoresis (CZE) has been used in chemical analysis for a long time (pharmaceutical, clinical, environmental, industrial). However, using CZE requires the use of complicated injection systems to introduce the sample into the capillary (electrokinetic injection, vacuum, hydrodynamic).

The isotachophoresis may be an alternative method thanks to the fact that the concentration of the tested sample between LE and TE occurs during the separation. Hence, ITP can be used as a preseparation and injection method for subsequent CZE analysis. It can be realized in any isotachophoretic equipment with a column coupling system. What is more, ITP-CZE enables the determination of analytes with a large quantitative excess of the matrix that is troublesome when other techniques are used. An example of the application of ITP-CZE is the analysis of the imazalil in fruits and vegetables [32]. The imazalil (IMZ) is a fungicide used for preserving citrus fruits, cucumbers, bananas, and other fruits during a long storage. In practice, the chromatographic techniques (HPLC, GC, TLC) with various detectors (UV, DAD, ECD, MS) are mainly used to the determination of IMZ in the plant material (Fig 6.11).

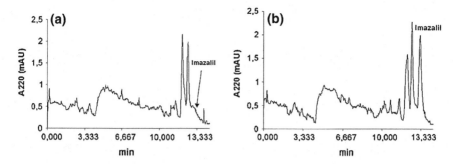

Fig. 6.11 Isotachophoregram of apple extract (**a** not exposed with IMZ; **b** exposed with IMZ). Reprinted with permission from Kvasnicka et al. [32]. Copyright 2003 Springer

ITP-CZE connection was also used to study disinfection by-products of the surface water conditioning for drinking purposes. Bromate content was determined (potential carcinogens) according to several 100-fold excess of chlorate [33].

6.4.2 Isotachophoresis-High Performance Liquid Chromatography (ITP-HPLC)

The offline connection (using preparative valve) and the online connection between the isotachophoresis and the high performance liquid chromatography is a very powerful equipment for study, for e.g., pesticide residues (bentazone) in water (Fig. 6.12). It turned out that the omission of the phase of enrichment in SPE columns reduces the limit of the detection and significantly reduces the analysis time [34].

The chromatogram of the determination of bentazone ($t_r = 19.1$ min) in the surface water (Rhine) by HPLC was shown in Fig. 6.12a. Figure 6.12b shows the same sample after ITP-HPLC analysis. The figures show that ITP-HPLC coupling significantly improves the result of the analysis by the removal of interfering components of the matrix and improves the peak symmetry which has a direct impact on the result of quantitative analysis.

6.4.3 Isotachophoresis-Mass Spectrometry (ITP-MS)

The mass spectrometer as a detector for separation techniques has been used for many years, mainly in the coupling of the gas chromatography and the liquid chromatography. The advantage of using GC–MS is the possibility of using the wide spectrum libraries (NIST Library or Wiley Library), there is a different

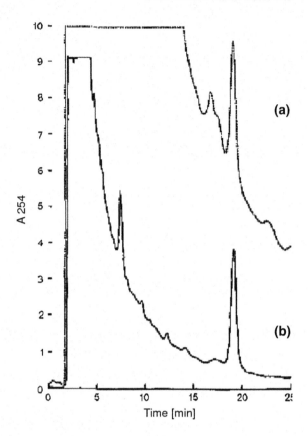

Fig. 6.12 Chromatograms of direct HPLC **a** and ITP-HPLC **b** analysis of 30 μl Rhine spiked sample. Detection UV, range = 0.005 AUFS. $t_r = 19.1$ min indicating Bentazone. Reprinted with permission from Hendriks et al. [34]. Copyright 1992 Springer

situation with LC–MS. The expensive equipment and the amortization of the equipment make that this technique is not as popular as GC–MS and therefore, there are no spectrum libraries. Each user of the system LC–MS is forced to create his own spectrum libraries using the pure standards. It is not a cheap solution and therefore, it is the further obstacle in the wide use [35]. A similar situation concerns ITP-MS coupling.

ITP-MS has been used to test antibiotic residues in the blood of calves so far, but the results are not repeatable. Tomas with co-workers [36] obtained significantly better results. He studied cytochrome C after enzymatic digestion under the cationic mode. The results (Fig. 6.13a) show that the method can be repeatable, which is very good in this case (overlapped three isotachophoregrams). While Fig. 6.13b shows the mass spectrum of six peptides which were separated after the digestion and identified. The advantage of the presented application is the primarily short analysis time and the ease of sample preparation for the analysis.

Fig. 6.13 **a** Overlay of three consecutive isotachophoretic analyses of cytochrome C tryptic digest. Injected volume 10 µl; **b** selected ion monitoring of the cytochrome C tryptic digest. Reprinted with permission from Tomas et al. [36]. Copyright 2010 Elsevier

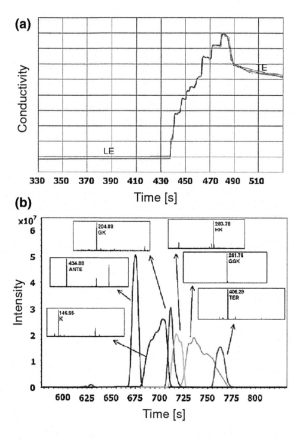

6.4.4 Isotachophoresis-Nuclear Magnetic Resonance (ITP-NMR)

The nuclear magnetic resonance (NMR) is a specific detector used mainly for the analysis of organic compounds and in connection with separation techniques (HPLC, CZE and ITP) gives many research opportunities. ITP-NMR coupling allows to:

- determine trace amounts of impurities towards the excess of the main ingredient (up to 1,000-fold excess of a matrix or other analyte);
- concentrate the sample before NMR measurement;
- use the measuring cells with the capacity from 5 nl to 1 µl;
- determine *online* the structure of the tested analytes;
- improve S/N ratio, i.e., the mass sensitivity;
- measure pH (pD) and temperature (*on-line*) during the process of the separation thanks to a detailed explanation of the mechanism of the separation being possible [37–39].

The measurement equipment is analogous to the set of CZE-NMR and consists of two basic elements: the separation part (ITP) and the detector (NMR).

An interesting use of isotachophoresis coupled with the nuclear resonance is the determination of atenolol (one of the main β-blocers used in cardiac drugs). During this determination the problem is an excess of about 1000-fold sucrose as one of the fillers of the pharmaceutical formulation. Typical concentrations of atenolol is about 200 μmol/L, while the concentration of sucrose is about 200 mmol/L (Fig. 6.14).

ITP-NMR coupling gives also the opportunity for full explaination of the mechanism of isotachophoretic separation. The conducted studies allow to determine online temperature changes (Joule's heat) and the pH and/or pD during the separation. Some linear relationships between the chemical shift δ and the

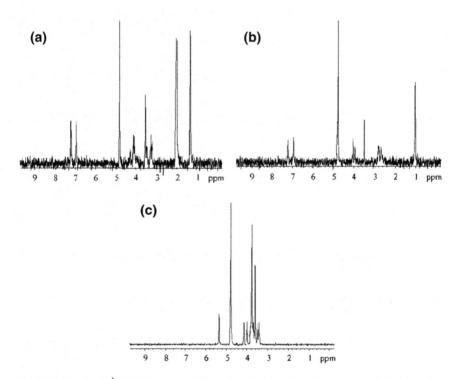

Fig. 6.14 Microcoil ^{1}H NMR spectra. All spectra obtained from the second coil, with identical data acquisition and processing parameters. Spectra **a** and **b** are plotted at the same scale; spectrum **c** is plotted at 1/4 scale. **a** on-flow cITP/NMR spectrum depicting the atenolol sample band at peak maximum during analysis of the trace impurity sample (200 μm atenolol and 200 mM sucrose in 50 % TE/D$_2$O solution). No sucrose peaks can be observed. Acetate peak visible at 2.0 ppm. S/N for atenolol methyl peak at 1.3 ppm, 34; **b** static NMR spectrum of 25 mM atenolol (direct injection) acquired after cITP/NMR analysis of the trace impurity sample. S/N for the atenolol methyl peak at 1.0 ppm, 21 (same moiety as in **a**, change in chemical shift due to pD difference); **c** stopped-flow cITP/NMR spectrum of sucrose during peak maximum from the trace impurity sample. Reprinted with permission from Wolters et al. [39]. Copyright 2002 American Chemical Society

temperature for half deuterated water and acetate ions were obtained during the separation [38].

6.5 Miniaturization

The development of modern analytical systems is closely related to miniaturization and nanotechnology. Chips for the capillary electrophoresis, electrochromatography have been used for many years in many industries for the quick control of the course of many processes. The clinical monitoring also uses chip systems [40]. The advantage of chips is: short analysis time (2–3 min), possibility of automation, low use of reagents, and low cost.

For several years, two research groups have been conducting the use of chip systems in isotachophoresis. One produces chips based on the silicon rubber [41], the other based on PMMA (polymethyl methacrylate) [42]. The diagram of this chip is shown in Fig. 6.15.

The measuring set consists of a chip in which separation occurs (with a channel volume of about 3,900 nl), of the system of 4 peristaltic micropumps (for filling the leading and terminating electrolytes and for sample introduction) and a

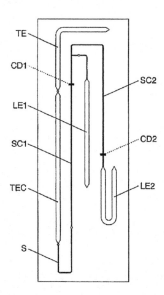

Fig. 6.15 A schematic of the CC chip, where: TE, terminating buffer reservoir, (8.8 μl); TEC, terminating channel (9 ml); S, a 0.9 ml sample injection channel [21.8 × 0.26 × 0.2 mm (length × width × depth)]; SC1, the first separation channel (a 2.1 μl volume; 51.8 × 0.26 × 0.2 mm) with a platinum conductivity sensor (CD1); SC2, the second separation channel (a 2.6 μl volume; 64.0 × 0.26 × 0.2 mm) with a platinum conductivity sensor (CD2); LE1, LE2, leading buffer reservoirs (each 8.8 μl); W, waste outlet. Reprinted with permission from Kaniansky et al. [42]. Copyright 2002 Wiley–VCH

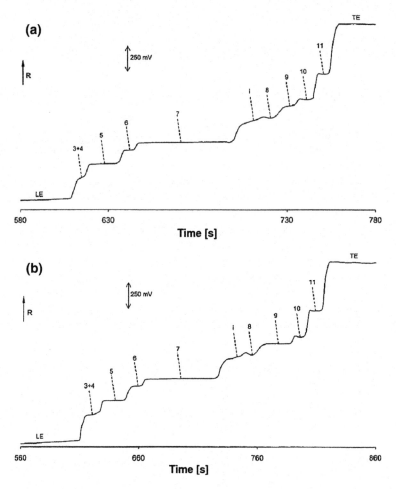

Fig. 6.16 An isotachophoreogram from the ITP separation of organic acids and inorganic anions present in a white wine **a** and red wine **b** samples (20 times diluted) on the CC chip. Zone assignments: LE, leading anion (chloride); *1** migration position of sulphate; *2* sulphite; *3* phosphate; *4* malonate; *5* tartrate; *6* citrate; *7* malate; *8* lactate; *9* gluconate; *10* aspartate; *11* succinate; *12* ascorbate; *13* acetate; *14* sorbate; TE, terminating anion (capronate). Reprinted with permission from Masar et al. [43]. Copyright 2001 Elsevier

computerized system of data collection and processing. Similar to the classical equipment for isotachophoresis, chip systems use the conductometric detector [42].

An example of an interesting use of the analysis using the chip is to study the authenticity of wine (white, red). It was noted that the quantitative and qualitative analysis and the relationship of concentrations of individual organic acids are typical for (like a "fingerprint") the proper species and the manufacturer. Figure 6.16 shows the results of the analysis of sample of white wine and red wine with the use of chip for isotachophoresis [43].

6.6 Summary

Thanks to the continuous development of isotachophoresis, it gives many unique research opportunities that other techniques do not have. The future of isotachophoresis is primarily miniaturization and automation, as it is in other branches of analytical chemistry. What is more, coupled techniques will be further developed.

isotachophoresis has several key advantages that distinguish it from other analytical techniques. The most important are:

• the ease of the sample preparation for the analysis, caused by the fact that separation takes place in an empty capillary tube of a large diameter 0.8 or 0.3 mm, just so that the filtration of the sample before analysis by filter paper or filter with a pore size of 0.45 μm is sufficient;
• ultrasonic extraction, mixing the sample with water, or wet digestion are the most commonly used methods for the sample preparations of the solids before isotachophoretic measurement;
• the volume of the sample for the analysis ranges from 30 μL to several nL;
• the low use of LE and TE and the fact that they are water solutions of inorganic and organic salts (i.e., biodegradable) causes that ITP belongs to the groups of "green chemistry techniques";
• the short analysis time coming up to 40 min, on average the analysis lasts 15 or 20 min;
• the low cost of single analysis and the amortization of the measurement equipment;
• the cost of the equipment is several times lower compared with ion chromatography (IC) or capillary electrophoresis (CZE).

The disadvantage of the analysis with isotachophoresis is a problem with selection of the appropriate LE and TE electrolytes and their proper preparation. Isotachophoresis requires the researcher to strictly and clearly define research objectives and/or analytes. Thanks to this fact the appropriate selection of electrolytes and the way of the sample preparation before the measurement will be possible.

References

1. P. Bocek, M. Deml, P. Gebauer, V. Dolnik, *Analytical Isotachophoresis* (VCH, Weinheim, 1988)
2. F.M. Everaerts, J.L. Beckers, T.P.E.M. Verheggen (eds.), *Isotachophoresis: Theory, Instrumentation and Applications. Journal of Chromatography Library*, vol. 6 (Elsevier, Amsterdam, 1976)
3. I. Medziak, A. Waksmundzki, Izotachoforeza Aparatura, zasada pomiaru i zastosowanie. Wiad. Chem. **2**, 71–92 (1978)
4. L. Krivankova, F. Foret, P. Gebauer, P. Bocek, Selection of electrolyte systems in isotachophoresis. J. Chromatogr. **390**, 3–16 (1987)

5. P. Gebauer, L. Krivankova, P. Bocek, Inverse electrolyte systems in isotachophoresis: Impact of the terminating electrolyte on the migrating zones in cationic analysis. J. Chromatogr. **470**, 3–20 (1989)
6. Villa Labeco Company Materials (Villa Labeco, Spisska Nova Ves, Slovakia)
7. E. Poboży, E. Wojasińska, M. Trojanowicz, Ion chromatographic speciation of chromium with diphenylcarbazide-based spectrophotometric detection. J. Chromatogr. **736**, 141–150 (1996)
8. S. Motelier, H. Pitsch, Determination of aluminum and ist fluoro complexes in natural waters by ion chromatography. J. Chromatogr. **660**, 211–217 (1994)
9. C. Ehrlin, U. Schmidt, H. Liebscher, Analysis of chromium (III)-fluoride-complexes by ion chromatography. Fresen. J. Anal. Chem. **354**, 870–873 (1996)
10. Z. Li, S. Mou, Z. Ni, J.M. Rivello, Sequential determination of arsenite and arsenate by ion chromatography. Anal. Chem. Acta. **307**, 79–87 (1995)
11. D. Connolly, B. Paull, Fast separation of UV absorbing anions using ion-interaction chromatography. J. Chromatogr. **917**, 353–359 (2001)
12. P.R. Haddad, P.E. Jackson, M.J. Shaw, Developments in suppressor technology for inorganic ion analysis by ion chromatography using conductivity detection. J. Chromatogr. **1000**, 725–742 (2003)
13. R.D. Rocklin, Detection in ion chromatography. J. Chromatogr. **546**, 175–187 (1991)
14. P.R. Haddad, Developments in detection methods for ion chromatography. Chromatographia **24**, 217–225 (1987)
15. W.W. Bucherberger, P.R. Haddad, Advances in detection techniques for ion chromatography. J. Chromatogr. **789**, 67–83 (1997)
16. W.W. Bucherberger, Detection techniques in ion chromatography of inorganic ions. TrAC **20**, 296–303 (2001)
17. H. Small, T.E. Miller, Indirect photometric chromatography. Anal. Chem. **54**, 462–468 (1982)
18. I. Nagyova, D. Kaniansky, Discrete spacers for photometric characterization of humic acids separated by capillary isotachophoresis. J. Chromatogr. A **916**, 191–200 (2001)
19. P. Kosobucki, B. Buszewski, Electromigration techniques as a modern tool to control work of sewage treatment plant. Tox. Environ. Chem. **19**, 109–116 (2001)
20. P. Kosobucki, B. Buszewski, The application of isotachophoresis to the compost analysis. Chem. Anal. (Warsaw) **48**, 555–565 (2003)
21. P. Kosobucki, B. Buszewski, Application of isotachophoresis for quality control of drinking and mineral waters. J. Liq. Chromatogr. Rel. Technol. **13**, 1951–1960 (2006)
22. P. Kosobucki, B. Buszewski, Isotachophoretic separation of selected imidazolium ionic liquids. Talanta **74**, 1670–1674 (2008)
23. P. Kosobucki, B. Buszewski, Determination of tetrafluoroborate and chloride anions by isotachophoresis and ion chromatography. Chem. Anal. (Warsaw) **53**, 895–903 (2008)
24. P. Gebauer, P. Bocek, Recent progress in capillary isotachophoresis. Electrophoresis **21**, 3898–3904 (2000)
25. P. Gebauer, P. Bocek, Recent progress in capillary isotachophoresis. Electrophoresis **23**, 3858–3864 (2002)
26. P. Gebauer, Z. Mala, P. Bocek, Recent progress in capillary ITP. Electrophoresis **28**, 26–32 (2007)
27. P. Gebauer, Z. Mala, P. Bocek, Recent progress in analytical capillary ITP. Electrophoresis **30**, 29–35 (2009)
28. F.I. Onuska, D. Kaniansky, K.D. Onuska, M.L. Lee, Isotachophoresis: trials, tribulations, and trends in trace analysis of organic and inorganic pollutants. J. Microcolumn Sep. **10**, 567–579 (1998)
29. STN 75 7430 (1997) Kvalita vody Izotachoforetické stanovenie chloridov, dusičnanov, síranov, dusitanov, fluoridov a fosforečnanov vo vodách
30. STN 75 7431 (1997) Kvalita vody Izotachoforetické stanovenie amoniaku, sodíka, draslíka, vápnika a horčíka vo vodách

31. M. Koval, D. Kaniansky, M. Hutta, R. Lacko, Analysis of saturated normal fatty acids in hydrocarbon matrices by capillary isotachophoresis. J. Chromatogr. **325**, 151–160 (1985)

32. F. Kvasnicka, J. Dobias, K. Klaudisova-Chudackova, Determination of Imazalil by on-line coupled capillary isotachophoresis with capillary zone electrophoresis (CITP-CZE). CEJC **1**, 91–97 (2003)

33. P. Praus, Determination of chlorite in drinking water by on-line coupling of capillary isotachophoresis and capillary zone electrophoresis. Talanta **62**, 977–982 (2004)

34. P.J.M. Hendriks, H.A. Classens, T.H.M. Noij, F.M. Everaerts, C.A. Cramers, On line isotachophoresis. A selective sample pretreatment prior to column liquid chromatographic analysis. Chromatographia **33**, 539–545 (1992)

35. R.E. Ardrey, *Liquid Chromatography—Mass Spectrometry: An Introduction* (Wiley, New York, 2003)

36. R. Tomas, M. Koval, F. Foret, Coupling of hydrodynamically closed large bore capillary isotachophoresis with electrospray mass spectrometry. J. Chromatogr. A **1217**, 4144–4149 (2010)

37. R.A. Kautz, M.E. Lacey, A.M. Wolters, F. Foret, A.G. Webb, B.L. Karger, J.V. Sweedler, Sample concentration and separation for Nanoliter-volume NMR spectroscopy using capillary isotachophoresis. J. Am. Chem. Soc. **123**, 3159–3160 (2001)

38. A.M. Wolters, D.A. Jayawickrama, C.K. Larive, J.V. Sweedler, Insights into the cITP process using on-line NMR spectroscopy. Anal. Chem. **74**, 4191–4197 (2002)

39. A.M. Wolters, D.A. Jayawickrama, C.K. Larive, J.V. Sweedler, Capillary isotachophoresis/ NMR: extension to trace impunity analysis and improved instrumental coupling. Anal. Chem. **74**, 2306–2313 (2002)

40. M. Szumski, B. Buszewski, State of the art in miniaturized separation techniques. Critical. Rev. Anal. Chem. **1**, 1–46 (2002)

41. S.J. Baldock, P.R. Fielden, N.J. Goddard, J.E. Prest, B.J. Treves Brown, Integrated moulded polymer electrodes for performing conductivity detection on isotachophoresis microdevices. J. Chromatogr. A **990**, 11–22 (2003)

42. R. Bodor, D. Kaniansky, M. Masar, K. Silleova, B. Stanislawski, Determination of bromate in drinking water by zone electrophoresis-isotachophoresis on a column-coupling chip with conductivity detection. Electrophoresis **23**, 3630–3637 (2002)

43. M. Masar, D. Kaniansky, R. Bodor, M. Johnick, Determination of organic acids and inorganic anions in wine by isotachophoresis on a planar chip. J. Chromatogr. A **916**, 167–174 (2001)

Chapter 7
Capillary Isoelectric Focusing

Michał J. Markuszewski, Renata Bujak and Emilia Daghir

Abstract Capillary isoelectric focusing (CIEF) is a widespread technique for the analysis of peptides and proteins in biological samples. CIEF is used to separate mixtures of compounds on the basis of differences in their isoelectric point. Aspects of sample preparation, capillary selection, zone mobilization procedures as well as various detection modes used have been described and discussed. Moreover CIEF, coupled to various types of detection techniques (MALDI or LIF), has increasingly been applied to the analysis of variety different high-molecular compounds. CIEF is considered as a highly specific analytical method which may be routinely used in the separation of rare hemoglobin variants. In addition, the application of CIEF in proteomic field have been discussed on the examples of analyses of glycoproteins and immunoglobins due to the meaning in clinical diagnostic.

7.1 Introduction

Capillary isoelectric focusing (CIEF) is an electrophoretic technique used to separate mixtures of compounds on the basis of differences in their isoelectric points (pI). CIEF is generally used to separate amphoteric compounds such as proteins, peptides, and amino acids. The principle of separation in CIEF is shown in Fig. 7.1.

M. J. Markuszewski (✉)
Faculty of Pharmacy, Medical University of Gdańsk, Gdańsk, Poland
e-mail: markusz@gumed.edu.pl

R. Bujak · E. Daghir
Faculty of Pharmacy, Ludwig Rydygier Collegium Medicum in Bydgoszcz, Nicolaus Copernicus University in Toruń, Toruń, Poland

B. Buszewski et al. (eds.), *Electromigration Techniques*, Springer Series
in Chemical Physics 105, DOI: 10.1007/978-3-642-35043-6_7,
© Springer-Verlag Berlin Heidelberg 2013

Fig. 7.1 Principle of separation in capillary isoelectric focusing

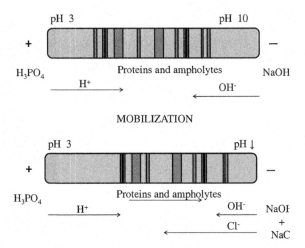

In CIEF, after filling the capillary with a mixture of solute and ampholytes, the pH gradient is formed. With a basic solution at the cathode and an acidic solution at the anode, upon application of the electric field the charged ampholytes and proteins migrate through the medium until they reach a region where they become uncharged (at pI).

There are two modes to perform capillary isoelectric focusing on:

- Multi-step CIEF: this mode of CIEF is characterized by presence of focusing and mobilization as two separated steps. Mobilization can be achieved by ion addition or applying a hydraulic force. This mode requires reduced electroosmotic flow (EOF) to a minimum.
- Single-step CIEF: in this mode focusing and mobilization steps occur simultaneously. The forces available to transport the zones are the same as for multi-step CIEF, but in this mode reduction of electroosmotic flow is not required.

Each of these modes need different instrument configurations and different methodologies for optimizing CIEF separation, and they will be discussed in the next parts of this chapter.

7.2 CIEF Methodology

7.2.1 Sample Preparation and Injection

Sample preparation for CIEF usually consists of three steps: adjustment of sample salt levels, selection of the proper ampholyte composition, and dilution of the sample to the appropriate analyte (e.g. protein) concentration. The ionic strength of the sample should be as low as possible, lower than 50 mM. To high ionic strength of sample due to the presence of buffer, salts or detergents will disrupt the isoelectric focusing process, causing long focusing times and peak broadening during mobilization. The high current and presence of salts can also cause precipitation of proteins in focused zones. To avoid this precipitation, samples with salt concentration higher than 50 mM should be desalted by one of the following methods: dilution, dialysis, gel filtration, or ultrafiltration.

The ampholyte composition should be selected based upon the desired separation range. During separation of complex samples containing proteins with broad range of isoelectric points or estimation of pI of unknown protein, a wide-range of ampholytes mixture should be selected, e.g., from 3 to 10 pH. The final ampholyte concentration should be between 1 and 2 % (v/v). To detect proteins at the basic end of the gradient during cathodic mobilization, it is required that the gradient span only the effective length of the capillary. When the total capillary length is greater than the effective length, sample components may focus on the "blind" zones, distal to the detection point and can be undetected during mobilization step [1].

The final protein concentration in the sample and ampholyte mixture depends upon method sensitivity and solubility of protein components during focusing step. The concentration of 0.5 mg/ml per protein should provide good sensitivity and proper focusing and mobilization performance. However, many proteins such as immunoglobulins, membrane proteins, high molecular weight, and hydrophobic proteins can precipitate during focusing at starting concentration. Therefore, for these kinds of compounds their concentrations during focusing should be lower, about 0.2 mg/ml. In such cases, use of very dilute protein solution may be necessary. Prior to injection, the prepared sample should be centrifuged to remove any particulate material and to degas the solution. The prepared sample and ampholyte mixture is introduced into the capillary. The pressure or vacuum injection can be applied. For good precision, appropriate volume (at least 3–5 volumes of mixture of sample and ampholytes) should be injected into the capillary to fill the tube by homogeneous mixture of sample.

7.2.2 Focusing

There are some approaches that can be used to achieve pH gradient. The pH gradient required to perform CIEF is achieved by the carrier ampholytes. If a capillary is filled with a solution containing an amphoteric substance flanked by an acidic solution at the anode and an alkaline solution at the cathode, while applying an electric field, all molecules with a net charge migrate towards the electrode of opposite charge. The focusing step starts with the immersion of the capillary in the anolyte and catholyte solutions, followed by application of high voltage. The most commonly used catholyte solution is 20–40 mM NaOH, and the anolyte is half the catholyte molarity, e.g., 10–20 mM phosphoric acid. After application of high voltage, the charged ampholytes migrate in the electric field. A pH gradient starts to develop, with low pH towards the anode and high pH towards the cathode.

The protein components in the sample migrate until their isoelectric point is reached. At this point, each protein is focused in a narrow zone. The focusing step is achieved rapidly and the exponential drop in current occurs. The current drop can be explained as follows: at the beginning of focusing step, when the electric field is applied, most components of sample are charged and act as current carrier. When ampholytes and analyzed compounds achieve the isoelectric point, their net charge is zero and the actual current decreases. The focusing step should be finished when the current has dropped to a level approximately 10 % of its initial value. It is not advisable to prolong focusing step beyond this point due to resistive heating and risk of protein precipitation. Moreover, loss of the ampholytes at the acidic or basic ends of the gradient can give rise to anodic or cathodic drift [2]. Anodic drift can be reduced by increasing the phosphoric acid concentration of the anolyte [3]. When the focusing step is completed, the final mobilization step begins with substitution of the anolyte or catholyte solutions with a suitable mobilization solution or by applying hydraulic force.

7.2.3 Two-Step CIEF

The focused zones are usually transported past the detection point by applying chemical or hydraulic mobilization [2–4].

7.2.3.1 Chemical Mobilization

After completing focusing step, high voltage is turned off and the anolyte or catholyte is replaced by mobilization solution. The selection between anodic or cathodic mobilization depends on isoelectric points of the analytes (proteins) and the aim of separation. Cathodic mobilization is more often used when the majority of proteins have isoelectric points between 6 and 9 (migration of focused zones towards cathode). The most common chemical mobilization mode is the addition of a neutral salt such as sodium chloride to the anolyte or catholyte. Sodium is a cation in anodic mobilization and chloride function as an anion in cathodic mobilization. At the beginning of mobilization step, current is on the low value level as at the end of focusing step, but begins to rise when the chloride ions enter the capillary. During mobilization step, when chloride is present in whole capillary, the rapid rise in current points to the end of mobilization. Movement of ions into the capillary causes a pH change at the capillary end, especially. The actual slope of the pH gradient through capillary is lower in the direction opposite to mobilization. Neutral and basic proteins are transported towards cathode in sodium chloride. Mobilization times of these proteins correlate with pI. In case of acidic proteins, mobilization is performed with low efficiency. Moreover, zone broadening may occur and acidic proteins can be undetected. Use of zwitterions is an alternative approach which provides more effective mobilization of protein zones across a wide pH gradient [5].

For example, mobilization of proteins with pI's in the range of 4.46–9.60 can be achieved by cathodic mobilization with a low-pI zwitterion. Effectiveness of zwitterions mobilization depends on the choice of proper mobilization solution. For example, zwitterion with pI between the pH of anolyte and the pI of the most acidic analyte's protein is needed in cathodic mobilization.

7.2.3.2 Hydraulic Mobilization

Positive pressure or negative pressure (vacuum) are the most commonly used forces that move the focused proteins zones to detection point in hydraulic mobilization step. During hydraulic mobilization, it is required to apply an electric field across the capillary in order to maintain focused protein zones [3].

- Pressure mobilization: CIEF with hydraulic mobilization was first described by Hjerten and Zhu [6]. Mobilization was performed by displacing focused zones from the capillary by pumping solution of anolyte into the capillary. This approach provides a proper flow rate in the capillary. Other forms of pressure, such as compressed gas or pressure achieved by height difference of liquid levels in reservoirs can be used in this mode.
- Gravity mobilization: focused protein zones can be transported towards the detection point by means of difference in levels of anolyte and catholyte contained in the reservoirs. Comparing with pressure or vacuum, force created by liquid height difference can be easily manipulated. The force using during this mode of mobilization can be regulated by adjusting the level difference of the reservoirs, by changing a liquid level in vials or raising or lowering the vials. From the analytical point of view, gravity mobilization is the simplest mode to transport focused protein zones to the detection point [7].
- Vacuum mobilization: hydraulic mobilization by vacuum and using online detection was described by Chen and Wiktorowicz [3]. The four-step vacuum-loading procedure was used in this approach. The following sequence: solution of catholyte (20 mM NaOH + 0.4 % methylcellulose), ampholytes and methylcellulose, sample solution and a final solution of ampholytes and methylcellulose was used to introduce these segments from the anodic end of the capillary. After loading of capillary, focusing step was carried out for 6 min. The mobilization of focused zones towards the cathode was performed by using vacuum at the capillary outlet with voltage simultaneously maintained to avoid distortion effects of laminar flow. The suppress EOF was obtained by applying a dimethylpolysiloxane (DB-1) coated capillary and addition of methylocellulose to the catholyte and ampholyte solutions. In this mode, mobility values of proteins were calculated by normalization of zone migration times of the catholyte-ampholyte and ampholyte-anolyte interfaces.

7.2.4 Single-Step CIEF

Single-step CIEF, so-called dynamic CIEF is a mode in which focusing and mobilization steps occur simultaneously. Mobilization using EOF is only used in single-step CIEF.

7.2.4.1 Single-Step CIEF in the Presence of EOF

In this approach standard capillaries with EOF are used. It is one-step process, with focusing occurring during mobilization of sample proteins towards detector by EOF. This mode has been used with both uncoated capillaries and with capillaries coated to reduce, but not eliminate EOF. Two approaches have been reported in this technique. In the first approach the sample and ampholyte mixture was introduced as a plug at the inlet of the capillary pre-filled with catholyte. In the second mode, whole capillary was pre-filled with a sample of ampholyte mixture [8]. In the first approach described by Thormann et al. [4], 75 μm i.d. × 90 cm uncoated capillaries were filled with catholyte (20 mM NaOH + 0.06–0.03 % hydroxypropylmethylcellulose [HPMC]) and a 5 cm segment of sample in 2.5–5 % ampholytes was gravitaly injected at the anodic end of the capillary. The addition of HPMC to the catholyte provides coating of the fused silica wall and leads to reduction of both EOF and protein adsorption on the inner surface of the capillary. The crucial thing in this approach is optimization of the HPMC concentration, ampholyte concentration and sample load.

7.2.4.2 Single-Step CIEF in the Presence of Hydraulic Forces

During this mode samples only partially fill up the capillary. The inlet and outlet reservoirs contain H_3PO_4 and NaOH, respectively. After sample injection, voltage is applied. While the zones focused, they are transported to the detector by means of hydraulic flow, which is created by the liquid height difference of the reservoirs. The gravity is the simplest mode, although other forces can also be used such as pressure or vacuum. This technique of single-step CIEF requires EOF elimination. The most important advantage of this approach is its simplicity, but the main disadvantage is that the capillary is not filled completely, therefore due to the small amount of sample, detection signal is reduced and sensitivity may be limited.

7.2.5 Capillary Choice

In order to obtain good resolution during CIEF performance with chemical mobilization it is necessary to reduce EOF to a very low level. The use of coated capillaries is required for this technique. A viscous polymeric coating is recommended to reduce EOF. Use of neutral, hydrophilic coating materials eliminates protein interactions. The low value of EOF in coated capillaries allows separation to be performed with short effective capillary length. The earlier work using chemical mobilization was performed using capillaries as short as 11 cm with internal diameters up to 200 μm [9]. Although 12–17 cm capillaries with internal diameters of 25–50 μm have been used more recently, in theory, resolution in CIEF does not depend on capillary length since ampholytes are the same and only their concentration might be changed. However, in practice the resolution is reduced by using too short capillaries or too small sample volume injection. In single-step CIEF, the length of capillary is essential, especially when the driving force is EOF. It is necessary to optimize the capillary length according to the size of the injection and the velocity of EOF in order to ensure that sample mixture will not reach detection point before end of focusing. There are many advantages of using large diameter capillaries, such as increase in detection signal, the potential of micropreparative analysis, and reduction of capillary plugging. CIEF was also performed in capillaries with the wall chemically modified by the attachment of a hydrophobic reagent. In such case, the high value of EOF was observed, especially when the pH is above 2–3. The probable explanation of this phenomena is that proteins and ampholytes react with the hydrophobic wall in different degrees.

7.2.6 Detection Methods in CIEF

UV/Vis spectrophotometry is the most common detection method used for analysis of focused protein zones in CIEF. Unfortunately, the strong absorbance of the ampholytes at wavelengths below 240 nm results in impractical detection of proteins in the low UV region. Therefore, 280 nm is usually used in CIEF technique with UV/Vis detection, but for some proteins such as hemoglobin and cytochromes detection can be in the visible range of the spectrum.

The other detection method used by Wu and Pawilszyn [10] in CIEF technique is fluorescence imaging detection. This detector incorporates laser beam (He-Ne) which detects deflections generated by the passage of substances with a refractive index different than that of background buffer. Universality and low cost are the most essential advantages of this detector. The detection system can be built to scan focused protein zones during detection without mobilization step. Concentration gradient detector provides fast analysis time (2 minutes) and detection limits in the 1–5 mg/ml range. It can be applied in detection of peptides without aromatic amino acids, which are necessary for UV detection in 280 nm. Detection modes which require derivatization step of the sample, are rarely used in CIEF techniques since during derivatization reaction, p*I* of analytes can be changed.

7.3 Application of CIEF

CIEF is a sensitive, high resolution and highly reproducible technique. CIEF is routinely used in the separation of amphoteric analytes such as aminoacids, peptides, and proteins, therefore it can be applicable in protein profiling. Coupling CIEF with modern analytical detection techniques makes it a useful analytical tool in the field of medicine.

7.3.1 CIEF in Hemoglobin Analysis

Hemoglobinopathies can be divided into two types. The first type presents structural abnormalities in the chemical structure of hemoglobin molecule, whereas the second type represents thalassemias in which the molecular structure of hemoglobin is normal but its quantity is reduced due to the congenital disorder of biosynthesis of globin chains.

The analysis of hemoglobin types is of primary meaning in clinical diagnostic due to over 600 of hemoglobins that have been identified so far and each type may differ in amino acid sequence. A single amino acid substitution can cause telomeres instability, reduced affinity to oxygen, or iron oxidation.

The separation of hemoglobins by capillary zone electrophoresis (CZE) is difficult and problematic. Utilizing CIEF one can obtain a baseline resolution of the most commonly observed human hemoglobin types: HbA, HbF, HbS, HbC [11]. All of them have almost identical isoelectric points. Their pIs are in the range of 7.0–7.4 pH units, and there is only a minimal difference between HbA and HbF pI's [12]. Jenkins et al. [11] adopted fused silica capillary with poliacrylamide coating (24 cm length, i.d. 50 μm) in order to identify hemoglobin different types. Haemolyzed erythrocytes were introduced into the poliacrylamide coated capillary and then isoelectric focusing under the voltage of 8 kV was performed. The separation of HbS and HbA was observed at 280 nm. The electrophoretic flow reduction optimization was done by the use of polymeric coatings based on methylcellulose, polyvinyl alcohol and polivinylpirylidon alcohol, and resulted in improved resolution and increased speed of electrophoretic separation. Also, Yao et al. [13] adopted fused silica capillary with methylcellulose coating for the separation of the same four hemoglobins (HbA, HbS, HbF, HbC) with high peak resolution.

Since coated capillaries may be quite unstable resulting after a certain time in the irreproducible migration times and the poor peak resolutions, the alternative way is the utility of dynamic coated capillaries. In order to obtain hemoglobin types separation (HbA1C, HbA, HbF, HbD, HbS, HbE) utilizing the dynamic capillary coating, one should add methylcellulose to the catholyte solution. The optimization of separation process is provided by the appropriate concentration of methylcellulose in background electrolyte (BGE). Moreover the way of mobilization affects the resolution of hemoglobin performance [14].

Fig. 7.2 Electropherogram
of the separation of Koln
hemoglobin blood sample (**a**),
and M-Saskatoon hemoglobin
blood sample (**b**), by the use
of CIEF technique. Reprinted
with permission from Wang
et al. [14]. Copyright 2006
Wiley-VCH

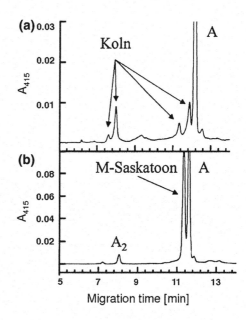

Another example of application CIEF in hemoglobin analysis is the rare hemoglobin variants separation: hemoglobin Koln and hemoglobin M-Saskatoon which both arise on the basis of gene mutation. Figure 7.2 presents electropherogram of Koln hemoglobin blood sample. Peaks raised from HbA and four other peaks characteristic of Koln hemoglobin are visible. Considering hemoglobin M-Saskatoon one characteristic peak of pI 7.003 is observed.

The ability of analyzing rare hemoglobin types makes CIEF a highly specific and sensitive analytical technique [15].

7.3.2 CIEF in Protein Analysis

CIEF has also been used for the analysis of human serum content. Hyphenation of CIEF to Matrix-Assisted Laser Desorption/Ionization Time of Flight Mass Spectrometry (MALDI-TOF-MS) is a very powerful tool in protein analysis. In case of such hyphenation at first the sample is focused and then it is eluted and detected with MALDI-TOF-MS.

Crowley et al. [16] performed complex analysis of human serum utilizing CIEF-MALDI-TOF technique in "off-line" mode. The results provided 160 peaks from ions separated during 40 minutes focusing under the voltage of 20 kV. CIEF-MALDI-TOF can be an interesting alternative against two-dimensional gel electrophoresis widely used in proteomics [16].

Another interesting application in protein analysis is the hyphenation of online solid phase microextraction (SPME) to CIEF with laser-induced fluorescence whole column imaging detection (LIF-WCID). This coupling was first time introduced by Liu and Pawliszyn [17] for analysis of labeled macromolecules. Two advantages of SPME-CIEF-LIF-WCID have been observed. Focusing and desorption are performed at the same time, whereas in conventional SPME-HPLC the separation step is performed after the desorption step. Moreover, long desorption time had no negative effect on peak diffusion. This method appeared to be a simple, fast, and effective way for protein analysis [17].

CIEF was suggested as an appropriate reference method for the determination of chronic alcohol abuse based on transferrin isoforms [18]. In comparison to chromatographic methods CIEF was able to successfully separate all of transferrin isoforms together with their genetic variants [18].

Capillary isoelectric focusing was applied in proteomics in studies on two cancer ovarian cell lines which were genetically and morphologically different from each other. Coupling CIEF with RP-HPLC-MS combined with bioinformatic data analysis enabled better understanding of pathogenesis of protein profiles of neoplastic cells [19].

In another proteomics study CIEF-ESI-MS/MS was introduced for protein analysis in yeast. The result of performed experiment was the identification of 55 % of yeast proteins which constitute 70 % of all proteins responsible for growth process [20].

7.3.3 CIEF in Glycoprotein and Immunoglobin Analysis

Ongay et al. [21] was the first who performed the analysis of glycoprotein $VEGF_{165}$ in insect and *Escherichia coli* cells using CIEF-MALDI-TOF technique. Seven peaks were observed in insect cells which indicated the presence of at least four $VEGF_{165}$ types as a result of post-translation protein modification (Fig. 7.3) [21].

Glycoprotein peptide hormone known as erythropoietin is often used as steroid due to its ability to stimulate erythropoesis. Dou et al. [22] introduced CIEF-WCID technique to α-erythropoietin (EPO-α) and β-erythropoietin (EPO-β) glycoforms separation. Simultaneously, the mobilization step was omitted due to its negative effect on shape and analysis of focused regions. The time of analysis was relatively short and high resolution was maintained throughout whole analysis, respectively [22]. During analysis of EPO-α and EPO-β (Fig. 7.4) seven and eight peaks were observed, respectively. Moreover, on EPO-α electropherogram one peak was not missing (no. 7) and in case of EPO-β the first peak was barely visible (Fig. 7.4).

The results showed that CIEF-WCID technique could be a promising tool for the identification of erythropoietin in biological fluids, and moreover could be used in determination of doping among sportsmen. Capillary isoelectric focusing can also be used in recombinant immunoglobulin analysis (rIgG). Tang et al. [23]

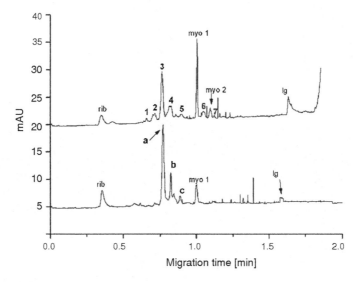

Fig. 7.3 Electropherogram of VEGF$_{165}$ glycoprotein expressed in insect cells (**a**), and *E. coli* (**b**), myo$_1$, myo$_2$—horse mioglobin, Ig-ß lactoglobulin, (*1–7*) VEGF$_{165}$ peaks expressed in insect cells, (*a–c*) VEGF$_{165}$ peaks expressed in *E. coli* cells. Reprinted with permission from Ongay et al. [21]. Copyright 2009 Wiley-VCH

Fig. 7.4 Electropherograms of α-erythropoietin (EPO-α) i β-erythropoietin (EPO-*β*) utilizing CIEF-WCID method. Reprinted with permission from Dou et al. [22]. Copyright 2008 Elsevier

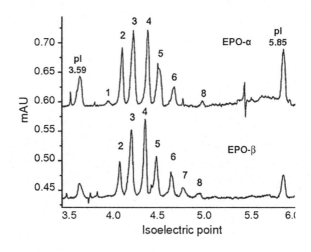

compared one- and two-step CIEF in rIgG analysis. Coated capillary was introduced and the separation was performed in 2 % ampholyte medium solution with 0.4 % methylcellulose addition. As a result it was proven that two-step CIEF provides better resolution and repeatability compared to one-step CIEF when considering routine analysis (Fig. 7.5) [23].

Janini et al. [24] introduced capillary isoelectric focusing for determination of murine monoclonal antibodies (MU-BS). The antibodies variation characterization

Fig. 7.5 Comparison of one-step and two-step CIEF in rIgG-1 analysis. Reprinted with permission from Tang et al. [23]. Copyright 1999 Elsevier

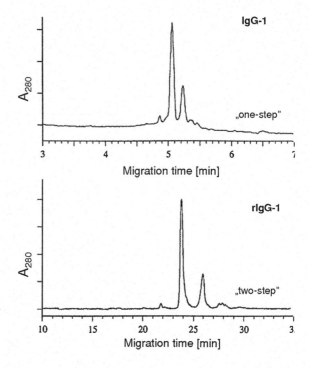

by CIEF provides the detection of chemical and post-translation modifications and antibodies purity determination, respectively. The results showed that this technique can be easily validated with migration time reproducibility of 0.5 % RSD. Nevertheless the preparative isolation of antibodies focused in capillary has many limitations such as possibility of contamination by ampholyte solution introduced into the capillary, fragmentation of peaks and complicated isolation procedures. Therefore the preparative analysis in conventional CIEF is not effective enough and does not provide products of high purity. Meert et al. [25] used pH gradient in immobilized capillary and succeeded in separation antibodies isoforms different from each other by charge in the range of 0.1 unit. Preparative isolation of antibodies under investigation provided the molecular characteristic of antibodies, moreover, it has been proven that immobilized pH gradient in capillary may enable high resolution and preparative isolation of the analytes [25].

References

1. R.A. Mosher, W. Thorman, Experimental and theoretical description of the isotachophoretic behavior of serum albumins. Electrophoresis **11**, 292–298 (1990)
2. J.R. Mazzeo, I.S. Krull, Improvements in the method developed for performing isoelectric focusing in uncoated capillaries. J. Chromatogr. A **606**, 291–296 (1992)

3. S.M. Chen, J.E. Wiktorowicz, Isoelectric focusing by free solution capillary electrophoresis. Anal. Biochem. **206**, 84–90 (1992)
4. W. Thorman, J. Caslavska, S. Molteni, Isoelectric focusing with electroosmotic zone displacement and on-column multichannel detection. J. Chromatogr. A **589**, 321–327 (1992)
5. M. Zhu, R. Rodriquez, D. Hansen, T. Wehr, Capillary electrophoresis of proteins under alkaline conditions. J. Chromatogr. A **516**, 123–131 (1990)
6. S. Hjertén, M. Zhu, Adaptation of the equipment for high-performance electrophoresis to isoelectric focusing. J. Chromatogr. A **346**, 265–270 (1985)
7. R. Rodriquez, C. Siebert (1994) *Poster presentation 6th international symposium on capillary electrophoresis* (San Diego, CA, 1994)
8. J.R. Mazzeo, I.S. Krull, Peptide mapping using EOF-driven capillary isoelectric focusing. Anal. Biochem. **208**, 323–329 (1993)
9. S. Hjertén, K. Kubo, A new type of pH-and detergent-stable coating for elimination of electroendosmosis and adsorption in (capillary) electrophoresis. Electrophoresis **14**, 390–395 (1993)
10. J. Wu, J. Pawliszyn, Fluorescence imaging detection for capillary isoelectric focusing. Electrophoresis **16**, 1474–1478 (1995)
11. M.S. Jenkins, Three methods of capillary electrophoresis compared with high-resolution a garowe gel electrophoresis for serum protein electrophoresis. J. Chromatogr. A **720**, 49–58 (1999)
12. X. Liu, Z. Sosic, I.S. Krull, Capillary isoelectric focusing as a toll in the examination of antibodies, peptides and proteins of pharmaceutical interest. J. Chromatogr. A **735**, 165–190 (1996)
13. X.W. Yao, D. Wu, F.E. Regnier, Manipulation of electroosmotic flow in capillary electrophoresis. J. Chromatogr. A **636**, 21–29 (1993)
14. J. Wang, S. Zhou, W. Huang, Y. Liu, C. Cheng, X. Lu, J. Cheng, CE-based analysis of hemoglobin and its applications in clinical analysis. Electrophoresis **27**, 3108–3124 (2006)
15. J.M. Hempe, R.D. Craver, Separation of hemoglobin variants with similar charge by capillary isoelectric focusing: value of isoelectric point for identification of common and uncommon hemoglobin variants. Electrophoresis **21**, 743–748 (2000)
16. T.A. Crowley, M.A. Hayes, Analysis of human blood serum using the off-line coupling of capillary isoelectric focusing to matrix-assisted laser desorption/ionization time of flight mass spectrometry. Proteomics **5**, 3798–3804 (2005)
17. Z. Liu, J. Pawliszyn, Coupling of solid-phase micro extraction and capillary isoelectric focusing with laser-induced fluorescence whole column imaging detection for protein analysis. Anal. Chem. **77**, 165–171 (2005)
18. R. Hackler, T. Arndt, A. Helwig-Rolig, J. Kropf, A. Steinmetz, J.R. Schaefer, Investigation by isoelectric focusing of the initial carbohydrate-deficient transferrin (CDT) and non-CDT transferrin isoform fractionation strep involved in determination of CDT by the CronAlcolD assay. Clin. Chem. **46**, 483–492 (2000)
19. L. Dai, C. Li, K.A. Shedden, D.E. Misek, D.M. Lubman, Comparative proteomic study of two closely related ovarian endometrioid adenocarcinoma cell lines using cIEF fractionation and pathway analysis. Electrophoresis **30**, 1119–1131 (2009)
20. W. Wang, T. Guo, T. Song, C.S. Lee, B.M. Balgley, Comprehensive yeast proteome analysis using a capillary isoelectric focusing-based multidimensional separation platform coupled with ESI-MS/MS. Proteomics **7**, 1178–1187 (2007)
21. S. Ongay, S. Puerta, J.C. Diez-Masa, J. Bergquist, M. Frutos, CIEF and MALDI-TOF MS methods for analyzing forms of the glycoprotein VEGF 165. Electrophoresis **30**, 1198–1205 (2009)
22. P. Dou, Z. Liu, J. He, J.J. Xu, H.Y. Chen, Rapid and high-resolution glycoform profiling of recombinant human erythropoietin by capillary isoelectric focusing with whole column imaging detection. J. Chromatogr. A **1190**, 372–376 (2008)

23. S. Tang, D.P. Nesta, L.R. Maneri, K.R. Anumula, A method for routine analysis of recombinant immunoglobulins (rIgGs) by capillary isoelectric focusing (CIEF). J. Pharm. Biomed. Anal. **19**, 569–583 (1999)

24. G. Janini, N. Saptharishi, M. Waselus, G. Soman, Element of a validation method for MU-BS monoclonal antibody using an imaging capillary isoelectric focusing system. Electrophoresis **23**, 1605–1611 (2002)

25. C.D. Meert, L.J. Brady, A. Guo, A. Balland, Characterization of antibody charge heterogeneity resolved by preparative immobilized pH gradients. Anal. Chem. **82**, 3510–3518 (2010)

Chapter 8
Two-dimensional Gel Electrophoresis (2DE)

Ewa Kłodzińska and Bogusław Buszewski

Abstract The chemical compounds, which are present in the environment, increasingly cause bad effects on health. The most serious effects are tumors and various mutations at the cellular level. Such compounds, from the analytical point of view, can serve the function of biomarkers, constituting measurable changes in the organism's cells and biochemical processes occurring therein. The challenge of the twenty-first century is therefore searching for effective and reliable methods of identification of biomarkers as well as understanding bodily functions, which occur in living organisms at the molecular level. The irreplaceable tool for these examinations is proteomics, which includes both quality and quantity analysis of proteins composition, and also makes it possible to learn their functions and expressions. The success of proteomics examinations lies in the usage of innovative analytical techniques, such as electromigration technique, two-dimensional electrophoresis in polyacrylamide gel (2D PAGE), liquid chromatography, together with high resolution mass spectrometry and bio-informatical data analysis. Proteomics joins together a number of techniques used for analysis of hundreds or thousands of proteins. Its main task is not the examination of proteins inside the particular tissue but searching for the differences in the proteins' profile between bad and healthy tissues. These differences can tell us a lot regarding the cause of the sickness as well as its consequences. For instance, using the proteomics analysis it is possible to find relatively fast new biomarkers of tumor diseases, which in the future will be used for both screening and foreseeing the course of illness. In this chapter we focus on two-dimensional electrophoresis because as it seems, it may be of enormous importance when searching for biomarkers of cancer diseases.

E. Kłodzińska · B. Buszewski (✉)
Faculty of Chemistry, Nicolaus Copernicus University, Toruń, Poland
e-mail: bbusz@chem.umk.pl; kojo@chem.uni.torun.pl

B. Buszewski et al. (eds.), *Electromigration Techniques*, Springer Series
in Chemical Physics 105, DOI: 10.1007/978-3-642-35043-6_8,
© Springer-Verlag Berlin Heidelberg 2013

8.1 Introduction

The chemical compounds, which are present in the environment, more often cause bad effects on health. The most serious effects are tumors and various mutations at the cellular level. Such compounds, from the analytical point of view, can serve the function of biomarkers, constituting measurable changes in the organism's cells and biochemical processes occurring therein. The challenge of the twenty-first century is therefore searching for effective and reliable methods of identification of biomarkers as well as understanding bodily functions, which occur in the living organisms at the molecular level. The irreplaceable tool for these examinations is proteomics, which includes both quality and quantity analysis of proteins composition, and also makes it possible to learn their functions and expressions. The success of proteomics examinations lies in the usage of innovative analytical techniques, such as electromigration technique, two-dimensional electrophoresis in polyacrylamide gel (2D PAGE), liquid chromatography, together with high resolution mass spectrometry and bioinformatical data analysis. Proteomics joins together a number of techniques used for analysis of hundreds or thousands of proteins. Its main task is not the examination of proteins inside the particular tissue but searching for the differences in the proteins' profile between bad and healthy tissues. These differences can tell us a lot regarding the cause of the sickness as well as its consequences. For instance, using the proteomics analysis it is possible to find relatively fast new biomarkers of tumor diseases, which in the future will be used for both screening and foreseeing the course of illness.

   The diagnostics used so far was based on one marker for diagnosing a numerous and heterogeneous group of patients, which causes lack of precision and peculiarity, whereas proteomic examinations enable to use several protein markers at the same time. Proteomics also enables more precise diagnosis of the disease, preparing new kinds of therapy, depending on individual patient's needs (medical testing by the patient's bed), and monitoring results of the treatment. Thanks to proteomics it is possible to learn more about pathogenic mechanisms, to prepare new medications and vaccines and also the detailed examination of pathogens. The examination of proteome can be divided into three primary stages: acquiring material and its prefatory treatment, the proper analysis of proteins' profile, and the analysis of the received data. Due to the fact that proteome is a very dynamic structure, properly sampled material and its storage is a very important stage of proteomic analysis. Some problem can be found during isolation, for instance from a fragment of tissue cells changed by the disease. For the purpose of very precise separation of healthy cells from sick ones there is a new method to be used—called LCM. This method consists in accurate preparation of tissue under microscope, using laser. There are various methods used for the examination of proteome, of which the most basic ones are:

- Two-dimensional electrophoresis, mass spectroscopy and X-ray crystallography.
- Two-dimensional electrophoresis which consists in separation of protein mix, based on isoelectric point and then their size.

- Mass spectrometry which enables to identify the proteins based on the differences in relation to their mass and charge.
- X-Ray Crystallography and NMR spectroscopy which are used to analyze three-dimensional structure of proteins, based on analysis of defractional images which came into being as a consequence of bombarding the compound with X-rays.

Proteomics examinations, therefore, include separation of proteins, its identification, quantity measuring, sequence analysis (bioinformatics), examination of structure and examination of mutual interactions, and modifications of proteins.

In this chapter we are focusing on two-dimensional electrophoresis because as it seems, it may be of enormous importance when searching for biomarkers of cancer diseases. Also, the basic theory of this technique is not described in detail in Polish, most of the publications on this topic are in English.

This technique is not a novelty in the field of science. It was first introduced in 1975 independently by Patrick O'Farrell from California University in San Francisco and Joachim Klose from Medical University of Berlin. It is a time-consuming technique, which requires a lot of patience but also has enormous value in the area of proteomics examinations [1, 2].

8.2 Theoretical Basis of 2DE

Two-dimensional gel electrophoresis (2DE) is a chemicophysical technique used for proteins separation, and which is combined of two or sometimes even three basic electrophoretic methods. Its scope is within three stages. The first stage is a procedure of separation of proteins by isoelectric concentration on the carrier strap in a horizontal system (called first direction—One dimension). The second stage is to transfer the separated proteins on the vertical polyacrylamide gel electrophoresis, where proteins are being separated based on their molecular mass. The final procedure consists in gel discoloring and obtaining two-dimensional map of the proteins' decomposition on the basis of the value of isoelectric point (Ip) and the weight of molecular protein (Fig. 8.1) [2].

In the standard procedure the first direction (dimension) isoelectric concentration was made in the polyacrylamide gel placed in a glass or plastic pipe. Inside the gel there were such substances as urea, detergents, reducing substances and ampholytes which give a possibility to create pH gradient value in the electrical field. The sample usually was introduced from the cathodic side—alkaline value of pH gradient. This method, called "O'Farrel" method [2] was used for over two decades with no major modifications. However, the potential of the method for creating a base of protein structures was discovered rather late, only in 1978 the procedure of creation and analysis of some gels with reproducible conditions was introduced and the "Human protein collection" was developed. Currently, after a few modifications of the method, the immobilization, deposited on the carrier strap pH gradients are being used. In such gels the pH gradient is formed by

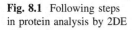

Fig. 8.1 Following steps in protein analysis by 2DE

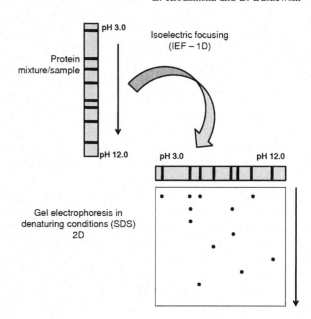

copolymerization of acrylamide monomers with acrylamide derivations containing carboxyl groups and third-line amine groups. This procedure was proposed in 1982 with the cooperation of three research groups. Angelika Gorg from Technical University of Munich was one of the members of those groups. She replaced 2 mm polyacrylamide gels with 2D much more thinner gels (<0.5 mm) immobilized on the stable carrier [3–13].

Many different factors can influence electrophoresis, for instance, isoelectric point (pI), pH value, ionic power, electrical parameters, and the temperature.

Each protein has a characteristic value of isoelectric point. It is such pH value with which the resultant protein charge is equal to zero. The higher the value of pH medium, the more negative the charge of protein. When the medium gets lower value the protein charge gets a positive value. The amphoteric nature of proteins depending on the environment conditions is visible here. With the anode the pH value is relatively low; it is increasing from the direction of anode to cathode. Gel pH by cathode has the highest value. If we add protein sample to the gel and apply voltage the protein will migrate to cathode or anode, depending on the charge. If the examined protein stops it means that pH value and isoelectric point were leveled, in which case protein charge has 0 value and the protein will not migrate under electrical field.

Another very important factor for the analysis is using electrolyte that has adequate value of ionic strength. If the electrolyte strength is too high we can observe a drop in ionic mobility due to friction between ions in the solution. When the strength value gets too low we can observe deficiency of ions which are carriers of electric charge in the solution [2].

The next parameter that should be controlled is temperature. During the process of acrylamide polymerization heat is emitted. This heat can destroy the gel and irreversibly denature protein structure (deactivating them), which significantly worsens the results of electrophoretic separation. It can also result in damaging glass plates and the equipment (deforming plastic elements) [2].

The last of the parameters that influence the process are electric parameters. The strength with which the field affects ion charge is proportional to the intensity of this field, whereas this value is proportional to the voltage put to the electrodes. Within the low values of ionic power there is a simple interrelation following from Ohm's law. It defines the dependence between voltage and electrical resistance of electrolyte. Power supply units of high voltage used for electrophoresis enable maintaining stable value of one of the parameters: field strength, field voltage, or power. Thus, if we establish for instance a stable value of voltage, for the remaining parameters we establish only upper range of value. Usually the protein analysis is performed with stable electric strength, nucleic acid with stable tension value, and the isoelectric focusing with stable electrical power [2].

8.3 Stages of Analytical Procedures Before the 2D Electrophoresis Gel Analysis

8.3.1 Sample Preparation

Proper sample preparation is a very important stage in every analytical procedure. Before the 2D analysis, the proteins in the sample must be submitted for denaturation, reduction, and appropriate solubilization. This procedure is performed in order to completely disrupt intermolecular interactions, so that each spot would be an individual polypeptide in the gel. The main problem when visualizing proteins lies in their heterogenic nature, thus the sample preparation for the analysis should be simple and reproducible. Any protein modifications during the whole process should be avoided, as it can provoke changes in the spot obtained after separation in the gel. So far no universal method for sample preparations has been established. What is needed is a one-stage procedure of extraction which would be simple in preparation and reproducible [2–13].

What should be considered is the process of extraction and solubilization of samples coming from microorganisms, seeds and leaves of plants, as well as from animal tissue. Cells and tissue must be submitted to preliminary fractionation (individually or in combination) using techniques such as grinding in mortar using liquid nitrogen, cutting or homogenization.

The first step in the sample preparation is to separate organelle. Cells can be fractionized when they have different shape, density, or other characteristics that can be marked or determined (charge, antigen, or enzyme presence). Cells that form part of complex tissue (for instance of liver or kidneys), first need to be disrupted from other cells. In some cases cells can be disrupted using chelate environment (removing Ca^{2+} and/or Mg^{2+}). In most cases cells will have to be disrupted mechanically or enzymatically. In Table 8.1 are presented the methods for sample preparation and receiving cell fractions. In Fig. 8.2 can be seen the standard procedure regarding isolation, identification, and characteristics of proteins [2–13].

The first step in cell fractionation is spinning (whirling), but it only allows separating those elements with different sizes. It is possible to achieve appropriate separation level by generating a thin layer of supernatant in the upper part of diluted salt solution which fills the test tube. That test tube is later put in the centrifuge [2–13].

Popular techniques of cells or tissue organelle isolation are techniques which use ultrasounds. Ultrasounds can be used for cells separation or deactivation of cell membrane and for isolating untouched organelle from the inside of the cell [2–13].

A process of disintegration is used for microorganism cells very often. It is a process which results in destruction of cell membrane and releasing protoplasts.

Cell disintegration can be caused by internal factors, related to the presence of incorrect genes, or external factors: chemical, physical, mechanic, and biological.

Table 8.1 Methods for protein and cellular fraction isolation before 2D analysis		
	Physical–mechanical	Homogenization
		French press
		Ultrasonic
		Centrifgation
		Ultrafiltration
	Physical–nonmechanical	Osmotic shock
		Decompression
		Stream bomb
		Freezing and drying
		Temperature decay (destruction)
	Biological	Autolyis
		Phage decay
		Ensymes decay
	Chemical	Antibiotics
		Chelating agents
		Chaotropic agents
		Salting out
		Dialysis
		Extraction
		Detergents
		Organic solvents
		Three-phase separation
		Ionic strength
		Polyetylene glycol
		Precipiation using diffrent dyes

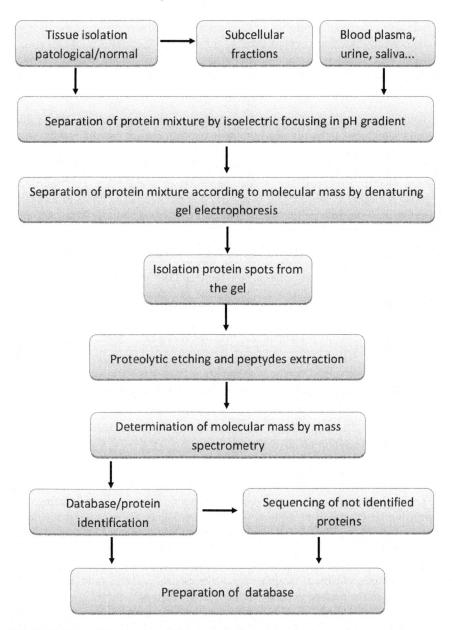

Fig. 8.2 Isolation, identification, and characterization of proteins

Apart from natural disintegration, especially in industrial laboratories, artificial disintegration is also used. Instrumental methods use special machines—disintegrators. The aim of disintegration process is not only to destroy cell wall and to isolate protoplast (which also means irreversible destruction of living cell), but also:

- Selective process of separation of anatomically and structurally faulty cells from the correctly developed population.
- Tightly located disintegration on the surface of cells which are in direct contact.
- Obtaining untouched organelle, which are later used for reconstructing cells that are in direct contact.

Another method used for sample preparation is homogenization. Homogenization techniques can be divided into those that can change osmotic media, from which the cell has been isolated, or those which require "physical strength" in order to achieve cell disintegration. Physical methods can be the following: mortar, blender, compression, or decompression or ultrasounds. For molecules separation mechanic blenders are used, they can be different in terms of technical advance, starting with very simple domestic ones and ending with very fast blenders, equipped with specially designed blades and chambers (for instance Virtis Tissue Homogenizer). Mechanical procedures of separation at this stage are completed with different kinds of organic solvents and/or detergents helpful with denaturation and molecule isolation (for instance DNS with histones). If we want to isolate particular molecules we need to make sure that they do not get destructed by strong enzymes (e.g. RNA during extraction with RNA). This can be achieved by cooling test tube or adding organic inhibitors [14].

After homogenization distracting factors should be removed from the cell (e.g. salt ions). For some of the components (blood, cell suspension) this can be achieved by using cells dropping with the gravitation–gravitational sedimentation. This procedure uses separation based on natural differences in size, shape, and density of the cell.

To sum up, the basic stages of sample preparation are:

- Cell lysis (disruption of the cell wall);
- Deactivation or removing disrupting substances;
- Preliminary protein solubilization.

Cell disruption can be achieved by: osmotic lysis, cyclical freezing and defrosting, detergent disintegration, (agents superficially active), enzymatic disintegration of cell wall (lysis), and ultrasounds. In addition, grinding with (or without) liquid nitrogen is used as well as high pressure (e.g. French press), homogenization or grinding with glass beads and rotating blend homogenizer. These methods have their advantages and disadvantages; they can be used individually or in combination. The choice of the method depends on the type of test tube.

Microbiological cells or plant tissue require drastic conditions of cell lysis due to the structure of wall cells, when more gentle methods can be used for animal tissue. Gentle process of cell disruption (for instance enigmatic lysis) is required for isolation of intact cell organelles (mitochondria) essential for further analysis [2, 9, 10].

During (or after) cell disruption interfering compounds must be removed (e.g. proteolytic enzymes, salt, lipids, polysaccharides, nucleic acids, or plant phenols). Further analysis is impeded by salt ions and protease—enzymes belonging to the group of hydrolyze. Protease should be deactivated in order to avoid presence of artifacts (additional trails on the protein solubilization map) or loss in molecular protein mass. To achieve this protein inhibitors are added to the sample, as they can modify protein and change its charge. Other means that can stop proteolysis are: warming the sample in SDS buffer (without urea) and protein deactivation by low pH (e.g., precipitating with ice-cold TCA). A very useful method is TCA/ acetone precipitation which minimizes protein degradation and removes interfering compounds, such as salt and polyphenols. It can also be used for enriching very alkaline ribosomal proteins from total cell lysates [2, 15, 16].

Attention should also be paid to protein loss caused by its incomplete precipitation or diverse solubilization.

Moreover, a completely different protein composition can be achieved by extraction with lysis buffer (depending on whether there was acid TCA used or not), which precedes TCA precipitation stage. On the other hand, this effect can be used for enrichment of very alkaline proteins such as ribosomal or nuclear protein coming from total cell disruption [2, 15, 16].

Salt ions can interfere in the electrophoretic separation and should be removed when their concentration is too high (>100 mM). Otherwise, it can cause horizontal or vertical trails in the gel. Salt also increases the conductivity of isoelectrofocusing gel—IEF, which prolongs the time needed for reaching the steady state. Salt can be removed by dialysis, protein precipitation from TCA or organic solvents (e.g., cold acetone). One of the alternatives is to use immobilized pH gradients on the carrier strap. In this case the sample is desalted in the gel by applying very low voltages (100 V). This procedure takes a few hours and paper beads beneath the electrodes should be replaced several times. In order to remove from the biological material (e.g. brain tissue) lipid compounds (fats) organic solvent extraction is used (cold acetone/ethanol). During this stage some of the proteins can be lost, due to the fact that some proteins are solubilized in the organic solvents, or that precipitated proteins do not always resolubilize. Alternatively high-speed centrifugation is used and subsequently lipid-layer is removed.

Polysaccharides and nucleic acids can interact with ampholytic carrier and wit proteins which raise the streaks on the map of protein decomposition. Moreover, these macro-molecules can increase the viscosity of the solution and can result in obstructing the pores in polyacrylamide gel. Other methods for removing nucleic acids are: digestion by a mixture of protease-free RNAses and DNAses or by ultracentrifugation [2, 15, 16].

Phenols present in the plant material (mostly in plant leaves) may interact with proteins provoking spots in the gel. Polyphenol compounds may be removed either by binding polyvinylpolypyrrolidone (PVPP) or by precipitation with TCA and other extractions with cold acetone. Sometimes, excessive protein amount in the sample might become a problem as it worsens the resolution and detection of low amount of protein by masking them and by limiting the amount on the gel. Albumins, which constitute over 60 % of protein components in the plasma, are the main problem as they can interact with other proteins in the solution [2, 15, 16].

8.3.2 Protein Solubilization

After cell disruption and removal of interfering compounds, single polypeptides must be submitted for denaturation and reduction in order to disrupt the intermolecular interferences. Another stage is protein solubilization to maintain proper charge on protein surface. Sample solubilization is frequently performed in buffer containing chaotropes (urea or thiourea) nonionic and amphoteric detergents (e.g., Tryton X-100 or CHAPS), and reducing factors. The most common sample solubilization buffer is based on O'Farrell's lysis buffer which can be modified (by 9 M urea, 2–4 % CHAPS, 1 % DTT). Lysis urea buffer is not the ideal for solubilization for all types of proteins, especially protein membrane or other highly hydrophobic proteins. Obvious improvement of solubilization of hydrophobic proteins comes from the use of thiourea and new amphoteric detergents, e.g., sulfobetains (NDSB) [2, 15].

8.3.3 Chaotropic Substances

Denaturation consists in changing the spatial structure (conformation) of proteins and leads to biological activity decay. It can be obtained in a chemical way: chaotropes, detergents, substances reducing disulfide bridge or by heating up. This process can be reversible (e.g., using chaotropes) or irreversible (e.g., thermic denaturation). Chaotrop substances used before 2DE analysis are the kind of substances which in water solutions strongly disturb hydrogen bonds between water molecules. They are effective only with high concentration (ionic strength)—for instance Urea 8 M. They disturb the structure of water causing that water ceases to interact with protein, which is why protein has more "freedom"— it is not stabilized by water molecules surrounding it.

Urea used for sample preparation effectively breaks hydrogen bonds which causes protein disruption and denaturation, however, thiourea (introduced by [17]) is better for breaking the mutual hydrophobic interactions. It is more soluble in concentrated urea solution [17]. Commonly, the best method for hydrophobic protein solubilization is the combination of: 5–7 M urea and c = 2 M thiourea in conjunction with appropriate detergents (e.g. CHAPS). The problem is that urea exists in equilibrium with ammonium isocyanate which reacts with α-amino group of protein and with rest of α-amino lysis group. This results in artificial connections such as blocking the peptide in free amino group and introduces non-homogeneous charge distribution on protein surface [2, 15–17].

8.3.4 Detergents

In order to prevent interactions between hydrophobic protein parts surfactants are added to the sample. Sodium Dodecyl Sulfate—SDS is the commonly used detergent. It is recommended to heat proteins in SDS solution in order to solubilize them. Before further analysis sample with protein should be diluted with 4-fold excess of (thiourea/urea) lysis buffer. If SDS concentration gets lower than 0.2 % additional streaks may appear on the protein map. A major problem in analysis of micropreparative protein with method 2-DE is to receive proper dilution as the amount of sample volume applied to IPG gel (strip) is limited. Currently nonionic or zwitterionic surfactants are preferred. The most popular nonionic detergents are: NP-40, Triton X-100, dodecyl maltoside. Unfortunately NP-40, Triton X-100 are not very effective with solubilization of very hydrophobic membrane proteins. On the other hand, amphoteric detergents such as CHAPS and sulfobetaines are more suitable and can be used with the interfering factors like urea and thiourea for solubilization of protein membrane. Moreover, SDS effectively masks the original protein charge present in a given electrolyte. SDS presence brings many advantages:

- Most of the proteins are solubilized in SDS electrolytes, especially after reduction of disulfide bridge;
- Protein separation occurs in accordance with their molecular mass;
- Coloring of protein-SDS complex is much more efficient than coloring protein only;
- SDS presence effectively eliminates enzymatic degradation of protein during separation.

Currently, there is no universal solvent which could be used for protein membrane solubilization. Most protein membranes cannot be solubilized in single nonionic or amphoteric detergents [2, 3, 6–13].

8.3.5 Reducing Agents

Reduction and prevention of oxidation of disulfide bonds is a pivotal step in sample preparation. In order to obtain a complete protein lysis it is necessary to use reducing agents for intra and intermolecular disulfide bonds reduction. The most popular reducing agents are: dichlorodiphenyltrichloroethane—DDT, and dithioerythritol—DTE, which are applied excessively, in concentration up to 100 mM. Unfortunately, these agents are weak acids, their pK values are between 8.5 and 9. This means that they will ionize in basic pH and they will run short in alkaline gel due to the migration to the anode during IEF. Moreover, DDT and DTE are not proper for the reduction and solubilization of protein containing high amount of cysteine such as wool keratins [2].

8.3.6 Protein Enrichment

In order to prevent loss of analytes, protein extract should not be diluted. Optimal protein concentration is 5–10 mg/ml, whereas minimal concentration should not get lower than 0.1 mg/ml. In diluted samples and with high salt concentration time of IEF analysis may be prolonged, which is why those samples should be done with no salt. To remove salt the protein is precipitated with mixture of TCA/ acetone. Diluted samples with low salt concentration can be directly applied on dry IPG strips, which again expand in sample solution. In this case stable urea, CHAPS and DTT are added to the sample till obtaining the required concentration.

It is possible to store the extract for a short time (few hours or one night) only in refrigerators (4 C). If a longer storage time is needed, samples should be kept in the freezers with the temperature −70 °C, in dry ice (−78) or in liquid nitrogen (−196). Samples should not be alternatingly frozen and defrosted. Small amount of the solution (100–200 µl) can be frozen only once [2].

8.3.7 Simplified Fractionation Procedure

Since no method for proteins amplification, similar to PCR technique has been introduced, often it is beneficial to perform the procedure of sample prefractionation. This procedure reduces the complexity of the sample. By performing this method we can obtain information regarding topology of the analyzed proteins.

This can be performed by:

- Isolation of types of cells from tissue: FACS and LCM;
- Isolation of cell or organelle fragments for instance by centrifugation in sucrose gradient or free flow electrophoresis (FFE);
- Selective precipitation of some types of proteins (e.g., precipitation with acetone/ TCA used for ribosome protein);
- Sequent procedures of extraction with increasing buffer solubilization, e.g., water buffer solution, organic solvents (ethanol, chloroform, methanol), extraction solutions based on detergents;
- Chromatographic and electro-kinetic separation methods, e.g., infinity chromatography, capillary electrophoresis (CZE), or isoelectric focusing (IEF) [2, 3, 9, 15, 16].

The main problem when analyzing very complex biological samples (for instance tumor cells) follows from their heterogenic nature. It is therefore extremely important to obtain proper samples, e.g., directly from tumor cells. In order to obtain proper cells for analysis some microanalytical techniques are used, such as: LCM— *Laser Capture Microdissection*, which allows isolation of pure cell population. The difficulty of this technique is the fact that it is time-consuming and requires considerable amounts of protein for analysis. Another method used for obtaining tumor cells form clinical samples is FACS—*Fluorescence Activating Cell Sorting*. Inside the cell, isolation of organelles such as mitochondria is made with centrifugation in accordance with the increase of sucrose gradient in the sample. Other less popular techniques of separation are EFE—*Free Flow Electrophoresis* or methods based on immunoaffinity. Those methods can be easily used for cells that do not have cell wall—mammalian cells; however, in most microorganisms access to organelles is more difficult due to the presence of cell wall. In that case use of lysis methods is required as they efficiently disrupt the cell wall and at the same time are gentle enough to leave organelles intact. Gentle methods of cell disruption can be achieved by mechanical treatment or using hypotonic solutions. However, the quality of sample prepared in this way may result insufficient. Sample loading capacity on the IEF gel is small and a whole cell lysate may not provide sufficient quantities of lower abundance proteins, which may affect their dyeing in the gel. Protein precipitation procedure allows increasing their amount in the sample, whereas the total mass and their charge are maintained. Precipitation of protein with the TCA/acetone is used in order to enrich the alkaline proteins such as ribosomal proteins from total cell lysates [2, 9, 10, 12].

Chromatographic methods are commonly used for enrichment of small amount of proteins. Chromatography based on hydroxyapatite $[Ca_{10}(PO_4)_6(OH)_2]$ and heparins is used as well as hydrophobic interactions chromatography. Anther approach is based on electrophoretic separation in accordance with isoelectric point in the liquid phase—EFE. In addition, rotating, multichamber device or multifunctional electrokinetic apparatus is used for preparation of samples, where samples are separated by charge or size. These techniques permit protein detection when their concentration in the solution is not too high, due to the fact that albumins, which interfered with other proteins, were removed [2].

Main disadvantage of those methods is that they require usage of advanced equipment for sample pre-fractionation and also require very qualified personnel. Figure 8.3 summarizing information regarding sample preparation for 2DE analysis [18].

8.4 First Dimension: Isoelectric Focusing

Isoelectric focusing of proteins is done with usage of pH value gradient. Proteins are amphoteric molecules and depending on the environment may exist as acidic forms or alkaline forms. In alkaline environment acidic groups present in the protein molecule are negatively charged, whereas in acidic environment alkaline groups are charged positively. Resultant charge of a given protein is a sum of all negative and positive charges present in amino-acid chain. For each protein a resultant curve of the charge can be drawn (titration curve) in the scale of pH value. The crossing of resultant curve with the x-axis makes isoelectric point—pH value, in which the resultant protein charges is zero. This situation is presented as a picture in Fig. 8.4.

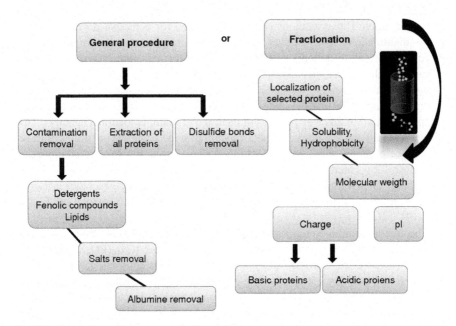

Fig. 8.3 Preparation of protein sample for analysis

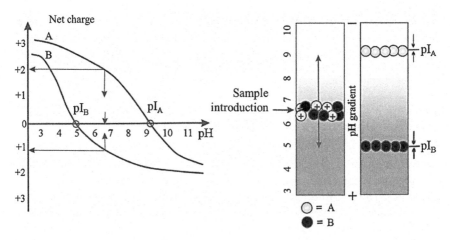

Fig. 8.4 Isoelectric focusing and exemplary curie of net charge for two model proteins, according to [2]

8.4.1 Forming pH Gradient

In practice there are two methods for pH gradient value determination in the gel:

• Gradient formed in electrical field in amphoteric buffer solutions—ampholyte;

- Immobilized pH gradient, in which buffering groups constitute an integral part of gel matrix (IPG straps).

In the first case the gradient rises due to the presence of carrier ampholytes, which are synthesized heterogenic mixture of polyelectrolytes. In this way a series of low molecular mass ampholytes are obtained and they have big buffer capacity in their isoelectric point. Leading ampholytes should meet the following conditions: have high buffer capacity, good conductivity in the isoelectric point, show no biological effects that could cause interferences, and have low molecular mass. Amino acids and peptides present in nature do not have sufficient buffering capacity in the isoelectric points, which is why they should not be used as leading ampholytes [2, 4].

pH gradient comes into being due to the presence of electrical field. Initially gel, together with leading ampholyte has an average pH value, and almost all ampholytes are charged: those of higher value—pI positive and those of lower value—pI negative. After applying electric field negatively charged ampholytes migrate in the direction of anode, and positively charged ampholytes migrate to cathode with their speed depending on their resultant charge. Ampholytes position themselves between the two electrodes as per their pI value and condition the pH of the environment. Ampholytes then lose some of their charge and gel conductivity is lowered. Electrode solutions between gel and electrodes are used in order to stabilize the obtained gradient. Acid solution is used from the anode side (anolyte) and alkaline solution is used from cathode side (catholyte). Furthermore, ampholytes facilitate protein solubilization [2, 4].

Isoelectric focusing is however, a slow process due to the fact that close to the isoelectric point proteins have low resultant charge and therefore low mobility. To obtain high definition a considerable separation segment is required, even with high electricity field power. In denaturing conditions matrix has a considerable viscosity which causes slow protein migration. Additionally, polypeptides after denaturation migrate much slower in the gel than native proteins. Due to the fact that ampholytes are in the solution, pH gradient remains unstable—it is called *gradient drift*. Distribution of spots and streaks that appear in the gel depends on the time; most of the alkaline proteins will be eluted together with alkaline part of the gradient. Another problem is the fact that proteins present in the sample behave like ampholytes, which also modifies the profile of pH gradient. Gradient, therefore, depends on the composition of sample that is analyzed [2, 4].

Problems with instability of gradient were solved after introduction of immobilized pH gradients produced by acrylamide derivatives which have buffering properties. These were submitted to copolymerization with gel matrix. The buffering groups are called immobiline chemicals. They are weak acids or alkaline with general formula: $CH_2 = CH–C0–NH–R$, where R is correspondingly amine or carboxyl group. Detailed chemical structure was proposed in 1990 by Righetti [2].

Substances with acidic features have stable dissociation between pK 0.8 do pK 4.6, and alkaline forms pK 6.2 do pK 12. To make the buffering effective two immobilines are used—one with acidic properties and the other with alkaline ones. In Fig. 8.5 polyacrylamide gel with polymerized immobiline is presented [2, 4].

pH value is defined by immobiline ratio in the polymerizing mixture. PH gradient can be easily controlled by changing the ratio of immobiline and pH can be defined with Handerson-Hasselbach equation [2].

$$pH = pK_B + log\frac{c_B - c_A}{c_A} \tag{8.1}$$

when buffering immobiline is alkali. C_A i C_B is molar concentration of acid and base of immobiline.

When buffering immobiline is acid, the equation is as follows:

$$pH = pK_B + log\frac{c_B}{c_{A-c_B}} \tag{8.2}$$

In the years 1986–1990 computer programs where introduced for designation and simulation of immobilized pH gradients and for designation of ionic strength and buffering capacity. However, this procedure should be compared to and based on the experiment.

Currently, for one-direction analysis fixed (ready) strips with immobilized pH gradient are used. This ensures reproducibility of the results. Dry strips with immobilines can be stored at temperature of −20 to −80 °C for months or even years. Producing strips for isoelectric focusing is completely reproducible, production in factories reduces possibility of making mistakes—which was happening when it was done by people, strips are easily stored, gradient is not modified by sample components, and stable gradients in alkaline environment allow repeatedly separate, even strongly alkaline proteins. Another advantage is variety in sample

Fig. 8.5 Buffer with
immobiline, according to [2]

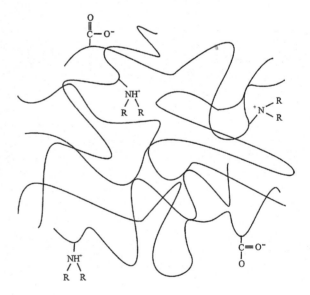

introduction, as dry strips can be submitted to direct rehydration together with sample solution. Also, some supplements or additions can be used with rehydration solution, such as detergents or reducing agents. Protein loss is also low due to the usage of balancing buffers before the 2D analysis. pH gradient can be freely designed on the prepared (ready) strips, depending on user's needs and furthermore, it is perfectly stable and constant.

Commercially available strips for isoelectric focusing after rehydration get around 3 mm thick and are accessible in various lengths—from 7 to 24 cm. Gradient that is included there might be between pH of 3 and 12, but also narrow range of pH value can be used. The lower the range the higher the resolution. The main disadvantage is the price of these kind of strips—the thinner pH gradient are the most expensive [2, 4].

Conditions of conducting the focusing process depend strictly on the properties and features of analyte and sample complexity. It is important to perform the focusing with precisely defined temperature, even if proteins were previously denatured. Depending on the temperature during the process, the position of spots in the gel can be changed. 20 °C is the recommended temperature as it prevents protein carbamylation and urea crystallization. Desalting and using adequate value of electric field are also important parts of the process. pH gradients immobilized on strips having low conductivity and current intensity with focusing process is limited to 50–70 μA per strip. In order to prevent aggregation and protein precipitation as well as heating the gel in some places of the strip the procedure starts with applying low voltage. It is later slowly increased up to maximum strength in the phase of protein focusing (concentration). Changes in the voltage that is applied are completely programmed in the respective phases or in the special ramp. Module of voltage is defined as Vh interval—the amount of applied voltage in time unit. For instance 5,000 Vh can be treated as 5,000 V in 1 h or 1,000 V in 5 h [2, 4].

Measurement of isoelectric point for a given protein on the strip cannot be defined from the surface of working electrodes for reasons of low conductivity. Defining pH gradient basing on the pK value and immobiline concentration is rather complicated due to the presence of urea. Adding isoelectric point patterns is not recommended either because it causes increasing additional proteins in the analyzed sample and might complicate analysis or provoke some mistakes in interpretation of protein solubilization map. One of the methods is drawing a curve for pH gradient. These kinds of works are commonly accessible on the Internet site: www.proteomics.amershambioscience.com and are defined in 20 °C in the presence of 8 mol/L urea. pH values are drawn relative to gel length indicated in percentages. Isoelectric point of protein is defined by comparing the position of that protein in the gel containing SDS to the original position on the focusing strip. This procedure is usually performed with help of computer programs for calibration and data analysis (1-D calibration).

The second method for defining pI is interpolation between marked proteins in the sample of a known value pI. Proteins that are characterized by their intensity, easily seen on every 2D map can be analyzed for identification and recognizing amino-acidic sequence. Theoretically, pI values can be used as values for defining pI in proteins. This procedure is also performed with computer program "2-D calibration".

Besides immobilized chemical compounds, gels contain a mixture of acrylamide and bisacrylamide (4 %T/3 % C, T—these are parameters describing properties of the gel). Parameters that characterize and describe gel properties are:

- Total strength of acrylamide which is counted with the pattern:

$$T[\%] = \frac{(acrylamide + bisacrylamide)\,[g]}{volume\,[ml]} \cdot 100, \qquad (8.3)$$

- Weight ratio of crosslinking agent:

$$C[\%] = \frac{bisacrylamide\,[g]}{(acrylamide + bisacrylamide)\,[g]} \cdot 100. \qquad (8.4)$$

When alkaline gels are used (with narrow range of pH 9–12 units) acrylamide can be replaced with N-dimethyloacrylamide, which guarantees better stability of gel matrix. After polymerization, IPG gel is immersed (washed) in deionized water and subsequently impregnated in 2 % glycerol and then dried. Dry gel is later cut into 3 mm width strips. With the increase of T value, average size of pores decreases [2, 4].

8.4.2 Rehydratation and Sample Loading

Before initiating the process of isoelectric focusing IPG strips should be rehydrated. It is possible to use set for focusing with thermally controlled program for strips rehydration or a separate program for rehydration. There are three ways of achieving this process: integral dehydration (integral phase of the method), separate rehydration (separate process), and rehydration performed outside the focusing set. This procedure can be conducted as active rehydration, protein sample is present in rehydration buffer, 50 V are applied to the strips to move samples further into the strips. Passive rehydration may occur in the presence of sample in hydration solution or without the buffer (cup loading). In this method the strips are rehydrated with no voltages. It takes around 12 h to get a complete process of rehydration.

For IPG strips rehydration in rehydration buffer it is recommended to use two types of buffering solutions. One is for rehydrating process with protein sample and the other is done without the sample. Performing rehydration with the sample minimizes problems connected to protein solubility as well as allows applying

Table 8.2 Buffers for protein separation in the first dimension

Rehydration buffer (without protein sample)	Rehydration buffer (with protein sample)
8–9.8 M urea	8–9.8 M urea
0.5 % CHAPS	1–1.4 % CHAPS
10 mM DTT (dithiothreitol)	15–100 mM DTT (dithiothreitol)
0–0.2 % (w/v) Biolyte (ampholyte)	0–0.2 % (w/v) Biolyte (ampholyte)
0.001 % orange G or bromophenol blue	0.001 % orange G or bromophenol blue

Table 8.3 Equalization buffers before separation in the second dimension

Balancing buffer I Step 1: 10–15 min	Balancing buffer II Step 2: 10–15 min
6 M urea	6 M urea
2 % SDS	2 % SDS
0.375 M Tri-HCl, pH 8.8	0.375 M Tri-HCl, pH 8.8
20 % glycerol	20 % glycerol
130 mM DTT	135 mM iodoacetamide

bigger amount of sample on the IPG strip. The amount of urea, CHAPS, DTT, and ampholyte solution depend on sample (protein) solubility. The amounts presented in Table 8.2 can be used as a hint (indication), however, optimal composition of buffering solution for each type of sample should be defined empirically. pH value gradient of ampholyte solution should be approximate to gradient on the IPG strips [18].

Rehydrating buffers should be prepared just before use, or can be stored in the temperature at −70 °C. The gel on the IPG strips is located in the lower part, in the grooves with no air access. These strips cannot be immobilized, they are covered with a special silicone or a liquid called "Dry Strip" and are rehydrated in the temperature of 20 °C. In case of higher temperature (>37 °C) protein carbamylation can happen. In temperature lower than 10 °C urea crystallization can appear on the strips. Gel hydration is not recommended when there are protein samples with big molecular mass, or with alkaline and strongly hydrophobic proteins. Rehydration of sample in the gel is a method less precise than "cup loading"—it is significant when amount analysis is performed [2]. Before starting a second part of analysis it is necessary to move IPG strips and to store them in proper buffering solutions (Table 8.3).

8.4.2.1 Rehydration With Protein Sample

This method is the easiest and enables using considerable amount of protein (up to 1 mg) and at the same time does not precipitate the protein. To apply the protein on the IPG strip, samples should be located in the rehydrating buffer. For dyeing with silver the required protein concentration is 5–100 μg, and for Coomassie blue 1 mg of protein can be used.

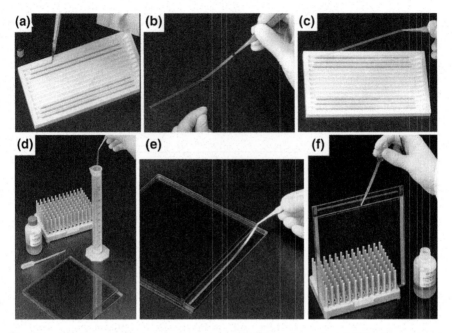

Fig. 8.6 Following steps in sample loading on IPG strip and transfer of IPG strip to second dimension: **a** protein sample loading in Buffet solution together with rehydration on focusing plate, **b** removing the cover sweet from the IPG, **c** loading IPG strip on the focusing plate (*tray*), **d** IPG Flashing IPG strips with Tris-glicine-SDS solution, **e** loading of IPG strips on the second, **f** introduction of agarose solution [18]

8.4.2.2 Rehydration Without Protein Sample

In case of proteins with no properties of IPG strips penetration, sample is applied on the IPG strip after rehydration process is done, just before starting isoelectric focusing. IPG strips should be located in the rehydrating solution; and protein samples should be applied after rehydration directly on the grooves of focusing tray. In Fig. 8.6 the respective stages of sample applying procedure are presented, together with moving it with the strip to the second dimension of electrophoresis [18].

8.5 Second Dimension: Gel Electrophoresis in Denaturizing Conditions

2D electrophoresis in the presence of SDS may be performed on horizontal or vertical systems [2]. SDS is ionic detergent which bonds the proteins, enables their separation based on their molecular mass. Small proteins move fast in the gel and bigger ones stop and remain in the upper part, where sample was applied. After adding SDS to the proteins we get the protein-SDS complex, its electric charge to the mass ratio is stable, and amounts 1 g of protein to 1.4 g SDS [2]. The horizontal setup is ideal for ready-made gels, whereas electrophoresis in vertical system is used for parallel analysis. The most commonly used buffers for the second dimension are nonlinear buffers in Laemmli system and their modifications. For special purposes other systems are used, like borate buffers, for separation of strongly glycoside proteins. The most recommended are gels of 20×25 cm^2 size and 1 mm thick [2].

8.5.1 Methods of Protein Dyeing (Staining): Detection

After finishing 2D procedure separated proteins must be stained (detection) with universal or specific methods of dyeing (staining) the spots (Fig. 8.7, Table 8.4). Concentration of proteins in a single cell can differ even by 6–7 orders of magnitude between rich protein cells and cells poor in protein material. This is a real challenge for almost all existing methods of protein detection. Methods of protein visualization (spot form) should be characterized by high sensitivity, reproducibility, low limit of detection, give a linear response to analyte, and enable subsequent protein identification, e.g., when using mass spectrometry (MS). Unfortunately, there are no universal methods for dyeing (staining) proteins which would meet all the requirements for proteomic analysis [2].

Methods for protein staining in the gel:

- Staining with anionic dyes (*Coomassie blue*);
- Negative staining (opposite) dyeing with metal cations and silver staining;
- Fluorescence staining;
- Radioactive isotopes—autoradiography, fluorography, or imaging with phosphor.

Most of these staining methods require protein solubilization for a few hours in ethanol/acetic acid/water (H$_2$O) to remove compounds which could interfere or interrupt detection [2].

Coomassie brilant blue (CBB) staining is commonly used for detection of proteins in the gels. Its advantage is good price, easy use, and conformity with later protein analysis and their identification (e.g. with mass spectrometry). The main limitations of CBB are related to insufficient sensitivity, which makes it unable to detect proteins present in the sample in a very small quantity (detection limits for CBB is 200–500 ng of proteins in the spot). The use of this method is compatible

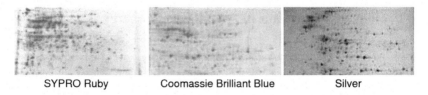

    SYPRO Ruby   Coomassie Brilliant Blue   Silver

Fig. 8.7 Dying methods for protein spots [18]

Table 8.4 Methods for dying of proteins [18]

Protein dyeing	Selectivity	Linear	Cost	Aparatus	Compatibility with MS analysis
Coomassie blue	10–25 ng	2 orders of magnitude	+	Densitometer	+++++
Silver dyeing	0.5–1 ng	1 order of magnitude	+++	Densitometer	–
SYPRO Ruby	0.5–1 ng	1 order of magnitude	+++++	Densitometer	+++

with subsequent identification by mass spectrometry as staining dyes is completely washed out from analyzed proteins. What is also very important is the purity of reagents what prevents sample pollution and additional signals on mass spectrometry. Usually staining the gel with CBB is recommended as well as analysis of the image of a given gel to obtain comparison of spots intensity in the gel. Proteins (spots) in the gel are later cut out with scalpel or automatic tools compatible with software for 2D, submitted to enzymatic cut, and finally analyzed by mass spectrometry. If the staining sensitivity with Coomassie is not sufficient, the gel can be stained again with silver and re-scanned. Then the interesting spots can be found and image can be analyzed again. If also this time the result is not satisfactory—electrophoresis for this sample can be done, using narrower range of pH value and applying bigger amount of sample.

With reverse staining protein bound metal cations (e.g. K, Cu and Zn) are usually less reactive than the free salt in the gel. Thus the speed of free or weakly bounded ions precipitation from insoluble salt is slower in the places occupied by proteins than in the protein-free spots. Due to zink or imidazole zink usage insolubilized salt is precipitated, whereas in the place where protein remains the precipitation process is much slower. This generates transparent spots with protein and dark non-transparent background where insoluble salt is. Silver staining is much more sensitive than CBB and negative (reverse?) staining. Detection limit is smaller than 0.1 ng protein in the spot. However, this method is not reproducible. When using aldehyde the sensitivity of this method increases. This, however, makes it impossible to perform further protein analysis, e.g., with mass spectrometry (MS).

Better and more reliable results as far as sensitivity and detection is concerned are obtained by marking the protein with radioactive isotopes or fluorescence compounds. Marking the protein is performed with the following isotopes: ^3H, ^{14}C, ^{32}P, ^{33}P, ^{35}S, ^{125}I or ^{131}I. After visualization is finished, a technique that allows identification and protein quantitation should be used, e.g., mass spectrometry [2].

8.6 Protein Identification by Mass Spectrometry

Prior to peptides and proteins identification with mass spectrometry proteins are cut out of the gel, dyed out (washing the staining agent from a given spot), submitted to enzymatic cut, and peptide extraction. It should be highlighted here that manual cutting out the spots from the gel is very laborious and inaccurate. It is recommended to cut the spots with the controlled digital camera.

Mass spectrometry is the most frequently used method of detection of proteins separated in the gel. It is an important analytical tool used in clinical proteomics, especially for detection and identification of some biomarkers of diseases. In this method molecules are ionized and later analyzed as ions. The created ions have mass (m) and charge (z). During the process of fragmentation ions, radicals and inert (neutral) molecules are created. Ions separation in mass analizator appears due to the ratio of mass to charge. Then ions translocate to the detector where ionic current signal is changed to electric signal, which with the use of register (recorder?) is converted to mass spectrum. In the last decade ESI and MALDI became much more useful for protein identification with mass spectrometry [2].

8.7 Computer in the Visualization of Analysis Results

Comparing complicated maps of protein spots is possible only with use of appropriate computer software. After staining the gel, places where proteins stopped are analyzed. For this purpose computer systems showing the results of analysis are used. The first step for visualization of achieved results is taking the photograph of the gel in digital format. For this purpose devices that have modified scanners, laser densitometry, CCD cameras, fluorescence cameras and projectors for phosphorus imaging. Next step is computerized processing of the photo. This stage is not fully automatic [2]. Densitometry device is an apparatus that facilitates moving digital photos of the gel to the computer's memory. It is a photo-electric device which measures density of optical materials (transparent and half-transparent). Image working involves reduction of spots' background and removing horizontal and vertical streaks. This procedure is fast and usually there is no need for analyst intervention. Individual spots on the map of protein are not identified, they are only defined by quantity—this step is automatic. Unfortunately most of the programs for image analysis do not identify all the spots correctly, especially when resolution is low. It is then necessary to manually correct it having as reference the originally stained gel. The obtained map of protein is compared to database which gives possibility to locate specific spots in both gels [19]. Attention should be paid that the pictures should be done in grey and have high resolution (definition). Also the densitometers used for scanning the images should be calibrated with adequate software.

Analysis of obtained protein map is used for:

- Comparing maps of proteins for healthy and bad (sick) cells;
- Constructing average gel images for few parallel analysis of the same sample;
- Detecting new or modified proteins;
- Defining location of protein in the gel for the later method of cutting the protein out;

- Defining and featuring families of similar proteins;
- Statistical analysis of experimental results;
- Database creation;
- Data bonding of 2D—mass spectrometry.

In order to compare few gels, the image of virtual gel used for reference should be done, which contains all possible for detection spots in a given experiment. These kinds of gel photos are available in the online databases (Fig. 8.8) [19].

Comparative analysis of obtained protein profiles is performed on samples from individual patients or on the material from a group of patients clinically diagnosed. Analysis of a group of a few hundred with controlling group should be done after performing inoculation tests for both of these groups for the chosen group to be representative. For this purpose the proteomic bio-informatics is used, a new discipline of science which enables processing algorithms and programs used for gathering and analysis of results of proteomic experiments. Most often with bio-informatics, analysis of matrix correlation is used. Analysis of main components, cluster analysis (grouping objects with similar features) or even neuron system is used [1].

8.7.1 2D Database

DynaPort 2D is and advanced online database with dynamic access to proteins and gels. The base of the gels has a function of reference for analysis performed in those gels, and also is used as a tool for navigation. It enables switching between experimental and anticipated data. Each spot identified in the gel is connected to complex information regarding given protein. Thus, we can obtain information regarding the protein that is of interest to us (e.g. pI, molecular mass, etc.). The majority of existing databases for two-dimensional electrophoresis is saved in Word-2DPAGE format. Database structure is based on programmed pages in

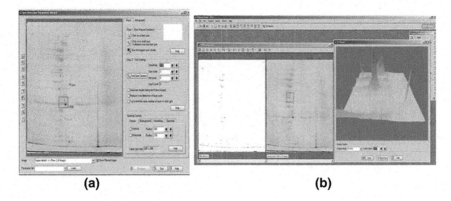

(a)   (b)

Fig. 8.8 Exemplary map for protein spots **a** and differential map for two gels (**b**)

HTML format. Information regarding each spot is connected to individual HTML page. A dynamic system of online databases relates proteins maps with gels used in 2D. Databases do not contain the information regarding every existing protein; they are continuously updated with information of newly found proteins [19].

8.8 Summary

Electromigration techniques became popular methods of giving possibility to separate given components from the analyzed sample. Among these techniques in the field of proteomics 2D gel electrophoresis has a special place, as it enables separation and analysis of individual protein fractions. After performing one- and two-dimensional electrophoresis a map of protein spots is obtained; and by using bio-informatics it can be compared to the map of, for instance, tumor changes in a given cell. On this ground a potential biomarker can be defined, which does not appear in the tissue of a healthy person, but does appears in the case of a sick person. 2D gel electrophoresis can be therefore used for diagnostics in medicine—especially for diagnosing protein markers causing different types of diseases. It is likely that in the near future this technique will give us the possibility to quickly diagnose tumor processes and to prepare the necessary medication which will save lives.

References

1. D.R. Baker, Capillary electrophoresis. (Wiley, New York, 1995)
2. T. Naven, R. Westermeier, Proteomics in practice: *A laboratory Manual of Proteome Analysis*. (Wiley, New York, 2002)
3. A. Görg, W. Postel, A. Domscheit, S. Gunther, Two-dimensional electrophoresis with immobilized pH gradients of leaf proteins from barley (Hordeum-Vulgare)—method, reproducibility and genetic-aspects. Electrophoresis **9**, 681–692 (1988)
4. A. Görg, W. Postel, S. Gunther, The current state of two-dimensional electrophoresis with immobilized pH gradients. Electrophoresis **9**, 531–546 (1988)
5. A. Görg, Two-dimensional electrophoresis. Nat. Biotechnol. **349**, 545–546 (1991)
6. A. Görg, Two-dimensional electrophoresis with im-mobilized pH gradients Current state. Faseb J. **11**, A1131 (1997)
7. A. Görg, G. Boguth, C. Obermaier, W. Weiss, Two-dimensional electrophoresis of proteins in an immobilized pH 4–12 gradient. Electrophoresis **19**, 1516–1519 (1998)
8. A. Görg, C. Obermaier, G. Boguth, W. Weiss, Recent developments in two-dimensional gel electrophoresis with immobilized pH gradients: wide pH gradients up to pH 12, longer separation distances and simplified procedures. Electrophoresis **20**, 712–717 (1999)
9. A. Görg, C. Obermaier, G. Boguth, A. Harder, B. Scheibe, R. Wildgruber, W. Weiss, The current state of two-dimensional electrophoresis with immobilized pH gradients. Electrophoresis **21**, 1037–1053 (2000)
10. A. Görg, W. Weiss, M.J. Dunn, Current two-dimensional electrophoresis technology for proteomics. Proteomics **4**, 3665–3685 (2004)

11. A. Görg, W. Weiss, M.J. Dunn, Current two dimensional electrophoresis technology for proteomics. Proteomics **5**, 826 (2005)

12. A. Görg, A. Klaus, C. Lueck, F. Weiland, W. Weiss, Two-dimensional electrophoresis for proteome analysis; outdated or still indispensable? Mol. Cell. Proteomics. **5**, S8 (2006)

13. A. Görg, O. Drews, C. Luck, F. Weiland, W. Weiss, 2-DE with IPGs. Electrophoresis **30**, S122–S132 (2009)

14. H.J. Issaq, T.P. Conrads, G.M. Janini, T.D. Veenstra, Methods for fractionation, separation and profiling of pro-teins and peptides. Electrophoresis **23**, 3048–3061 (2002)

15. C. Dennison, *A guide to protein isolation* (Kluwer Academic Publishers, New York, 2002)

16. T.P. Cutler, *Methods in Molecular Bi-ology: Protein Purification Protocols*, vol 244, 2nd edn. (Humana Press Inc, New Jercy, 2003)

17. T. Rabilloud, Use of thiourea to increase the solubility of membrane proteins in two-dimensional electrophoresis. Electrophoresis **19**, 758–760 (1998)

18. V. Piljac, G. Piljac, P. Maricic, *Electrofocusing in polyacrylamide or agarose gels containing ampholine carrier ampholytes*, In: Genetic Engineering. Centrifugation and electrophoresis, (TIZ "Zrinski" Cakovec, 1986)

19. V. Piljac, G. Piljac (eds.), Genetic Engineering. Centrifugation and electrophoresis, (TIZ "Zrinski" Cakovec, 1986)

Chapter 9
Electrochromatographic Methods: Capillary Electrochromatograpy

Michał Szumski

Abstract Capilary electrochromatography is a separation technique that brings together advantages of liquid chromatographic selectivity with high efficiency of CE provided by flat flow profile of the electroosmosis. This chapter provides the basic knowledge on generation of the EOF in capillary electrochromatography and parameters that influence it. Furthermore, attention is paid to the methods used for gradient elution in CEC, which have been one of the problems that restrict the wide use of CEC in chemical laboratories. The chapter also describes the problem of bubble formation in CEC and provides a step-by-step guide of how to perform CEC separation. Moreover, methods of preparation of CEC columns are discussed, including preparation of packed and monolithic silica and polymeric beds. The position of CEC among other contemporary separation methods is also discussed.

Capillary electrochromatography (CEC) is a liquid phase separation method using electroosmotic flow to increase separation efficiency in comparison to liquid chromatography. In numerous papers CEC is regarded as a hybrid of capillary electrophoresis and liquid chromatography (HPLC, TLC), which however, is not totally correct. A common characteristic of CE is using one of the electrokinetic phenomena—electroosmosis (EOF)—to move a mobile phase across the stationary phase. It is well known that this phenomenon is not responsible for electrophoretic separation mechanism and it can, but not necessarily, accompany it. A common feature of liquid chromatography is using separation mechanism; however, it has been pointed out in some works that the electric field present during separation may have some influence on the division of the analytes between mobile and stationary phases; hence, the separation mechanisms in CEC and HPLC are not

M. Szumski (✉)
Faculty of Chemistry, Nicolaus Copernicus University, Toruń, Poland
e-mail: michu@chem.umk.pl

B. Buszewski et al. (eds.), *Electromigration Techniques*, Springer Series in Chemical Physics 105, DOI: 10.1007/978-3-642-35043-6_9, © Springer-Verlag Berlin Heidelberg 2013

identical. Therefore, electrochromatography can be regarded as an independent separation technique.

The first reports on possibilities of using electroosmosis were given by Strain (1939—electrokinetic separation of dyes) [1] and Mould (1952—"electrokinetic filtration" of polysaccharides on alumina bed) [2]. The concept of fast liquid chromatography using electroosmosis for "pumping" of eluent through a micro-column ($d_c = 1$ mm) packed with an adsorbent or along the chromatographic plate (TLC) was proposed by Pretorius [3]. The obtained efficiencies were worse than in LC counterpart, but it was so likely due to poor Joule heat dissipation and problems with construction of the setup. It was in 1981 that Jorgenson and Lukacs first performed CEC separation, using packed 170 μm i.d. capillary, in a manner that is followed till the present times[4]. The authors indicated remarkably high efficiencies of the column. Later, Knox and Grant [5] described the theoretical aspects of application of CEC. Particular attention was paid to obtaining high efficiencies in comparison to analogous LC separation and possibility of using stationary phases of smaller diameters ($d_p < 3$ μm) with traditional or longer column lengths.

At present, two most important electrochromatographic techniques can be distinguished: capillary electrochromatography (CEC) and planar electrochroma-tography (PEC), a method relatively new and still being developed. As the two techniques differ significantly in equipment construction and technical solutions, CEC and PEC will be described in separate sections below.

9.1 Theory

A basic difference between LC and CEC is the way of inducing a mobile phase flow along the column filled with the stationary phase. In LC the flow is provided by pressure generated by precisely controlled mechanical pumps, but in CEC it is a result of the applied voltage. These differences determine both the instrumental requirements and the stationary phase properties. In contrast to LC the stationary phase in CEC plays a double role: on the one hand its properties should provide interactions responsible for the separation of the analytes, and on the other hand it should provide generation of the EOF of practical significance [6].

In CEC a stable and precisely controlled high voltage power supply and a stationary phase able to induce EOF (for example possessing ionizable groups) in the range of required mobile phase compositions and pH values are necessary. The first prerequisite of generation of electroosmosis is creation of electrical double layer on the border of two phases: liquid and solid. The electrical double layer is created spontaneously after contact of the electrolyte (a mobile phase) with the solid phase surface, which is able to possess the charge. Such ability may be a result of dissociation of the surface functional groups or adsorption of the ions from the solution. The surface of fused silica, the most important material to manufacture capillaries and a silica gel (so far an important support for LC stationary phases) can be negatively charged at pH > 2 because of dissociation of silanol groups.

$$\equiv Si - OH \Leftrightarrow \equiv Si - O^- + H^+$$

In the liquid phase a strongly bound to the surface (Coulomb forces) counter ion layer is then created. Such a layer, however, is not able to neutralize a surface charge completely, so another layer is formed which is significantly diffused. In both layers, together with the counter ions, there are also present non-dissociated polarized mobile phase constituents and common ions from the buffer. Such a layer, formed in the electrolyte is called a diffusion layer.

A *plane of shear* is also formed between strongly (compact) and weakly bound layers. The ions of the compact layer (also known as a Helmholtz or Stern layer) do not take a part in electrokinetic phenomena, whereas the ions of diffuse layer (also known as Gouy-Chapman layer) can be continuously exchanged with the ions from the bulk solution. In the first approaches such a system was considered a model capacitor in which a potential drops with the distance between a solid phase surface and a diffusion layer. Such a potential is called a ζ (zeta) potential, and a distance on which its value drops by e^{-1} is called a double layer thickness δ. It is schematically presented in Fig. 9.1 [7].

According to von Smoluchowski equation a linear velocity of the electroosmotic flow (v_{EOF}) is proportional to the ζ potential, which, in turn is proportional to the double layer thickness δ:

$$v_{EOF} = \mu_{EOF} \cdot E = \frac{\varepsilon_0 \varepsilon \zeta E}{\eta} \qquad (9.1)$$

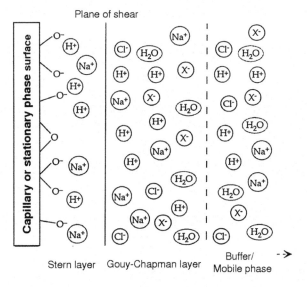

Fig. 9.1 Graphical representation of the electrical double layer

$$\zeta = \frac{\sigma \delta}{\varepsilon_0 \varepsilon} \qquad (9.2)$$

where μ_{EOF} is electroosmotic mobility, ε is relative permittivity of the medium, ε_0 is permittivity of the vacuum, η is mobile phase viscosity, and σ is charge density.

The thickness of the double layer decreases with increasing of the electrolyte concentration. In 0.1 mol/L buffer solution the thickness is ca. 1 nm, while in 0.01 and 0.001 mol/L solutions δ is ca. 3.1 and 10 nm respectively.

When voltage is applied an electric field gradient, E, is created:

$$E = \frac{V}{L_{tot}} \qquad (9.3)$$

where V is applied voltage and L_{tot} is total column length.

As a result of the application of the voltage, a loosely bound ion layer moves towards the electrode of opposite sign. As the ions are solvated a mobile phase moves also with them. Such a flow is called electroosmotic flow. Thus, the electroosmosis (electroendoosmosis) is a movement of a liquid in a capillary structure in a direction parallel to the direction of the applied electric field [7].

The profile of electroosmotic flow is nearly flat, and its velocity is independent of the channel width only if it is larger than 10–40 δ. Such a criterion is easily fulfilled in open channel capillaries, however, in packed CEC columns double layers of the particles and capillary wall can touch each other which make their double layers overlap. According to theory it should reduce the flow significantly, but in practice application of the stationary phases of particle diameter smaller than 1 μm with very diluted buffers (thick double layer) is possible. Hence it seems that the only problem is preparation of the columns using small particles. In CEC the electroosmotic flow is provided by particles' surface in all the places where they do not stick to each other and all the pores in the double layers are not overlapped. In comparison to laminar flow, in EOF the flow velocities in inter-particle and intraparticle channels are approximately the same ($v_2 \approx v_1$), while in pressure-induced flow the ratio of these two velocities is $v_2/v_1 = (d_2/d_1)^2$ so that it means that $v_2 > v_1$ [8]. In packed CEC columns the velocity of EOF depends particularly on stationary phase and not significantly on electroosmosis generated by the capillary wall (despite that EOF in empty capillary is several times faster than in a packed capillary column) [9] (Fig. 9.2).

Because of the friction forces the EOF is slower at the channel wall (a channel can be a capillary, interparticle space, pores in the stationary phase particles, flow-through pores, and pores in the monolithic bed, channel of the chip device, etc.) but its profile is flat. The flat flow profile makes all the analyte molecules move at the same velocity and are eluted as narrow bands which result in sharp peaks of high efficiency (Fig. 9.3, Table 9.1). In turn—a laminar flow, induced by the pressure is characterized by a parabolic profile. As a result the chromatographic band is wider and the separation efficiency drops.

Fig. 9.2 Schematic representation of interparticle and intraparticle EOF in a column packed with porous particles. Reprinted with permission from Rathore [8]. Copyright 2002 Wiley–VCH

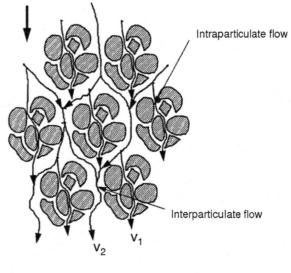

Intraparticulate flow

Interparticulate flow

v_2 v_1

Fig. 9.3 Flat and parabolic flow profiles and corresponding peak shapes

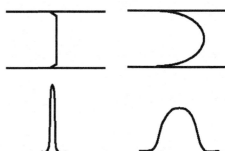

A linear velocity of the mobile phase in the column packed with the particles is given by the following equation:

$$v = \frac{d_p^2 \Delta p}{\Phi \eta L} \qquad (9.4)$$

where Δp is pressure drop across the column, Φ is flow resistance factor, L is column length, and η is mobile phase viscosity.

From the comparison of Eqs. (9.1) and (9.4) one can deduce that linear mobile phase velocity of EOF does not depend on the channel diameter. It suggests that one can use particles of very small diameters. For example utilization of adsorbents characterized by d_p between 0.5 and 3 μm has not been possible in LC (if only typical $L = 25$ cm is assumed) since very high pressures would be necessary to generate mobile phase flow (Table 9.2). The columns packed with very small particles have been available in the past due to UHPLC development. However, it is possible to use such adsorbents in CEC in columns of up to several dozens of centimeters in length.

Table 9.1 Comparison of the efficiencies under pressure and electroosmotic flow conditions of the capillary columns packed with silica particles, according to [65]

d_p [μm]	Pressure flow		Electroosmotic flow	
	L [cm]	N [plates/column]	L [cm]	N [plates/column]
5	50	45,000	50	90,000
3	30	50,000	50	150,000
1.5	15	33,000	50	210,000

Table 9.2 Comparison of capillary column parameters under liquid chromatographic and electrochromatographic conditions, according to [65]

	Pressure flow		Electroosmotic flow	
Particle size, d_p [μm]	3	1.5	3	1.5
Column length, L [cm]	66	18	35	11
Elution time, t_0 [min]	33	–	18	6
Pressure Δp [MPa]	40	120	0	0

Assumptions: L, t_0 and Δp given for a column affording 50,000 theoretical plates at a mobile phase velocity of 2 mm/s

In CEC the electroosmotic mobility can be a function of the following parameters:

- Stationary phase properties:

  – kind of the supporting material (silica, titania, etc.),
  – pore size,
  – kind of chemically bonded ligands (presence of ionizable grous),
  – coverage density,
  – end-capping.

- Mobile phase properties:

  – buffer pH,
  – buffer concentration (ionic strength),
  – kind and concentration of organic modifier,
  – additives.

The EOF/pH relationship is often connected with the kind of support and chemically bonded phase (Fig. 9.4). Several types of stationary phases can be distinguished. Silica-based materials show the relationship no. 2. At low pH values EOF is slow and then speeds up as pH increases. Since the surface charge of silica is negative the mobile phase flows from the anode (positive electrode at capillary inlet) to the cathode (negative electrode at the capillary outlet)—such EOF is called "cathodic" and the electroosmotic mobility, μ_{EOF}, has the sign "+". Sigmoidal shape of the μ_{EOF} versus pH relationship proves the gradual dissociation of the silanol groups. In a case of modification of silica gel support with amine groups (they are of weak base character, so they will be protonated, which results in

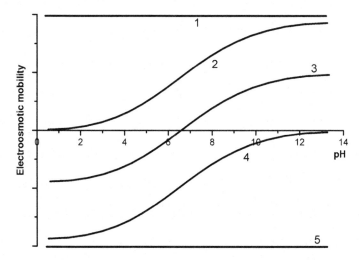

Fig. 9.4 Schematic representation of EOF versus μ_{EOF} relationships for different types of CEC stationary phases

positive charge on the surface) the direction of the EOF may reverse (negative values of μ_{EOF})—the exemplary curves are presented on plots 3 and 4 in Fig. 9.4. The resulting direction of the flow depends on the ratio of the surface functional groups giving negative (silanols) or positive (amine) charge. The end-capped (blocking of free silanols) aminopropyl stationary phase may show only anodic EOF (curve 4 in Fig. 9.4) which is fastest at low pH values. It is advantageous sometimes to introduce such functional groups which provide the same charge within all pH ranges. For example plot 1 is typical for the stationary phase containing sulfonic groups of negative value (dissociated at almost all pH values), while plot 5 characterizes material with ternary amine groups (positive charge at all pH values).

The curves presented in Fig. 9.4 are of course idealized ones. In practice it is often difficult to predict the direction and magnitude of EOF having known the stationary phase parameters. The general observations are as follows [10, 11]:

- stationary phases of higher surface area have faster μ_{EOF} (Fig. 9.5),
- end-capping influences the silanol activity, but it does not influence μ_{EOF} significantly,
- organic carbon content and coverage density do not affect μ_{EOF} significantly,
- in some stationary phases one can observe the influence of so-called "hydrolytic pillow" phenomenon which is related to preferential adsorption of water at the "base" of the chemically bonded stationary phase (e.g., alkylamide phases show such effect),
- introduction of ionized functional groups (SAX or SCX, strong cation/anion exchanger, phases), for e.g., sulfonic phases increase EOF significantly and make it pH independent [9].

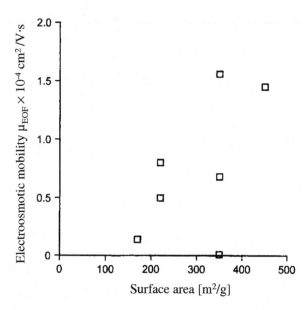

Fig. 9.5 Relationship between electroosmotic mobility and surface area of the stationary phase. The shown phases are: Nucleosil 5 C18 (350 m²/g, EC), LiChrospher RP-18 (450 m²/g), Spherisorb Diol (220 m²/g), Zorbax BP ODS (350 m²/g, EC), Spherisorb S5 ODS2 (220 m²/g, EC), Hypersil ODS (170 m²/g, EC). Measurement conditions: 70/30 ACN/CAPSO $c = 50$ mmol/L pH 9, 53, t_{EOF} marker—thiourea; EC—end-capped stationary phase. Reprinted with permission from Zimina et al. [11]. Copyright 1997 Elsevier

The presence of surface charged groups, or on the other hand, using the electrolyte solution are nor necessary prerequisites to generate the electroosmotic flow. There have been literature data which reported application of modified capillaries or non-charged stationary phases. For example, relatively fast electroosmosis can be observed when using polymeric styrene–divinylbenzene (STY-DVB) monoliths, copolymers of naphthyl methacrylate phenylene diacrylate or copolymers of octadecyl acrylate and PETA (pentaerythritol triacrylate). In such cases the EOF results from adsorption of buffer constituents on the polymeric stationary phase, which in turn is related to surface hydroxyl groups (PETA) or π-electrons of aromatic rings (STY, DVB and PDA).

It can also be derived from Eq. (9.1) that the electroosmotic flow velocity depends (except from zeta potential) on dielectric constant to viscosity, ε/η, ratio. When using liquids of high ε/η one can induce EOF high enough to be of practical significance. Table 9.3 presents properties of such solvents, while some μ_{EOF} experimental data are shown in Fig. 9.6.

Such phenomena are of great practical significance [12–15]. Excepts from traditional electrochromatography in which mixtures of organic solvents with aqueous buffers are used as mobile phases, non-aqueous electrochromatogtography has also been developed [12–21]. There are several reasons for advantageous usage of non-aqueous mobile phases:

Table 9.3 Properties of some solvents applicable as mobile phases in CEC

Solvent	Boiling point (°C)	η [MPa s]	ε	pK_{auto}	ε/η
Methanol (MeOH)	64.7	0.545	32.70	17.20	60.0
Ethanol (EtOH)	78.3	1.078	24.5	18.88	22.77
1-propanol (1-PrOH)	97.2	1.956	20.33	19.43	10.39
2-propanol (2-PrOH)	82.3	2.073	19.92	20.80	9.61
1-butanol (1-BuOH)	117.7	2.593	17.51	21.56	6.75
Acetonitrile (ACN)	81.6	0.341	37.5	>33.3	109.9
Formamide (FA)	210.5	3.30	111.0	16.8	33.63
N-methylformamide (NMF)	~180	1.65	182.4	10.74	110.5
N,N-dimethylformamide (DMF)	153.0	0.802	36.71	29.4	45.77
Dimetylosulfoxide (DMSO)	189.0	1.996	46.68	33.3	23.38
Tetrahydrofuran (THF)	66.0	0.460	7.58	–	16.47
Water	100.0	0.890	78.39	14.00	88.07

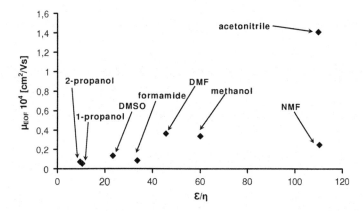

Fig. 9.6 Electroosmotic mobilities measured for pure solvents

- electrophoresis and electrochromatography of the analytes hardly soluble in aqueous or organo-aqueous solvents;
- selectivity adjustment by mixing of solvents of different properties;
- "preparative" CE/CEC—possibility of using capillaries of relatively large internal diameters because of much lower currents (and Joule heating!) in comparison to aqueous electrolytes;
- employing polar interactions during the separation—such interactions are practically non-existent in aqueous solutions due to leveling effect of water. It might be important in, for e.g., chiral separations where all subtle polar interactions between enantiomers and chiral selector can be fully exploited.

Some buffer substances soluble in organic solvents can be used in non-aqueous CEC. The most frequently used are: ammonium acetate, Tris, oxalic acid, ammonium oxalate, and acetic acid/sodium acetate.

9.2 CEC Equipment

CEC instrument is similar to CE instrument and consists of: high voltage power supply (up to 30–50 kV) connected to inlet and outlet electrodes, which are submerged in a mobile phase placed in the inlet and outlet vials, a sample vial, thermostated capillary column and a detector (Fig. 9.7). A significant difference is that high pressure (ca. 12 bar) used to be applied to the inlet and outlet vials during CEC separation which prevents from bubble formation in the outlet of the column (this phenomenon is described later in this chapter). Nowadays most of the commercially available CE instruments have such a feature. The pressure is usually provided by the application of an inert gas (nitrogen or helium) from the external tank. However, it must be emphasized here that most of the instruments are only partially designed as CEC instruments and they do not allow to fully exploit the potential of CEC. Presently, the most important problem in development and wider use of CEC is practical shortage of the instruments capable of gradient elution. Only two companies, namely Microtech Scientific, Inc. and Unimicro Technologies, Inc. have designed and launched multi-purpose setups for micro-HPLC, CE, CEC, and pCEC (pressurized CEC). In 2012, only the latter company still sells the instrument called TriSep®.

Many scientific groups developed their own CEC gradient setups that employed different approaches to change the mobile phase composition:

- step gradient,
- gradient provided by an external HPLC pump,
- gradient generated by electroosmotic pumping,
- gradient generated by increase of voltage.

Fig. 9.7 Schematic representation of a CEC equipment

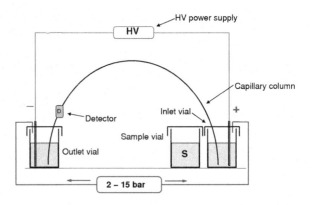

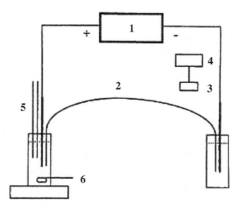

Fig. 9.8 Setup for gradient elution in CEC performed by continuous addition of the mobile phase component characterized by stronger elution power to the inlet vial. *1* power supply, *2* capillary column, *3* and *4* detector and integrator, *5* outlet of the HPLC pump, *6* magnetic stirrer. Reprinted with permission from Steiner and Scherer [22]. Copyright 2000 Elsevier

Step gradient is the easiest to perform in instruments available in the market—separation is stopped for a moment, then the inlet vial is exchanged for a new one that contains a mobile phase of stronger elution power. The process can be performed repeatedly to obtain the desired gradient profile.

In another very simple approach for generating gradient elution the external HPLC pump adds the stronger component of the mobile phase (like acetonitrile) to the inlet vial initially containing the weaker mobile phase. Figure 9.8 presents an example of such a setup [22].

A more sophisticated instrument, based on CE and HPLC components is presented in Fig. 9.9 [23]. Such a device allows for gradient CEC separation at high pressure conditions and sample introduction using HPLC injection valve.

Yan et al. [24–26] developed a gradient CEC chromatograph based on electroosmotic pumping. The instrument, schematically presented in Fig. 9.10, consisted of two independently controlled HVPSs connected to the electrodes placed in mobile phase components. The solvents, via capillaries 1 and 2 were pumped to the mixer (a T-piece) to which a capillary column was connected. Column outlet was placed in a grounded outlet vial. The sample was electrokinetically injected, by detaching column inlet from the mixer and putting it to the sample vial. Gradient elution was completed by generating of two independent electroosmotic flows through the capillaries 1 and 2 in such a way that a mobile phase of desired composition could be created in the mixer. Due to the independent control of the power supplies the mobile phase composition could be adjusted dynamically. Exemplary separation of 16 PAHs is presented in Fig. 9.11 [25].

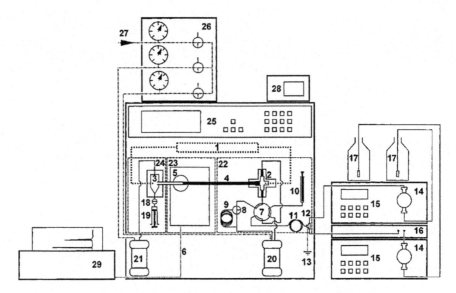

Fig. 9.9 Gradient electrochromatograph. The key parts are: *1* power supply, *2* inlet vial with the electrode, *3* outlet vial with the electrode, *4* packed capillary column, *5* detector, *7, 8, 9* injection valve with splitter, *15* HPLC pumps, *17* bottles with mobile phase components, *26* gas pressure control panel, *27* gas inlet (nitrogen at 1.4 MPa). Reprinted with permission from Huber et al. [23]. Copyright 1997 American Chemical Society

Fig. 9.10 System for gradient elution CEC employing electroosmotic pumping. Detection—LIF. Reprinted with permission from Yan et al. [25]. Copyright 1996 American Chemical Society

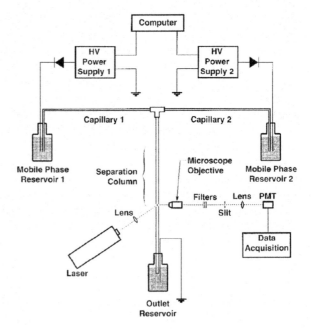

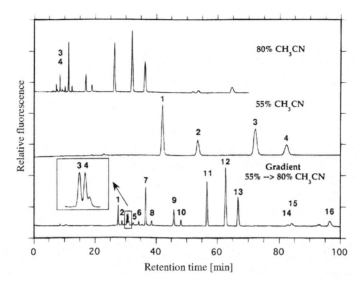

Fig. 9.11 Separation of 16 PAHs performed of the set-up shown in Fig. 9.10. During isocratic elutions the voltage was 20 kV. Reprinted with permission from Yan et al. [25]. Copyright 1996 American Chemical Society

9.3 Problem of Bubble Formation in CEC

Drying of the outlet part of the packed bed or bubble formation just close to the outlet frit are serious problems of electrochromatographic separations performed in packed capillary columns. These two phenomena disadvantageously influence the separation process by:

- problems with detection if the bubbles are smaller than internal diameter of the capillary;
- breaking of the electrical circuit if the bubbles are larger—this stops the mobile phase flow and breaks the separation.

Several theories have been proposed to explain the formation of bubbles. Knox and Grant [27] believed that the main reason for it is heating of the capillary after the voltage is applied. There are, however two drawbacks of such explanation. First, during CEC process the current is much lower than in CE separation preformed in a capillary of the same dimensions. This is so because in CEC the capillary is partially filled with the stationary phase and the mobile phase buffers are of lower concentrations than in CE to achieve higher electroosmotic mobility. Secondly, the above-mentioned phenomena affect column of small internal diameter as well, so Joule heating is well dissipated. Moreover, efficiently thermostated columns (it can be checked by plotting $I = f(V)$ in wide range of voltages) also show such effects.

The second theory indicates the differences between electroosmotic mobilities of packed (lower mobility) and empty (higher mobility) parts of the CEC column (see section CEC columns). Such difference generates a local pressure drop (effect of "suction") in the point of contact of these two sections, i.e., at the outlet frit [28].

The third theory says that the bubbles are generated due to the presence and properties of the frits themselves, for e.g., due to the difference between zeta potentials (and hence—the electroosmotic mobility) of the frit and the stationary phase. It may be a result of the sintering process (which require heating up to 450–500 °C) and thus decomposition of the chemically bound phase [28]. It was also shown that some negative effect can be attributed to the permeability of the frits [29] and their length [30].

The problem of bubble formation can be minimized by careful degassing of the mobile phase, performing separation under high pressure conditions or deactivate the outlet frit by secondary modification with, for e.g., octadecylsilane, to restore stationary phase properties. Also, to stabilize the EOF Seifar et al. [31] and Bailey and Yan [32] applied anionic surfactant (sodium dodecyl sulfate) as a mobile phase additive below its critical micelle concentration. Ye et al. [33] used cationic surfactant cetyltrimethylammonium bromide (CTAB) for dynamic modification of the silica gel used as a stationary phase. After CTAB ions were adsorbed on the silica, their hydrophobic parts allowed for reversed phase separation.

9.4 Performing CEC Separation

In general, performing CEC separation is similar to CE. However, design of the instrument and its functions like way of injection or providing of a gradient elution may have some influence on the individual stages. A general scheme may be as follows:

- Preparation of a mobile phase. In CEC water-organic and non-aqueous phases can be used, both types used to contain buffering electrolyte. To prepare a water-organic phase an aqueous buffer (of desired concentration and pH, sometimes neutral salts are added to increase ionic strength) is prepared separately and then mixed with an organic solvent, usually acetonitrile or methanol, in a proper ratio.
- Column preparation. Before use the column should be conditioned with a properly degassed (sonication + vacuum) and filtered mobile phase. With respect to the column permeability (hydrodynamic flow resistance) and its stability, it can be done either with a manually (or using a special screw) pressed syringe or just with an HPLC pump (when using analytical HPLC pump one should remember to use a proper splitter or work in constant-pressure mode to provide low flow rates!). The column should be conditioned as long as the bubbles stop to appear in the empty part and detection window.

- Setting the capillary in the cassette. Just before setting the capillary in the cassette the detection window should be wiped with a tissue or a filter paper wetted with isopropanol or methanol. The window should be examined for presence of particles or fingerprints. When installing the capillary care should be taken to set the window properly in a UV–Vis or FLD capillary holder, not to break it, and to check whether polyimide coating is not visible in a slit of the detection cell.
- Mounting of the cassette in a CEC instrument.
- Setting the separation parameters: temperature in the cassette, voltage, polarity (usually up to ±30 kV), way of application of the voltage (it is better to use a voltage ramp), setting the detection parameters (wavelength, bandwidth, etc.) separation time, applied high pressure and injection parameters. Injection can be done using HPLC valve, but most often it is performed electrokinetically. Electrokinetic injection relies on generation of EOF for a short period (several seconds) while dipping the capillary inlet in a sample vial. Such injection is described by applied voltage in kilovolts and time in seconds. For highly permeable columns hydrodynamic injection (like in CE) can be used.
- Start the separation. It is vital to keep track on current plot—rapid rises or falls may indicate instability of the column (damage to the frits, settlement of the bed, leakage of the stationary phase, bubble formation, cracking of the capillary under polyimide coating, breaking of the detection window).

9.5 CEC Columns

Developing of liquid chromatography towards miniaturized techniquesis connected with elaboration of columns of smaller and smaller internal diameters [34]. At the early stages the microcolumns were manufactured from stainless steel, Teflon, PEEK, or glass (by drawing the tube filled with stationary phase). Presently, fused silica is regarded as the best material for microcolumns of internal diameter smaller than 0.5 mm. Fused silica capillaries have been used in micro liquid chromatography, and electromigration techniques including capillary electrochromatography. As CEC columns are often regarded as unified (they can be used in LC, SFC, and CEC), the following fused silica features are considered most important:

- high mechanical strength, which is important during high pressure column packing (fused silica capillaries can withstand pressure at least 80 MPa);
- smooth inner surface which provides efficient packing of the adsorbent and allows to avoid "wall effects";
- flexibility provided by polyimide coating;
- facility, durability, and universality of connections with other parts of the system (ferrules, tubings, connections);
- fused silica allows to generate EOF;

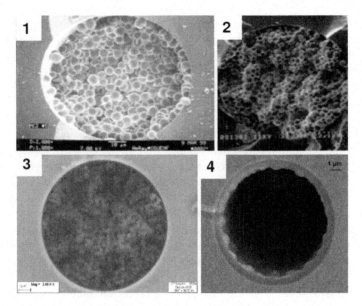

Fig. 9.12 Cross sections of different types of capillary columns: *1* packed [47], *2* fritless [71], *3* monolithic, *4* open tubular [72]. Reprinted with permission from: Chirica and Remcho [47], Dulay et al. [71] and Yue et al. [72]. Copyright American Chemical Society

- fused silica is transparent in UV–Vis range, which allows absorbance and fluorescence-based detection.

The fused silica capillaries are manufactured in a wide range of internal diameters and are used in such fields as gas chromatography, supercritical fluid extraction, and chromatography (as columns and restrictors), electromigration techniques and capillary liquid chromatography. It should be mentioned here that, despite their high mechanical strength, utilization of some thin-walled fused silica capillary columns under high pressure (for example during column packing) may result in formation of micro-cracks under polyimide coating which finally lead the column to rend. Moreover, packed capillary columns tend to dry quickly soon after disconnection from the source of liquid which may irreversibly influence the column efficiency. Also, some chromatographic organic solvents, like acetonitrile, damage the polyimide coating which weaken both of the column ends. Hence, sometimes it is necessary to keep the capillary columns ends dipped in water.

CEC column is a capillary containing a stationary phase in one of the following forms (see Fig. 9.12):

- a particulate-based bed held by the porous frits—packed column,
- monolithic bed,
- fritless bed,
- a layer of stationary phase on the inner surface of the capillary—open tubular column.

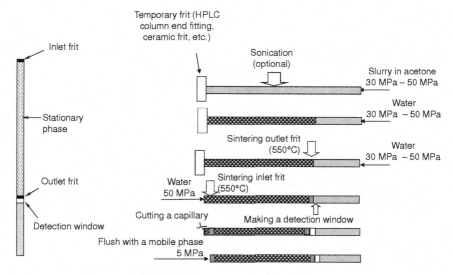

Fig. 9.13 Construction of typical packed capillary column and a scheme of a packing process

9.5.1 Packed Columns

Packed column shown in Fig. 9.13 is a result of the evolution of columns for micro liquid chromatography. In such a column one can distinguish the following parts: a stationary phase bed, inlet and outlet frits (which hold the bed) and empty outlet part with the detection window. The detection window is placed just after the outlet frit in order to minimize extra column volumes, which, in the case of capillary columns are in the range of nanoliters. In CEC the empty part is placed in the outlet vial to close the circuit (see Fig. 9.7).

Efficient packing of the capillary column requires that the analyst have some experience and special equipment. In the literature the following packing methods can be found:

- high pressure packing using liquids and a slurry of stationary phase prepared in a proper (usually low viscosity) solvent (see Fig. 9.13);
- supercritical fluid packing using carbon dioxide. Using SC carbon dioxide allows for packing of longer columns [35–39] due to its low viscosity, changeable density, lack of surface tension. It was reported that crucial parameters for obtaining a good column are using sonication, pressure ramp, proper restrictor length, and rate of decompression of the system after packing;
- packing using centripetal forces—in this method no pressure is used, and it is relatively fast (it takes ca. 15 min to pack one column). A rotating system (3,000 rev/min) is used which crowds the slurry into the capillary [40];

Fig. 9.14 Scheme of an apparatus for electrokinetic packing of capillary columns. Reprinted with permission from Yan [41]. Copyright 1995 Yan C

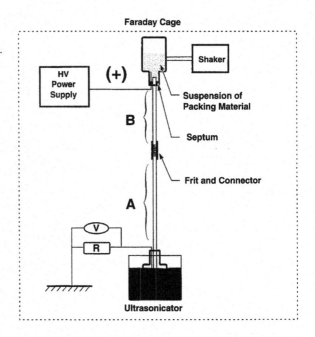

- electrokinetic packing, developed by Yan et al. is another pressure-free method [41]. The method can be characterized by:

  - Lack of pressure during packing due to electroosmotic or electrophoretic movement of the adsorbent. The method is suitable for packing very small particles.
  - It is possible to pack several columns simultaneously.
  - To some extent the electrokinetic packing should result in more homogeneous beds due to flat profile of electroosmotic flow in contrast to pressure-driven flow with parabolic profile.
  - To achieve good results and exploit the advantages of the method the adsorbent of narrow particle size distribution should be used. Particles of different sizes can be characterized by different electrophoretic mobilities which may affect bed homogeneity.

The electrokinetic packing can be performed in the apparatus schematically presented in Fig. 9.14. The columns packed with such a method are characterized by very high efficiency, for e.g., 102,000 plates/m for the stationary phase of 3 μm particles under CEC conditions [24]. Using electrokinetic packing technique Dadoo et al. packed the capillary column with 1.5 μm particle which resulted in column efficiency higher than 700,000 plates/m [26]—see Fig. 9.15.

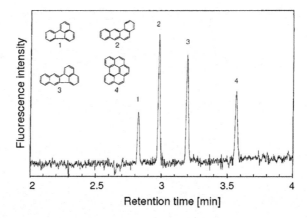

Fig. 9.15 Electrochromatographic separation of four PAHs on C18 stationary phase of 1.5 μm particles. Column properties: L_{eff} = 28 cm, L_{tot} = 36 cm, d_c = 100 μm. Mobile phase: 70/30 ACN/sodium tetraborate c = 4 mmol/L, V = 20 kV, electrokinetic injection 2 kV for 2s. Detection: in-column (through the stationary phase to avoid decrease of efficiency after the passage of the analytes through the outlet frit) LIF (argon laser, λ_{ex} = 257 nm). Number of theoretical plates: 600,000–750,000 plates/m. Reprinted with permission from Dadoo et al. [26]. Copyright 1998 American Chemical Society

9.5.2 Fritless Columns

Fritless columns ("pseudo-monolithic") are capillary columns in which the stationary phase bed is not retained by inlet and outlet frits. They are prepared by packing the capillary with the stationary phase and fixing the particles using several techniques:

- Heating (360 °C) of the column with the stationary phase in the presence of sodium carbonate solution. Under such conditions the particles are sintered with each other, however, the chemically bound stationary phase is then degraded [42].
- Filling the packed column with tetraethoxysilane solution and performing sol–gel reaction (at 100 °C) to fix the particles in the porous silica matrix [43, 44].
- Filling the packed column with potassium silicate (Kasil) solution in Formamide and heating it to 160 °C [45]. Such a method is and adaptation of Cortes method of preparation of ceramic frits for capillary columns [46].
- Filling of the packed capillary with the polymerization mixture based on, for example, methacrylates and completing the thermal or photopolymerization [47, 48]. According to the authors, such treatment does not negatively affect the chemically bound stationary phase (Fig. 9.16).

Fig. 9.16 Particles of the stationary phases (C18, 5 μm) embedded in a polymeric matrix. Reprinted with permission from Chirica and Remcho [47]. Copyright 2000 American Chemical Society

9.5.3 Monolithic Columns

Monolithic bed is defined as a continuous, uniform porous structure, in the form of skeleton of interconnected units, prepared by polymerization, polycondensation or fixing by other methods of the material inside a capillary (sintering, compression), which structure possesses chromatographic properties.

The first monolithic beds were polyurethane foams, which were attempted to be used in chromatography at the end of the 1960s and beginning of the 1970s. Due to unacceptable low mechanical stability and poor separation performance they did not become popular in separation methods. One of the forerunners in monolithic columns was Stellan Hjerten, who in 1989 used polyacrylamide gel compressed in a column for separation of proteins. "Monolithic approach" in preparation of stationary phases for chromatography turned out to be useful particularly in capillary columns—the synthesis can be easily performed in the entire length (basically—any length) of the capillary or in a chosen place. The synthesis is possible to be performed in classical columns as well as in capillaries of 10–20 μm i.d. and chip channels. The following types of the monolithic beds can be distinguished:

- organic—polymeric,
- inorganic—silica, titania based,
- mixed mode,
- carbon (not applied in CEC).

9.5.4 Polymeric Monoliths

One of the advantages of polymeric monolithic columns is their simple, sketchily of course, way of preparation. The scheme of their synthesis is shown in Fig. 9.17 and the detailed description is given below:

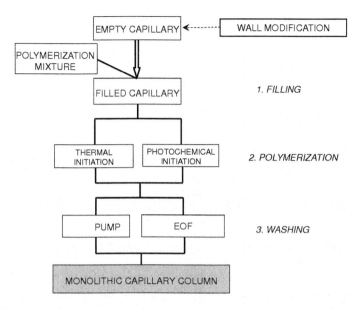

Fig. 9.17 Scheme of the preparation of capillary polymeric monolithic columns

Step 1—modification of inner capillary wall. The aim of the modification is introduction of the vinyl groups to the capillary inner surface, which will later serve as anchoring points for the polymer. It has key influence on the mechanical stability of the bed and prevents from creation of voids between the polymer and the capillary wall. Etching is a first step of the capillary pretreatment and relies on flushing it with 1 mol/L NaOH or KOH solution. The procedures vary and some of the authors leave the filled capillary in the oven (100–120 °C), while others just flush it for 20–30 min at room temperature. However, it was found that higher temperature increases the roughness of the inner capillary surface which helps to keep the monolith in the capillary [49]. Then the capillary is flushed with water, acetone and dyed in a stream of nitrogen for 0.5 to 1 h. The next step is filling with solution of bifunctional reagent, which can react with surface silanol groups and contains vinyl groups as well. The most frequently used reagents are 3-(trimethoxysilyl) propyl methacrylate (γ-MAPS) (Fig. 9.18) or glycidyl methacrylate use as 10–50 % solutions in acetone, toluene, or methanol. Such solution is left in the capillary for 2–3 h (for toluene) or 3–12 h (for acetone solutions), and the capillary is flushed with acetone and dried. The efficiency of the reaction can be quantitatively evaluated by measuring the wetting angle theta (θ). Modification gives rise to the inner surface hydrophobicity, so the wetting angle increases in comparison to unmodified capillary; it was assumed in the literature that $\theta > 70°$ proves good coverage with γ-MAPS.

Step 2—preparation of the polymerization mixture and filling the capillary. The polymerization mixture usually consists of three main components: monomers,

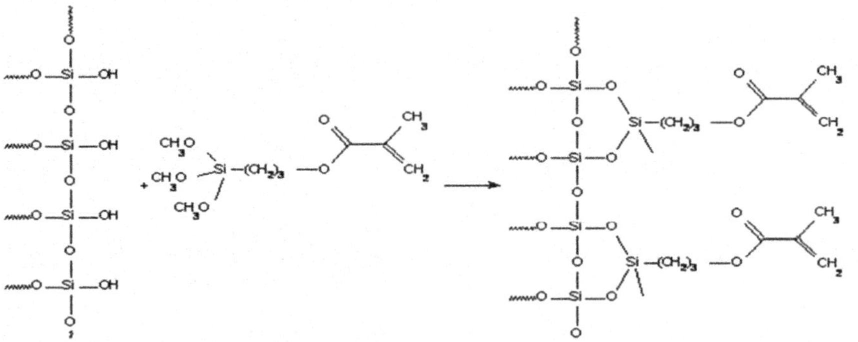

Fig. 9.18 Modification of inner capillary surface with γ-MAPS

porogen solvent, and initiator. The monomers can play different roles: there are monomers containing functional groups which provide interaction with the analytes, crosslinking reagents (provides rigidity of the porous polymer and influences the porosity), and optional polar (or ionizable) monomers (provide generation of the electroosmotic flow). The porogen may be either a single or multicomponent solvent. Different organic solvents have been used, water or supercritical fluid (e.g. CO_2). After mixing all the components the mixture is sparged with nitrogen or helium for 10–30 min to remove dissolved oxygen. Then the capillary is filled using gravitation or a syringe using appropriate connectors. Both ends of the capillary are plugged with pieces of silicone rubber (for e.g., pieces of GC septum).

Step 3—polymerization. Polymerization is initiated thermally (50–80 °C by placing the capillary in a water bath or an oven) or photochemically (most frequently radiation of $\lambda = 365$ nm is used). The time of the polymerization is 8–24 h for thermal process and 10 min to 2 h for photochemical process. It is also possible to initiate the polymerization by γ-radiation, then using of the initiator is not necessary.

Step 4—flushing. After polymerization is completed both ends are cut and all unreacted compounds are flushed away using an appropriate (e.g., HPLC) pump or EOF. In this step the detection window can also be created.

A wide range of commercially available monomers can be used during synthesis which allows for controlling of the surface properties of the monolithic stationary phases. In general the following polymeric monoliths have been described:

- Acrylamide;
- Styrene–divinylbenzene and their derivatives;
- Acrylic and methacrylic.

Most polymeric monolithic columns can be synthesized in a single-step procedure (Fig. 9.17), however, it must be remembered that physicochemical differences of the components of interest may not allow full miscibility and then obtaining of the homogeneous mixture may not be possible. In such cases, to achieve a desired chromatographic properties a grafting technique can be applied. In a first step a basic monolithic skeleton is polymerized to achieve a support of desired hydrodynamic properties and porosity. Trimethylolpropane trimethacrylate (TRIM) is often used as it gives hard polymers and leaves many free vinyl groups for further modification via grafting. In a second step a grafting polymerization mixture (containing monomers of needed properties dissolved in a porogen) is introduced. The monomers react with free vinyl groups of TRIM and cover its surface. Grafting is particularly useful in preparation of molecularly imprinted polymers in fused silica capillaries.

9.5.5 Monolithic Silica Beds

In contrast to polymeric monolithic columns the preparation of silica structures is a more complicated process, however, their chromatographic properties (much higher permeability) are advantageous in comparison to silica particulate-based stationary phase. For example, silica monoliths can be derivatized using procedures developed for traditional stationary phases, they also generate electroosmotic flow.

Silica monoliths are synthesized from alkoxysilanes according to reactions presented in Fig. 9.19 [50, 51]. The first published procedures of preparation of silica monoliths performed in a "macro" scale, included flushing of the bed with ammonium hydroxide solutions of increasing concentrations (starting from 10^{-4} up to 0.1 mol/L) and heating at 600 °C. Such conditions cannot be applied to capillary columns as polyimide coating would be pyrolized. In later procedures (Fig. 9.20) urea was added to the starting mixture as well as heating temperature

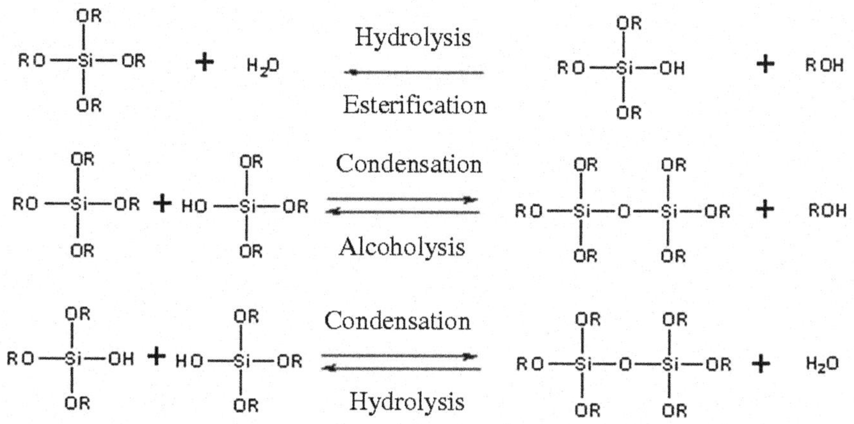

Fig. 9.19 Sol–gel reactions in preparation of silica monoliths

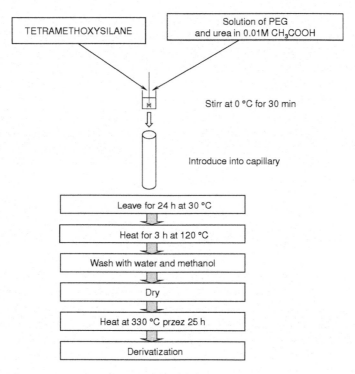

Fig. 9.20 Scheme of the preparation of monolithic silica columns

was limited to 330 °C. Here, urea decomposes at initial heating step (120 °C) and thus produces ammonia etched the pores in the silica skeleton. Heating treatment at 330 °C helps to get rid of residual organic compounds by their decomposition. Finally, the monolith can be derivatized by simply flushing it with, for e.g., octadecyldimethyl-N,N-diethylaminosilane solution at 60 °C for 3 h. The column quality (porosity) is influenced by accuracy in preparation of solutions and reproducibility of the performed actions. Exemplary compositions of the polymerization mixtures as well as the properties of the obtained silica monoliths are presented in Table 9.4.

9.5.6 Organic–Inorganic Mixed Mode Monoliths (PSG)

Photopolymerized sol–gel (PSG) monoliths can be synthesized during a combined sol–gel and photopolymerization processes [52–55]. The starting monomer is γ-MAPS (used normally for capillary wall modification) molecule which contains trimethoxysilyl group and methacrylic group. The process of PSG preparation comprises three steps:

Table 9.4 Exemplary compositions of polymerization mixtures and structural parameters of the obtained bed [50]

Column symbol	TMOS (ml)	PEG (g)	Urea (g)	CH$_3$COOH (ml)	Skeleton size (μm)	Flow-through pore size (μm)	Domain size (μm)	Permeability K (×10^{-14} m^2)
MS(100)-T1.0-A[a]	40	12.4	9.0	100	1.5	2.1	3.6	11.1
MS(100)-T1.4-A[b]	56	11.8	9.0	100	1.0	1.6	2.6	5.4
MS(100)-T1.6-A[a]	64	10.4	9.0	100	1.3	1.5	2.8	4.1
MS(100)-T1.8-A[b]	72	8.4	9.0	100	1.2	1.3	2.5	1.3
MS(100)-T1.0-B[a]	40	12.8	9.0	100	1.2	1.8	3.0	7.4
MS(100)-T1.4-BI[b]	56	11.7	9.0	100	1.3	1.8	3.1	10.1
MS(100)-T1.4-BII[b]	56	11.8	9.0	100	1.0	1.6	2.6	5.9
MS(100)-T1.4-BIII[**]	56	11.9	9.0	100	0.9	1.3	2.2	4.7

Preparation temperature: [a] 30 °C, [b] 25 °C, domain size—skeleton size + flow-through pore size)
Reprinted with permission from Hara et al. [50]. Copyright 2006 American Chemical Society

- Step 1—sol–gel reactions. A solution of monomer is prepared using the following (exemplary) chemicals: 750 μL of γ-MAPS, 22.5 μL of HCl (0.1–5 mol/L) and 225 μL of water. The solution is stirred at room temperature in darkness for 30 min to complete sol–gel process (see reactions shown in Fig. 9.19) and to form a "colloidal precursor".
- Step 2—making a polymerization mixture. Part of the colloidal precursor mixture is mixed with toluene (porogen solvent) in 10/90 to 20/80 ratio. Amounts of precursor less than 10 % result in formation of a thin layer of polymer on the inner capillary wall; on the other hand amounts greater than 20 % give impermeable polymer. Irgacure 1800 has been used as and initiator during UV-initiated photopolymerization or Irgacure 784 and onium salts (for example diphenyliodonium chloride—DPI) when visible light was used. Such prepared mixture is stirred for another 2.5 h and then it is introduced to the capillary.
- Step 3—photopolymerization. The process is initiated with UV ($\lambda = 365$ nm) or visible ($\lambda = 460$ nm) light. When UV light is used the protective polyimide coating should be removed from the FS capillary which adversely affects its durability. The polymerization time should be experimentally determined for a particular system but usually it is ca. 15 min.

The key parameters influencing PSG synthesis are: HCL concentration in the initial step (it should not be lower than 0.1 mol/L), and precursor/porogen ratio.

The PSG monoliths are of hydrophobic character and can be used in reversed phase separation system without further surface modifications. However, hydrophobicity of the surface properties can be increased and changed by modifying the monolith with chloro-, ethoxy- or methoxysilanes (for e.g., n-octadecyl dimethyl chlorosilane) in the same way as silica particles for HPLC. Because of the presence of silanol groups the PSGs provide relatively fast EOF. Also, the advantage of PSG is that during preparation no wall modification is required.

9.5.7 Open-Tubular Columns

In open tubular CEC (OT-CEC) columns the stationary phase is a layer of defined thickness on the inner capillary surface, while the cross section of the capillary remains open. Such a system resembles the GC capillary column. First open tubular columns for liquid separation techniques were described by Tsuda et al. [56]. Such columns were prepared by drawing of a glass tube, then etching the inner surface, which was eventually derivatized with silaned to provide hydrophobic character. The following are the advantages and drawbacks of using open tubular columns in HPLC and CEC:

- It is possible to use relatively long columns.
- High efficiencies can be obtained.

- No retaining frits are used.
- OT columns contain very small amounts of the stationary phase (unfavorable phase ratio) which make them prone to overloading. On the other hand, their low volume allows to analyze extremely low volume samples (injection volume is in the range of 1–5 nL) which may be useful in many fields.
- As the diffusion coefficient in liquids is lower than in gases, the internal diameter of the open tubular LC or CEC column must be much lower than in GC. Capillaries of internal diameter not larger than 50 μm must be used, and most frequently their diameter dos not exceed 25 μm. Working with such capillaries requires keeping the workplace, tools, and hands clean because of risk of clogging the column.

Low value of phase ratio (it is defined as stationary phase volume to mobile phase volume ratio in the column) in OT-CEC columns can be increased by increasing the porosity of the stationary phase layer, which can be done by etching of the inner capillary wall with NH_4HF_2 [57–61], multilayer coating, or making the porous silica [62] or polymeric layers [63]. Also, some nanomaterials can be used here as they are characterized by favorable surface area to volume ratio. Exemplary nanomaterials that were used in OT-CEC are: latex, lipoproteins, silica, titania, gold, and carbon nanotubes [64–72].

Preparation of polymeric and silica open tubular columns is very similar to preparation of monoliths. The significant difference is lower amount of monomers in the polymerization mixture while porogen solvent is in excess. Eeltink et al. suggest that the capillary should be rotated during polymerization (ca. 100 rev/min) to provide obtaining of the uniform layer of the stationary phase on the capillary wall (Fig. 9.21) [63].

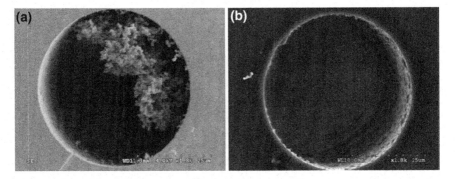

Fig. 9.21 Open tubular polymeric columns ($d_c = 50$ μm) synthesized by photopolymerization. **a** static synthesis; **b** column rotated during the polymerization. Polymerization mixture: butyl methacrylate 2 %, EDMA 2 %, 1-octanol 96 %, initiator: DMPA 0.1 % of monomers; UV ($\lambda = 254$ nm) irradiation performed at 30 °C for A 30 min, B 10 min. Reprinted with permission from Eeltink et al. [63]. Copyright 2006 Wiley–VCH

9.6 CEC Among Contemporary Separation Techniques

In the early 1990s CEC was anticipated to become a separation technique of a great significance, or that it could even displace CE and HPLC. Such predictions have never come true. Several factors affected the fate of CEC:

- As mentioned earlier there is lack in the market of CEC-dedicated instruments which provide gradient elution.
- So far, no new and efficient methods of packed capillary columns have been elaborated. Here, the low stability of the frits is a bottleneck. In fact, none of the manufactures will guarantee a stability of the frits, that is—a column lifetime. There are few CEC column types in the market and they are relatively expensive (you pay for manufacturer's know-how and experience, not for the amount of the stationary phase or capillary used). Most researchers prepare the columns (and very often—the stationary phases as well) themselves for particular analytical tasks. Monolithic polymeric capillary columns are much better from such point of view, however, their chromatographic performance rarely equals classical silica materials. Unpredictable durability and stability of packed capillary columns and lack of dedicated equipment are probably the most important reasons of low popularity of CEC in routine use.
- Despite many CEC-dedicated stationary phases (for e.g., those that provide fast and stable EOF) elaborated, companies do not manufacture them because of low popularity of the technique.
- There is no deep knowledge of the technique among analysts; it is taught as one of the less important CE variants—as a consequence, knowledge of the advantages and potential applications is poor. It should be noted that, virtually, CEC should become more and more significant for economic and ecological reasons; in comparison to HPLC stationary and mobile phases, consumption is very low.
- There is probably another "human factor"—GC and HPLC are such well established techniques that it would be difficult for many analysts to switch to miniaturized techniques.

References

1. H.H. Strain, On the combination of electrophoretic and chromatographic adsorption methods. J. Am. Chem. Soc. **61**(5), 1292–1293 (1939)
2. D.L. Mould, R.L.M. Synge, Electrokinetic ultrafiltration analysis of polysaccharides a new approach to the chromatography of large molecules. Analyst **77**(921), 964–969 (1952)
3. V. Pretorius, B.J. Hopkins, J.D. Schieke, Electro-osmosis: A new concept for high-speed liquid chromatography. J. Chromatogr. A **99**, 23–30 (1974)
4. J.W. Jorgenson, K.D. Lukacs, High-resolution separations based on electrophoresis and electroosmosis. J. Chromatogr. A **218**, 209–216 (1981)
5. J. Knox, I. Grant, Miniaturisation in pressure and electroendosmotically driven liquid chromatography: Some theoretical considerations. Chromatographia **24**(1), 135–143 (1987)

6. M. Szumski, Miniaturyzacja technik chromatograficznych—preparatyka i zastosowanie uniwersalnych mikrokolumn w analityce. Dissertation, Nicolaus Copernicus University, Toruń (2003)
7. S. Suś, (2002) Kolumny monolityczne. Dissertation, Nicolaus Copernicus University, Toruń
8. A.S. Rathore, Theory of electroosmotic flow, retention and separation efficiency in capillary electrochromatography. Electrophoresis 23(22–23), 3827–3846 (2002)
9. N. Smith, M.B. Evans, Comparison of the electroosmotic flow profiles and selectivity of stationary phases used in capillary electrochromatography. J. Chromatogr. A 832(1–2), 41–54 (1999)
10. M. Szumski, B. Buszewski, Study of electroosmotic flow in packed capillary columns. J. Chromatogr. A 1032(1–2), 141–148 (2004)
11. T.M. Zimina, R.M. Smith, P. Myers, Comparison of ODS-modified silica gels as stationary phases for electrochromatography in packed capillaries. J. Chromatogr. A 758(2), 191–197 (1997)
12. D. Belder, K. Elke, H. Husmann, Influence of pH*-value of methanolic electrolytes on electroosmotic flow in hydrophilic coated capillaries. J. Chromatogr. A 868(1), 63–71 (2000)
13. M. Girod, B. Chankvetadze, G. Blaschke, Enantioseparations in non-aqueous capillary electrochromatography using polysaccharide type chiral stationary phases. J. Chromatogr. A 887(1–2), 439–455 (2000)
14. J. Tjornelund, S.H. Hansen, Non-aqueous capillary electrophoresis of drugs: properties and application of selected solvents. J. Biochem. Biophys. Methods 38(2), 139–153 (1999)
15. M.-L. Riekkola, Recent advances in nonaqueous capillary electrophoresis. Electrophoresis 23(22–23), 3865–3883 (2002)
16. E. Tobler, M. Lämmerhofer, W. Lindner, Investigation of an enantioselective non-aqueous capillary electrochromatography system applied to the separation of chiral acids. J. Chromatogr. A 875(1–2), 341–352 (2000)
17. N.W. Smith, Comparison of aqueous and non-aqueous capillary electrochromatography for the separation of basic solutes. J. Chromatogr. A 887(1–2), 233–243 (2000)
18. S.H. Hansen, Z.A. Sheribah, Comparison of CZE, MEKC, MEEKC and non-aqueous capillary electrophoresis for the determination of impurities in bromazepam. J. Pharm. Biomed. Anal. 39(1–2), 322–327 (2005)
19. I. Bjornsdottir, S. HonoréHansen, Comparison of separation selectivity in aqueous and non-aqueous capillary electrophoresis. J. Chromatogr. A 711(2), 313–322 (1995)
20. B. Chankvetadze, G. Blaschke, Enantioseparations using capillary electromigration techniques in nonaqueous buffers. Electrophoresis 21(18), 4159–4178 (2000)
21. L. Roed, E. Lundanes, T. Greibrokk, Nonaqueous electrochromatography on continuous bed columns of sol–gel bonded large-pore C30 material: Separation of retinyl esters. J. Microcol. Sep. 12(11), 561–567 (2000)
22. F. Steiner, B. Scherer, Instrumentation for capillary electrochromatography. J. Chromatogr. A 887(1–2), 55–83 (2000)
23. C.G. Huber, G. Choudhary, C. Horvath, Capillary electrochromatography with gradient elution. Anal. Chem. 69(21), 4429–4436 (1997)
24. C. Yan, R. Dadoo, H. Zhao, R.N. Zare, D.J. Rakestraw, Capillary electrochromatography: Analysis of polycyclic aromatic hydrocarbons. Anal. Chem. 67(13), 2026–2029 (1995)
25. C. Yan, R. Dadoo, R.N. Zare, D.J. Rakestraw, D.S. Anex, Gradient elution in capillary electrochromatography. Anal. Chem. 68(17), 2726–2730 (1996)
26. R. Dadoo, R.N. Zare, C. Yan, D.S. Anex, Advances in capillary electrochromatography: Rapid and high-efficiency separations of PAHs. Anal. Chem. 70(22), 4787–4792 (1998)
27. J. Knox, I. Grant, Electrochromatography in packed tubes using 15–50 μm silica gels and ODS bonded silica gels. Chromatographia 32(7), 317–328 (1991)
28. A.S. Rathore, C. Horváth, Axial nonuniformities and flow in columns for capillary electrochromatography. Anal. Chem. 70(14), 3069–3077 (1998)
29. H. Rebscher, U. Pyell, In-column versus on-column fluorescence detection in capillary electrochromatography. J. Chromatogr. A 737(2), 171–180 (1996)

30. R.A. Carney, M.M. Robson, K.D. Bartle, P. Myers, Investigation into the formation of bubbles in capillary electrochromatography. J. High Resolut. Chromatogr. **22**(1), 29–32 (1999)
31. R. Seifar, W. Kok, J. Kraak, H. Poppe, Capillary electrochromatography with 15 μm ODS-modified non-porous silica spheres. Chromatographia **46**(3), 131–136 (1997)
32. C.G. Bailey, C. Yan, Separation of explosives using capillary electrochromatography. Anal. Chem. **70**(15), 3275–3279 (1998)
33. M. Ye, H. Zou, Z. Liu, J. Ni, Y. Zhang, Capillary electrochromatography with a silica column with a dynamically modified cationic surfactant. J. Chromatogr. A **855**(1), 137–145 (1999)
34. M. Szumski, B. Buszewski, State of the art in miniaturized separation techniques. Crit. Rev. Anal. Chem. **32**, 1–46 (2002)
35. D. Tong, K.D. Bartle, A.A. Clifford, A.M. Edge, Theoretical studies of the preparation of packed capillary columns for chromatography. J. Microcol. Sep. **7**(3), 265–278 (1995)
36. D. Tong, K.D. Bartle, A.A. Clifford, Preparation and evaluation of supercritical carbon dioxide-packed capillary columns for HPLC and SFC. J. Microcol. Sep. **6**(3), 249–255 (1994)
37. W. Li, A. Malik, M.L. Lee, Fused silica packed capillary columns in supercritical fluid chromatography. J. Microcol. Sep. **6**(6), 557–563 (1994)
38. A. Malik, W. Li, M.L. Lee, Preparation of long packed capillary columns using carbon dioxide slurries. J. Microcol. Sep. **5**(4), 361–369 (1993)
39. P. Koivisto, R. Danielsson, K.E. Markides factors affecting the preparation of packed capillary columns in supercritical carbon dioxide media. J. Microcol. Sep. **9**(2), 97–103 (1997)
40. T.D. Maloney, L.A. Colón, A drying step in the protocol to pack capillary columns by centripetal forces for capillary electrochromatography. Electrophoresis **20**(12), 2360–2365 (1999)
41. C. Yan, Electrokinetic packing of capillary columns. US Patent No 5453163 09/26/1995 (1995)
42. R. Asiaie, X. Huang, D. Farnan, C. Horváth, Sintered octadecylsilica as monolithic column packing in capillary electrochromatography and micro high-performance liquid chromatography. J. Chromatogr. A **806**, 251–263 (1998)
43. M.T. Dulay, R.P. Kulkarni, R.N. Zare, Preparation and characterization of monolithic porous capillary columns loaded with chromatographic particles. Anal. Chem. **70**(23), 5103–5107 (1998)
44. C.K. Ratnayake, C.S. Oh, M.P. Henry, Characteristics of particle-loaded monolithic sol–gel columns for capillary electrochromatography. J. Chromatogr. A **887**, 277–285 (2000)
45. G.S. Chirica, V.T. Remcho, Silicate entrapped columns—new columns designed for capillary electrochromatography. Electrophoresis **20**, 50–56 (1999)
46. H.J. Cortes, C.D. Pfeiffer, B.E. Richter, T.S. Stevens, Porous ceramic bed supports for fused silica packed capillary columns used in liquid chromatography. J. High Resolut. Chromatogr. **10**(8), 446–448 (1987)
47. G.S. Chirica, V.T. Remcho, Fritless capillary columns for HPLC and CEC prepared by immobilizing the stationary phase in an organic polymer matrix. Anal. Chem. **72**(15), 3605–3610 (2000)
48. R. Xie, R. Oleschuk, Photoinduced polymerization for entrapping of octadecylsilane microsphere columns for capillary electrochromatography. Anal. Chem. **79**(4), 1529–1535 (2007)
49. J. Courtois, M. Szumski, E. Byström, A. Iwasiewicz, A. Shchukarev, K. Irgum, A study of surface modification and anchoring techniques used in the preparation of monolithic microcolumns in fused silica capillaries. J. Sep. Sci. **29**(2), 325 (2006)
50. T. Hara, H. Kobayashi, T. Ikegami, K. Nakanishi, N. Tanaka, Performance of monolithic silica capillary columns with increased phase ratios and small-sized domains. Anal. Chem. **78**(22), 7632–7642 (2006)
51. N. Tanaka, H. Nagayama, H. Kobayashi, T. Ikegami, K. Hosoya, N. Ishizuka, H. Minakuchi, K. Nakanishi, K. Cabrera, D. Lubda, Monolithic silica columns for HPLC, micro-HPLC, and CEC. J. High Resolut. Chromatogr. **23**(1), 111–116 (2000)

52. M.T. Dulay, H.N. Choi, R.N. Zare, Visible light-induced photopolymerization of an in situ macroporous sol–gel monolith. J. Sep. Sci. **30**(17), 2979–2985 (2007)
53. M. Kato, K. Sakai-Kato, T. Toyo'oka, M.T. Dulay, J.P. Quirino, B.D. Bennett, R.N. Zare, Effect of preparatory conditions on the performance of photopolymerized sol-gel monoliths for capillary electrochromatography. J. Chromatogr. A **961**(1), 45–51 (2002)
54. M.T. Dulay, J.P. Quirino, B.D. Bennett, R.N. Zare, Bonded-phase photopolymerized sol–gel monoliths for reversed phase capillary electrochromatography. J. Sep. Sci. **25**(1–2), 3–9 (2002)
55. M.T. Dulay, J.P. Quirino, B.D. Bennett, M. Kato, R.N. Zare, Photopolymerized sol-gel monoliths for capillary electrochromatography. Anal. Chem. **73**(16), 3921–3926 (2001)
56. T. Tsuda, K. Nomura, G. Nakagawa, Open-tubular microcapillary liquid chromatography with electro-osmosis flow using a UV detector. J. Chromatogr **248**(2), 241–247 (1982)
57. J.-L. Chen, Etched succinate-functionalized silica hydride stationary phase for open-tubular CEC. Electrophoresis **30**(22), 3855–3862 (2009)
58. J.J. Pesek, M.T. Matyska, S. Sentellas, M.T. Galceran, M. Chiari, G. Pirri, Multimodal open-tubular capillary electrochromatographic analysis of amines and peptides. Electrophoresis **23**(17), 2982–2989 (2002)
59. J.J. Pesek, M.T. Matyska, Column technology in capillary electrophoresis and capillary electrochromatography. Electrophoresis **18**(12–13), 2228–2238 (1997)
60. J.J. Pesek, M.T. Matyska, G.B. Dawson, J.I.C. Chen, R.I. Boysen, M.T.W Hearn, Open-tubular electrochromatographic characterization of synthetic peptides. Electrophoresis **25**(9), 1211–1218 (2004)
61. M.T. Matyska, J.J. Pesek, I. Boysen, T.W. Hearn, Characterization and applications of etched chemically modified capillaries for open-tubular capillary electrochromatography. Electrophoresis **22**(12), 2620–2628 (2001)
62. A. Malik, Advances in sol-gel based columns for capillary electrochromatography: Sol-gel open-tubular columns. Electrophoresis **23**(22–23), 3973–3992 (2002)
63. S. Eeltink, F. Svec, J.M.J. Fréchet, Open-tubular capillary columns with a porous layer of monolithic polymer for highly efficient and fast separations in electrochromatography. Electrophoresis **27**(21), 4249–4256 (2006)
64. J.L. Chen, T.L. Lu, Y.C. Lin, Multi-walled carbon nanotube composites with polyacrylate prepared for open-tubular capillary electrochromatography. Electrophoresis **31**(19), 3217–3226 (2010)
65. F. Svec, Capillary electrochromatography: A rapidly emerging separation method, in *Modern Advances in Chromatography*, ed. by R. Freitag (Springer, Berlin, 2002)
66. Y.-L. Hsieh, T.-H. Chen, C.-P. Liu, C.-Y. Liu, Titanium dioxide nanoparticles-coated column for capillary electrochromatographic separation of oligopeptides. Electrophoresis **26**(21), 4089–4097 (2005)
67. C. Fujimoto, Titanium dioxide coated surfaces for capillary electrophoresis and capillary electrochromatography. Electrophoresis **23**(17), 2929–2937 (2002)
68. C.P. Kapnissi-Christodoulou, X. Zhu, I.M. Warner, Analytical separations in open-tubular capillary electrochromatography. Electrophoresis **24**(22–23), 3917–3934 (2003)
69. J.L. Chen, K.H. Hsieh, Polyacrylamide grafted on multi-walled carbon nanotubes for open-tubular capillary electrochromatography: Comparison with silica hydride and polyacrylate phase matrices. Electrophoresis **31**(23–24), 3937–3948 (2010)
70. J.L. Chen, Multi-wall carbon nanotubes bonding on silica-hydride surfaces for open-tubular capillary electrochromatography. J. Chromatogr. A **1217**(5), 715–721 (2010)
71. M.T. Dulay, R.P. Kulkarni, R.N. Zare, Preparation and Characterization of monolithic porous capillary columns loaded with chromatographic particles. Anal.Chem. **70**, 5103–5107 (1998)
72. G. Yue, Q. Luo, J. Zhang, S.-L. Wu, B.L. Karger, Ultratrace Lc/MS proteomic analysis using 10um-i.d. porous layer open tubular poly(styrenedivinylbenzene) capillary columns. Anal.Chem. **79**, 938–946 (2007)

Chapter 10
Electrochromatography Methods: Planar Electrochromatography

Adam Chomicki, Tadeusz H. Dzido, Paweł Płocharz and Beata Polak

Abstract Planar electrochromatography is a technique in which mixture components are separated in adsorbent layer of a chromatographic plate placed in electric field. In such separation system a mobile phase movement stems from electroosmosis phenomenon. Partition and electrophoresis mechanisms are involved in separation of mixture components with this technique. Two principal modes of planar electrochromatography are described: planar electrochromatography in an open system (PEC) and planar electrochromatography in a closed system (pressurized planar electrochromatography, PPEC). The development of both modes is presented beginning with the first paper on electrochromatography by Pretorius et al. in 1974 and finishing with the last papers by Dzido et al. in 2010. Constructional development of equipment to planar electrochromatography is provided and influence of operating variables on separation efficiency as well. The advantages and challenges of PPEC technique are especially discussed.

10.1 Introduction

Planar electrochromatography is a technique in which mixture components are separated in adsorbent layer of a chromatographic plate placed in electric field. In such separation system mobile phase movement stems from electroosmosis phenomenon. There are two techniques of planar electrochromatography: planar electrochromatography in open system (PEC, *Planar Electrochromatography*) and planar electrochromatography in closed system (PPEC, *Pressurized Planar*

A. Chomicki (✉) · T. H. Dzido · P. Płocharz · B. Polak
Faculty of Pharmacy, Medical University of Lublin, Lublin, Poland
e-mail: adam.chomicki@umlub.pl

B. Buszewski et al. (eds.), *Electromigration Techniques*, Springer Series
in Chemical Physics 105, DOI: 10.1007/978-3-642-35043-6_10,
© Springer-Verlag Berlin Heidelberg 2013

Electrochromatography). Chromatographic plates with functionalized silica gel layer are the most often applied in the latter mode. Electroosmotic flow of the mobile phase in systems with C-18 functionalized silica gel is generated due to electrical double layer formed at the mobile phase—stationary phase interface. Dissociation of residual silanol groups on the stationary phase and adsorption of ions from the mobile phase influence on charge of the electrical double layer. Separation mechanism in PPEC systems is dependent on two main effects: electrophoresis and partition of solutes between mobile and stationary phases. Molecular structure, charge of ions, and types of the mobile and stationary phases are involved in migration distance of the solutes. Hence, molecular interactions and electrokinetic phenomena influence on solute separation. Solute molecules as ions can migrate in the direction of electroosmotic flow or in the opposite direction. The main advantages of planar electrochromatography are: considerably shorter separation time relative to conventional planar (thin-layer) chromatography (TLC), high efficiency (plate height of PPEC system is comparable to that of HPLC one), and different separation selectivity in comparison to liquid chromatography and electrophoresis.

10.2 Development of the Method

At the breakthrough of the forties and fifties of the twentieth century some authors claimed possible application of electroosmotic flow to thin-layer chromatography separations [1, 2]. In 1974 Pretorius et al. published the first paper on this subject [3]. The results obtained by the authors were promising because of considerable shortening (fifteen times) of separation time in comparison to that with conventional planar chromatography. The method was named as high-speed thin layer chromatography (HSTLC). However, the method was not acknowledged to be a separation technique for a long time in spite of the unquestionable advantages mentioned. This technique was rediscovered 20 years later [4]. The authors proposed a method of chromatogram development in electric field using plates initially dry. The electrochromatograms were developed from the two opposite sides of the plates (from anode and cathode) simultaneously. The authors have named the mode as planar electrochromatography. They found that migration distance of the mobile phase was dependent both on electric field and chromatographic plate side from which it was fed with mobile phase solution, whereas solute migration was dependent on polarity of mobile and stationary phases and electrophoretic mobility of separated components as well. In 1999, Howard et al. [5] and in 2000, Nurok et al. [6] undertook an attempt to follow the experiments presented by Pretorius et al. [3]. The authors revealed 2.5–4.5 times increase in efficiency of planar electrochromatography system in comparison to the conventional TLC system. In addition, Nurok et al. found flow velocity of the mobile phase to be dependent on electrolyte concentration [7], and its optimal value advantageous for separation of components of the mixture investigated [8]. The authors also mentioned inconveniences of planar electrochromatography such as decrease of migration velocity of solute bands during PEC process, evaporation of mobile phase from the plate (caused by generation of Joule heat), and flux of the mobile phase to the surface of adsorbent layer. In the subsequent paper, Howard et al. [9] proposed a new device for PEC that featured restriction of mobile phase evaporation from the adsorbent layer. They also confirmed that mobile phase velocity increased with potential increase applied to electrodes. The authors also demonstrated that migration velocity of mobile phase front was dependent on composition of the mobile phase which was concerned with various physicochemical properties of solvents used in the experiments.

Subsequent development of PEC mode was concerned with design of new developing chambers in which some drawbacks mentioned above were eliminated or considerably reduced [10, 11]. Adaptation of the commercially available Horizontal DS Chamber, used for conventional development of thin-layer chromatograms, to PEC is an example. In this case the authors revealed that flow velocity of the mobile phase is approximately proportional to the potential applied to electrodes, and solvent evaporation from the adsorbent layer enhanced velocity of the mobile phase. Evaporation of solvents from the adsorbent layer was caused by Joule heat generated during PEC process and was dependent on buffer concentration in the mobile phase and chamber saturation with eluent vapors. This evaporation can be restricted by cooling the chromatographic plate in PEC chamber [12]. Additionally, prewetting the chromatographic plate should be carefully performed before electrochromatogram development.

Planar electrochromatography under open conditions (in open system) seemed to be an interesting method. However, the disadvantages mentioned above did not enable its broad application in laboratory practice. Based on the experience mentioned above a new technique has been worked out in that a gas phase was eliminated. This technique was named as the pressurized planar electrochromatography (PPEC). The first paper on PPEC was published by Nurok et al. in 2004 [13]. Preliminary results of investigations revealed that PPEC system featured considerably higher efficiency than conventional TLC one and enabled to obtain higher repeatability of solute migration distances in comparison to PEC [14]. Other groups also carried out investigations on PPEC development. Initially, in our research modified Horizontal DS Chamber for TLC, equipped with special cover of the chromatographic plate, was applied [15]. We claimed that prewetting time of the adsorbent layer of chromatographic plate with mobile phase solution influenced the migration distance of solutes, repeatability of its value, and shape of solute zones. Tate and Dorsey reported that attainment of equilibrium between mobile and stationary phases was decisive effect of repeatability and stability of separation process with PPEC [16, 17]. In 2006, our group published a paper in which a device with control of electroosmotic flow of the mobile phase was presented [18]. This device enabled to directly measure and adjust electroosmotic flow of the mobile phase in PPEC system for the first time. Our group presented very fast separation (45 s) of six component dye mixture. In the subsequent papers we presented the influence of various operating variables and parameters on separation efficiency of PPEC systems such as stationary phase type, mode of sample application [19], and flow velocity of the mobile phase [20, 21]. We also demonstrated advantages of PPEC and TLC combination in two-dimensional (2D) separation system [22].

10.3 Chambers for Pressurized Planar Electrochromatography

There are two main types of PPEC devices in contemporary publications. These device types differ in construction of PPEC chamber in which a chromatographic plate is placed. Nurok et al. demonstrated the first device type as it is mentioned above. In this device the chromatographic plate is placed into the PPEC chamber in vertical position, see Fig. 10.1. In the second type, worked out by our group, the chromatographic plate in the PPEC chamber is placed in horizontal position, Fig. 10.2.

In the device worked out by Nurok et al. [13] the chromatographic plate is placed between two metal blocks, which press, by means of hydraulic press, a Teflon foil and ceramic sheet to an adsorbent layer of the chromatographic plate. In this way gas phase is eliminated from the separating system, Fig. 10.3. The subsequent stage of development of the authors' device was concerned with introduction of temperature control of the adsorbent layer in the PPEC chamber. This modification consisted in channels made in metal blocks for circulating liquid of constant temperature in the range 3–60 °C.

Our group has presented another device with PPEC chamber for plates placed horizontally [15, 18, 23] as mentioned above. The modified Horizontal DS

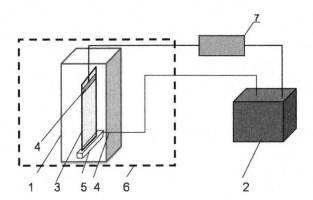

Fig. 10.1 Schematic view of the device for PPEC with chromatographic plate vertically placed in the PPEC chamber: *1* PPEC chamber, *2* high-voltage DC supply, *3* chromatographic plate, *4* electrodes, *5* mobile phase container, *6* casing of PPEC chamber, *7* ammeter. Reprinted with permission from Płocharz et al. [24]. Copyright 2010 Polskie Towarzystwo Chemiczne

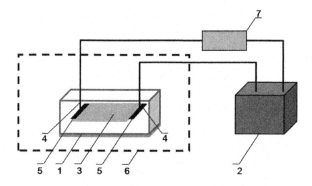

Fig. 10.2 Schematic view of the device for PPEC with chromatographic plate horizontally placed in the PPEC chamber: *1* PPEC chamber, *2* high-voltage DC supply, *3* chromatographic plate, *4* electrodes, *5* mobile phase container, *6* casing of PPEC chamber, *7* ammeter. Reprinted with permission from Płocharz et al. [24]. Copyright 2010 Polskie Towarzystwo Chemiczne

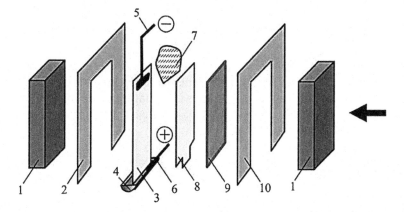

Fig. 10.3 Elements of PPEC chamber for chromatographic plate: *1* metal block, *2* frame, *3* chromatographic plate, *4* mobile phase solution, *5* cathode, *6* anode, *7* blotting paper, *8* Teflon foil, *9* ceramic sheet, *10* frame; adapted from. Reprinted with permission from Nurok et al. [13]. Copyright 2004 American Chemical Society

Chamber was applied as our first PPEC chamber [15]. In this chamber some part of adsorbent layer of the chromatographic plate extended out the chromatographic plate cover. Then volatile components of the mobile phase could evaporate somewhat from this extended adsorbent layer. However, the evaporation was particularly enhanced when higher electric field was applied during separation process. In the next development stage of our PPEC device we introduced considerable constructional changes [18, 23]. In the redesigned device the chromatographic plate was completely covered by special cover. In this way evaporation of the mobile phase was prevented and no gas phase was present in the separation system. In addition, the new device enabled to control flow velocity of

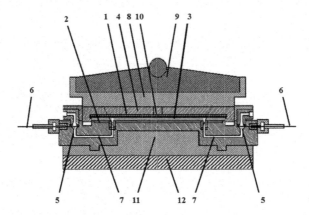

Fig. 10.4 Longitudinal section of PPEC chamber with chromatographic plate horizontally placed: *1* chromatographic plate, *2* elastic pad, *3* Teflon foil, *4* adaptor of the chromatographic plate, *5* electrode compartment, *6* platinum electrode, *7* channels, *8* poly-acetal part of the cover, *9* stainless steel part of the cover, *10* base for the chromatographic plate, *11* body of the chamber, *12* stainless steel chamber base. Reprinted with permission from Chomicki et al. [22]. Copyright 2009 Wiley–VCH

the mobile phase. This device was also modified [22]. The constructional changes were characterized by new shape of electrode compartments and additional flow of the mobile phase solution was generated with syringe pump through these compartments, Fig. 10.4. This additional flow was to remove gas bubbles formed close to electrodes during separation process.

It is worth quoting a device for PPEC described by Tate and Dorsey [16, 17]. This device was equipped with a grid of electrodes to research potential drop across adsorbent layer during separation process. Based on the experiments the authors concluded about the conditions of equilibrium of a mobile–stationary phase system.

10.4 Advantages and Disadvantages of Pressurized Planar Electrochromatography

Pressurized planar electrochromatography is characterized by various advantages and disadvantages similarl to other separation methods. Shortening of separation time seems to be the most sophisticated feature of this technique in comparison to conventional planar chromatography. In pressurized planar electrochromatography electroosmotic flow of the mobile phase is generated by electric field. According to Smoluchowski's equation this flow is directly proportional to potential drop between electrodes. Influence of potential drop on migration time of the mobile phase front for the definite distance is presented in Table 10.1 [24]. Analogous values of migration time of mobile phase front in conventional thin-layer chromatography are also presented in this table. As it can be seen in PPEC technique

Table 10.1 Migration times of the mobile phase in TLC and PPEC techniques; chromatographic plate: HPTLC RP-RP18 W (Merck), and mobile phase: 80 % acetonitrile in citric acid buffer (pH 5.0) [24].

Mobile phase migration distance [mm]	Time (min)			
	TLC	PPEC (1.5 kV)	PPEC (3.0 kV)	PPEC (4.5 kV)
15	1.08	0.94	0.42	0.20
30	2.67	1.87	0.84	0.39
45	5.25	2.81	1.26	0.59
60	9.00	3.75	1.68	0.78
75	14.00	4.68	2.10	0.98
90	19.00	5.62	2.53	1.17

Reprinted with permission from Płocharz et al. [24]. Copyright 2010 Polish Chemical Society

Fig. 10.5 a TLC, and **b** PPEC chromatograms of three component sample mixture: *1* propyphenazone, *2* caffeine, and *3* acetaminophen; TLC separation time—7.5 min, PPEC separation time—0.5 min, polarization voltage—3.5 kV, chromatographic plate: HPTLC RP-18 W (Merck), mobile phase: 20 % acetonitrile in buffer (pH 4.1). Reprinted with permission from Płocharz et al. [24]. Copyright 2010 Polskie Towarzystwo Chemiczne

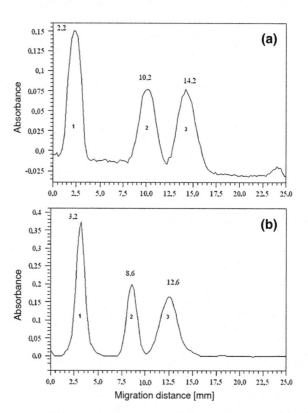

mobile phase migration time is considerably shorter than that in TLC. Typical separation time in TLC is in the range of a few to several tens of minutes. Hence PPEC technique enables considerable shortening of separation time in comparison to TLC. In Fig. 10.5 separation examples of three-component mixture (propyphenasone, caffeine and acetaminophen) with TLC (7.5 min) and PPEC (30 s) are presented. Analogous shortening of separation time with PPEC technique was reported in the papers [13, 14, 25].

Owing to the fact that profile of electroosmotic flow of the mobile phase is flat, lower dispersion of solute zones is observed in PPEC systems in comparison to liquid chromatography techniques (HPLC) and especially to TLC. This means that the PPEC technique is characterized by low values of theoretical plate height. The minimum value of plate height approaches double diameter of stationary phase particles whicht reflects the high performance of separating system [14, 18, 21]. In Fig. 10.6 chromatograms of test steroid mixture obtained with PPEC and HPTLC (high performance thin-layer chromatography) are demonstrated. These chromatograms clearly reveal that the former technique shows considerably higher separation efficiency than the latter [21].

Different separation selectivity is the next advantage of PPEC in relation to liquid chromatography and electrophoresis techniques. If sample mixture comprises solutes of both types, dissociated and non-dissociated, then separation selectivity in PPEC can be quite different from that in liquid chromatography and/or electrophoresis. Under such conditions components to be separated are present in the mobile phase solution as anions, cations, and neutral molecules. In electrochromatography systems with silica-based stationary phases electrophoretic migration of anions is reverse to electroosmotic flow. However, electrophoretic mobility of anions, and cations as well, is usually lower than electroosmotic mobility of the

Fig. 10.6 a PPEC, and **b** HPTLC chromatograms of steroid test mixture: *1* testosterone isobutyrate, *2* testosterone acetate, *3* methandienone, *4* hydrocortisone acetate, *5* 16-dehydropregnenolone acetate, *6* prednisolone succinate; mobile phase: 80 % acetonitrile in buffer (1.96 mM citric acid, 4.08 mM disodium hydrogen phosphate, pH 5.0), chromatographic plate: HPTLC RP-18 W (Merck), polarization voltage 2.5 kV (PPEC), separation time: PPEC—5 min, HPTLC—6.5 min. Reprinted with permission from Płocharz et al. [21]. Copyright 2010 Elsevier

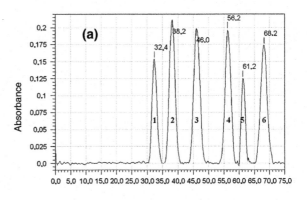

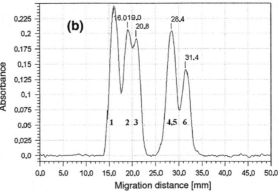

mobile phase. Then all mixture components including anions migrate towards cathode, i.e., according to electroosmotic flow of the mobile phase. On rare occasions when electrophoretic mobility is higher than electroosmotic mobilty, then anion zone can migrate towards anode, and its peak is registered before place of sample application, Fig. 10.7 [26]. Direction of electrophoretic mobility of cations is in accordance with electroosmotic mobility which implies cation zones showing higher migration distances relative to those of liquid chromatography systems. Migration distance of neutral solutes is dependent on electroosmotic flow of the mobile phase and their partition between the phases of a separating system. With respect to the discussion above increase in migration distance of solutes in pressurized planar electrochromatography can be ordered as follows: anions, neutral solutes, and cations. This order can be changed if the solute types mentioned undergo differentiation of partition between stationary and mobile phases.

PPEC has been successfully applied to two-dimensional separations (2D) in combination with liquid chromatography techniques especially thin-layer chromatography. It has been especially promising due to different separation selectivity demonstrated by the former technique relative to the latter as discussed above. Such separation process, 2D TLC/PPEC, is usually carried out in two stages. In the

Fig. 10.7 a PPEC, and b TLC chromatograms of: *1* acetylsalicylic acid, *2* caffeine, and *3* acetaminophen; chromatographic plate: TLC RP-18 (Merck), mobile phase: 60 % acetonitrile in buffer (pH 3.8), polarization voltage (PPEC) 2.0 kV (black arrow indicates application point), PPEC separation time—7 min, TLC separation time—24 min. Reprinted with permission from Hałka et al. [26]. Copyright 2010 Akadémiai Kiadó

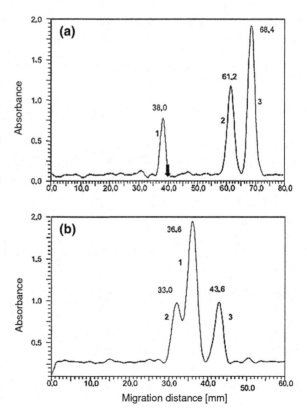

first stage thin-layer chromatography is performed, and in the second—pressurized planar electrochromatography. It is possible to proceed reverse order of these separation stages. Promising examples of application of such procedures were reported for separation of foodstuff dyes [22] and amino acids [27].

It seems that the discussed advantages of PPEC will be further revealed when more papers on combinations of this technique with mass spectrometry will be published in the near future [28, 29].

At the contemporary development stage of PPEC equipments for this technique are characterized by some inconveniences in operation. When these are eliminated, the mode will be applied in laboratory practice. Too high Joule heat generated during PPEC process constitutes the main impediment of this technique. This disadvantageous effect can be eliminated or considerably restricted by heat absorption as it has been proposed in paper [14]. Another important inconvenience of PPEC technique is related to the chromatographic plates used. Unfortunately, these are dedicated to conventional thin-layer chromatography. Adsorbent layer thickness of such plates is close to 0.2 mm, and it seems to be too thick for PPEC. Diminution of the adsorbent layer thickness could lead to lowering of Joule heat generation during PPEC separation process. In addition the adsorbent layers should generate high electroosmotic flow of the mobile phase. Ordinary chromatographic plates do not show this property. A last paper on this subject indicates that production of new stationary phases is very promising especially for separation of molecules of biological interest [30]. Another relatively weak point of pressurized planar electrochromatography is lower repeatability of migration distance values in comparison to liquid chromatography techniques. However, this feature is the property of all electromigration techniques [31].

In the course of PPEC development various improvements of devices have been emerging, however, these are not fully familiar in operation [18, 22]. There are no commercially available PPEC equipments so far with respect to this aspect. Hence, participation of researchers in launching a new PPEC device on the market also constitutes a challenge in this regard.

References

1. R. Consden, A.H. Gordon, A.J.P. Martin, Ionophorsis in silica jelly- a method for the separation of amino acids and peptides. Biochem. J. **40**, 33–41 (1946)
2. D.L. Mould, R.L.M. Synge, Electrokinetic ultrafiltration analysis of polysaccharides. A new approach to the chromatography of large molecules. Analyst **77**, 964–969 (1952)
3. V. Pretorius, B.J. Hopkins, J.D. Schieke, A new concept for high-speed liquid chromatography. J. Chromatogr. **99**, 23–30 (1974)
4. M. Pukl, R. Prosek, E. Kaiser, Planar electrochromatography part 1. Planar electrochromatography on non-wetted thin layers. Chromatographia **38**, 83–87 (1994)
5. T. Shafik, A.G. Howard, F. Moffatt, I.D. Wilson, Evaporation induced solvent migration in electrically-driven thin layer chromatography. J. Chromatogr. A **841**, 127–132 (1999)
6. D. Nurok, M.C. Frost, D.M. Chenoweth, Separation using planar chromatography with electroosmotic flow. J. Chromatogr. A **903**, 211–217 (2000)

7. D. Nurok, J.M. Koers, D.A. Nyman, W.M. Liao, Variables that affect performance in planar chromatography with electroosmotic flow. J Planar Chromatogr. **14**, 409–414 (2001)
8. D. Nurok, J.M. Koers, M. Carmichael, A role of buffer concentration and applied voltage in obtaining a good separation in planar electrochromatography. J. Chromatogr. A **983**, 247–253 (2003)
9. A.G. Howard, T. Shafik, F. Moffatt, I.D. Wilson, Electroosmotically driven thin layer electrochromatography on silica media. J. Chromatogr. A **844**, 333–340 (1999)
10. D. Nurok, J.M. Koers, M.A. Carmichael, T.H. Dzido, The performance of planar electrochromatography. J. Planar Chromatogr. **15**, 320–323 (2002)
11. T.H. Dzido, R. Majewski, B. Polak, W. Gołkiewicz, E. Soczewiński, Application of a horizontal DS chamber to planar electrochromatography. J. Planar Chromatogr. **16**, 176–182 (2003)
12. T.H. Dzido, R. Majewski, in *Proceedings of the International Symposium on Planar Separations Budapest*, Planar electrochromatography in horizontal chamber with cooling of the chromatographic plate (2003), pp. 129–138
13. D. Nurok, J.M. Koers, A.L. Novotny, M.A. Carmichael, J.J. Kosiba, R.E. Santini, G.L. Hawkins, R.W. Replogle, Apparatus and initial results for pressurized planar electrochromatography. Anal. Chem. **76**, 1690–1695 (2004)
14. A.L. Novotny, D. Nurok, R.W. Replogle, G.L. Hawkins, R.E. Santini, Results with an apparatus for pressurized planar electrochromatography. Anal. Chem. **78**, 2823–2831 (2006)
15. T.H. Dzido, J. Mróz, G.W. Jóźwiak, in *Planar Chromatography Proceedings of the International Symposium on Planar Separations Visegrad*, Application of a horizontal chamber for separation in opened and closed systems of planar electrochromatography (2004), pp. 19–28
16. P.A. Tate, J.G. Dorsey, Characterization of flow and voltage profiles in planar electrochromatography. J. Chromatogr. A **1079**, 317–327 (2005)
17. P.A. Tate, J.G. Dorsey, Linear voltage profiles and flow homogeneity in pressurized planar electrochromatography. J. Chromatogr. A **1103**, 150–157 (2006)
18. T.H. Dzido, P.W. Płocharz, P. Ślązak, Apparatus for pressurized planar electrochromatography in a completely closed system. Anal. Chem. **78**, 4713–4721 (2006)
19. P.W. Płocharz, T.H. Dzido, P. Ślązak, G.W. Jóźwiak, A. Torbicz, Influence of sample application mode on performance of pressurized planar electrochromatography in completely closed system. J. Chromatogr. A **1170**, 91–100 (2007)
20. T.H. Dzido, P.W. Płocharz, A. Klimek-Turek, A. Torbicz, B. Buszewski, Pressurized planar electrochromatography as the mode for determination of solvent composition-retention relationships in reversed-phases systems. J. Planar Chromatogr. **21**, 295–298 (2008)
21. P.W. Płocharz, A. Klimek-Turek, T.H. Dzido, Pressurized planar electrochromatography, high performance thin layer chromatography and high performance liquid chromatography—comparison of performance. J. Chromatogr. A **1217**, 4868–4872 (2010)
22. A. Chomicki, P. Ślązak, T.H. Dzido, Preliminary results for 2-D separation with high performance thin-layer chromatography and pressurized planar electrochromatography. Electrophoresis **30**, 3718–3725 (2009)
23. T.H. Dzido, P.W. Płocharz, Planar electrochromatography in a closed system under pressure-pressurized planar electrochromatography. J. Liq. Chromatogr. RT **30**, 2651–2667 (2007)
24. P.W. Płocharz, P. Ślązak, A. Hałka-Grysińska, A. Chomicki, T.H. Dzido, Elektrochromatografia planarna w układzie zamkniętym. Wiadomości Chemiczne **64**, 61–80 (2010)
25. B. Polak, A. Hałka, T.H. Dzido, Pressurized planar electrochromatographic separation of the enantiomers of tryptophan and valine. J. Planar Chromatogr. **21**, 33–37 (2008)
26. A. Hałka, P. Płocharz, A. Torbicz, T.H. Dzido, Reversed-phase pressurized planar electrochromatography and planar chromatography of acetylsalicylic acid, caffeine and acetaminophen. J. Planar Chromatogr. **23**, 420–425 (2010)

27. A. Chomicki, K. Kloc, T.H. Dzido, Two-dimensional separation of some amino acids with high performance thin-layer chromatography and pressurized planar electrochromatography. J. Planar Chromatogr. **24**, 6–9 (2011)
28. D.J. Janecki, A.L. Novotny, S.D. Woodward, J.M. Wiseman, D. Nurok, A preliminary study of the coupling of desorption electrospray ionization-mass spectrometry with pressurized planar electrochromatography. J. Planar Chromatogr. **21**, 11–14 (2008)
29. V. Panchagnula, A. Mikulskis, L. Song, Y. Wang, M. Wang, T. Knubovets, E. Scrivener, E. Golenko, I.S. Krull, M. Schultz, H.E. Hauck, W.F. Patton, Phosphopeptide analysis by directly coupling two-dimensional planar electrochromatography/thin-layer chromatography with matrix assisted laser desorption/ionization time of-flight mass spectrometry. J. Chromatogr. A **1155**, 112–123 (2007)
30. S.D. Woodward, I. Urbanova, D. Nurok, F. Svec, Separation of peptides and oligonucleotides using a monolithic polymer layer and pressurized planar electrophoresis and electrochromatography. Anal. Chem. **82**, 3348–3445 (2010)
31. K.D. Altria, N.W. Smith, C.H. Turnbull, Analysis of acidic compounds using capillary electrochromatography. J. Chromatogr. B **717**, 341–353 (1998)

Chapter 11
Non-Aqueous Capillary Electrophoresis

Michał Szumski and Bogusław Buszewski

Abstract Non-aqueous capillary electrophoresis and capillary electrochromatography are special variants of these techniques. Here, organic solvents or their mixtures with or without dissolved electrolytes are used as separation buffer or mobile phase, respectively. The most important features of non-aqueous systems are: better solubility of more hydrophobic ionic substances (many natural products) than in water, much less current and Joule heating allows for using highly concentrated buffers and/or larger capillary internal diameters, polar interactions are enhanced in organic solvents which is often highly advantageous in chiral separation systems. This chapter presents most frequently used solvents, their properties, as well as shows pH* scale which is often used in non-aqueous systems.

In classical CEC and CEC aqueous or hydroorganic solvents are used but it is different in non-aqueous CE (NACE) and CEC (NACEC) which employ a wide range of organic solvents, without any addition of water. Despite non-aqueous approach was described as far as in the 1980s (Walbroehl and Jorgenson) [1, 2], a systematic increase in number of published papers has been dated since 1994 [3, 4]. There are several reasons for using NACE:

- Improvement in selectivity.
- Electrophoresis in large bore capillaries—"preparative" CE.
- Electrophoresis of compounds insoluble or slightly soluble in water.
- Employing during some separations (for example chiral compounds) polar interactions which are not present in water because of its "leveling effect".
- In some cases an increase of fluorescence intensity may be observed when switching to NACE, which positively affects LOD.

M. Szumski (✉) · B. Buszewski
Faculty of Chemistry, Nicolaus Copernicus University, Toruń, Poland
e-mail: michu@chem.umk.pl

B. Buszewski et al. (eds.), *Electromigration Techniques*, Springer Series
in Chemical Physics 105, DOI: 10.1007/978-3-642-35043-6_11,
© Springer-Verlag Berlin Heidelberg 2013

11.1 Solvents

When compared to aqueous systems, application of a wide range of organic solvents and their mixtures, characterized by different properties opens new possibilities in controlling selectivity. For example, pK_a values in organic solvents may differ significantly from those determined in water, which may allow for separation of analytes impossible to be separated in aqueous conditions. This is so, because the solvents used in NACE can belong to amphiprotic (like methanol) or typical proton acceptors (for example formamide) (Table 11.1).

It may be quite important to pay attention to purity of the solvents used, particularly to possible water content. Sometimes when important data, not only analytical but also thermodynamic (for example dissociation constants), have to be obtained it might be necessary to determine water content, which may be done using Karl-Fischer method [5, 6].

In many pure solvents the electroosmotic flow may be observed during CE or CEC. EOF depends on dielectric constant to viscosity ratio and may vary from one solvent to another—some exemplary data are shown in Fig. 9.6 in Sect. 9.1. For practical reasons the solvents should be characterized by low viscosity (which influences not only EOF but electrophoretic mobility as well), low vapor pressure, and should not be toxic. Absorption of UV–V is radiation should also be taken into account, and if any problems occur, other detection methods need to be considered. Pure solvents or their mixtures as well as solutions of some electrolytes can be used as a running electrolytes. Acids and their ammonium (e.g., ammonium acetate, ammonium oxalate) or sodium salts are most popular as well as some organic buffers like Tris. Their cations and anions may have significant influence

Table 11.1 Classification of organic solvents in terms of their acid–base properties (Brønsted theory) according to Kolthoff [7, 8]

Properties		Relative acidity	Relative basicity	Example
Amphiprotic	Neutral	+	+	Methanol, glycerol, phenol, tert-butyl alcohol
	Protogenic	+	−	Sulfonic acid, formic acid, acetic acid
	Protophilic	−	+	Liquid ammonia, formamide, N-methylacetamide, NMF
Aprotic	Dipolar protophilic	−	+	DMSO, DMF, THF, 1,4-dioxane, pyridine
	Dipolar protophobic	−	−	Acetonitrile, acetone, nitrobenzene, sulfolan, propylene carbonate
	Neutral	−	−	Aliphatic hydrocarbons, benzene, 1,2-dichloroethane, carbon tetrachloride

+ or − indicate whether given type of solvent is stronger or weaker acid or base than water

Table 11.2 pH* values of methanolic electrolytes, according to [10]

Electrolyte composition	pH*
2 mol/L TFA, 20 mmol/L NH$_4$OAc	2.90
1 mol/L TFA, 20 mmol/L NH$_4$OAc	3.15
10 mmol/L oxalic acid, 10 mmol/L ammonium hydrogen oxalate	5.79
10 mmol/L succinic acid, 10 mmol/L lithium hydrogen succinate	8.75
20 mmol/L NH$_4$OAc	10.51
50 mmol/L NaOCH$_3$, 20 mmol/L NH$_4$OAc	13.50
0,1 mol/L NaOCH$_3$, 20 mmol/L NH$_4$OAc	15.24

on resolution in NACE due to not impaired (compared to water conditions) electrostatic interactions with analytes [3, 4, 9].

When electrolytes solutions in organic solvents are employed as background electrolytes their pH* values are sometimes used to characterize them. The values presented in Table 11.2 have been obtained with a pH-meter with glass electrode calibrated with the following methanolic buffers: oxalic acid/ammonium hydrogen oxalate and succinic acid/Lithium hydrogen succinate of pH* 5.79 and 8.75, respectively [10]. Exemplary plots of dependency of EOF on pH and pH* are presented in Fig. 11.1 [10].

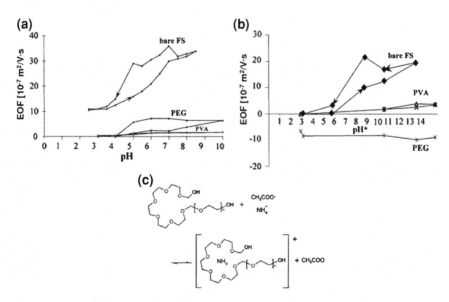

Fig. 11.1 Comparison of EOF in capillaries (coated with hydrophilic polymers) in aqueous and non-aqueous conditions and different pH and pH*. Bottom picture explains the reversed EOF: ammonium ions are complexed by PEG in non-aqueous conditions, which results in positive net charge on the inner capillary wall. Reprinted with permission from Belder et al. [10]. Copyright 2000 Elsevier

11.2 Separations in Wide Bore Capillaries

In classical CE the capillaries are characterized by rather small internal diameters
(25–75 µm) as the Joule heat needs to be efficiently dissipated. Joule heating is
much lower in NACE due to much less conductivity of electrolytes in organic
solvents, which allows to use high electrolyte concentrations, wide bore capillaries
(150 or even 530 µm), and higher voltages [11, 12]. Contrary to separations
performed in aqueous solutions, lower Joule heat generated in organic solvents
give linear plot I = f(V) in a broad range of voltages and capillary diameters
(Fig. 11.2).

Larger capillary diameter may in turn allow for larger volumes to be injected
which make "preparative" but with high resolution CE separations possible.
However, when using large bore capillaries a siphoning (i.e., flowing of the liquid
from the vial of higher liquid level to another vial of lower liquid level) effect may be
quite significant. Such phenomenon may adversely influence band width and resolution. It can be minimized by using capillaries tapered at both ends (to restrict
hydrodynamic flow) or by compensating the change of level of liquid by controlled
rise of the vial in which the level becomes lower due to the EOF—Fig. 11.3 [12, 13].

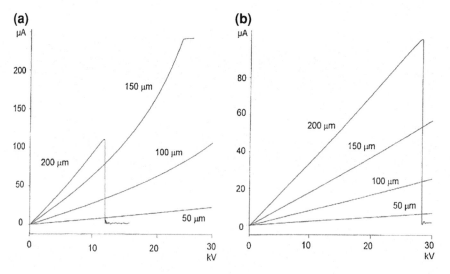

Fig. 11.2 Dependency of current on the applied voltage for capillaries of $L_{tot} = 58.5$ cm and
different internal diameters. Electrolytes: **a** 20 mmol/L ammonium acetate in water/acetic acid
(99/1, v/v), **b** 20 mmol/L ammonium acetate in ethanol/acetonitrile/acetic acid (50/49/1, v/v).
Temperature: 20 °C. Breakdown on the curves indicate the conditions when the medium starts to
boil which breaks the electric circuit. Reprinted with permission from Valkó et al. [20]. Copyright
1998 Elsevier

Fig. 11.3 NACE separation of diuretics (1—bumetanide, 2—furosemide, 3—ethacrynic acid) in wide-bore capillaries: 320 μm and total length of 60 cm with controlled vial lifting. Separation medium: ethanol/acetonitrile 50/50 + electrolyte: **a** 1 mmol/L KAc, lifting 0.72 mm/min, no siphoning effect, **b** 5 mmol/L NH$_4$Ac, lifting 0.84 mm/min, no siphoning effect, **c** 5 mmol/L NH$_4$Ac, no lifting, significant siphoning effect. Reprinted with permission from Jussila et al. [13]. Copyright 2000 Wiley-VCH

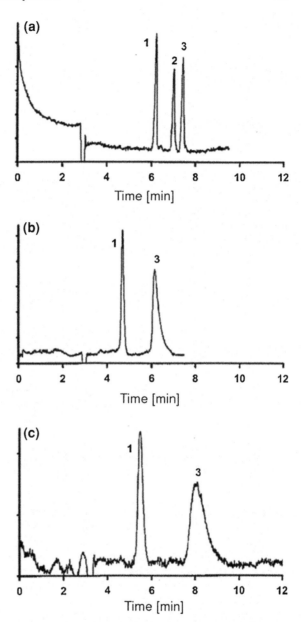

11.3 Separation of Large Molecules Insoluble in Water

Despite their ionic character many molecules of natural of anthropogenic origin are not soluble in water because of their high molecular weight, hydrophobicity or instability in water solutions [14]. Application of NACE may be the only way to

Fig. 11.4 Separation of fatty acids in the following mixtures: 2.5 mmol/L antraquinone-2-carboxylic acid, 40 mmol/L Tris in N-methylformamide/dioxane 3/1 (v/v). Separation conditions: V = 20 kV, L_{tot} = 67 cm, L_{eff} = 46 cm, d_c = 75 μm, indirect detection at λ = 264 nm. The migration order is: n-C_{26}, n-C_{24}, n-C_{22}, n-$C_{22\text{-}OH}$, n-C_{20}, n-C_{18}, n-C_{16}, n-$C_{16\text{-}OH}$, n-C_{14}. Reprinted with permission from Drange and Lundanes [15]. Copyright 1997 Elsevier

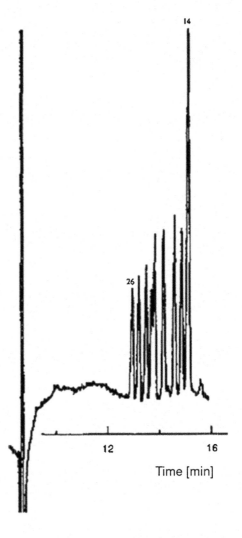

separate them provided their solubility in organic solvents. For example, it is possible to separate long chain fatty acids with NACE. Such separations used to be performed with GC, HPLC, or SFC. Drange and Lundanes [15] used 40 mmol/L Tris in NMF/dioxane 3/1 as a separation medium. Due to lack of chromophores in fatty acids, indirect UV detection (λ = 264 nm) was employed in the presence of 2.5 mmol/L antraquinone-2-carboxylic acid (Fig. 11.4).

Xu et al. [16] applied NACE for determination of cholesterol in egg yolk and milk. When compared to classical spectrophotometric analysis, NACE turned out to be much faster and provided good correlation of quantitative data. The authors used 100 mmol/L sodium acetate/acetic acid (99/1) methanolic solution as a background electrolyte. The other exemplary analytes that can be separated with

NACE are phospholipids, polymers, dyes, organomercury compounds, hydrophobic polypeptides, phosphoorganic pesticides, hydrophobic pharmaceuticals, steroids, and alkaloids [9].

11.4 Separation of Uncharged Compounds

Electrophoretic separation in NACE can be completed if analytes are charged in an organic solvent or if non-charged analytes interact with charged additive, introduced to the system for that purpose [17]. Under aqueous conditions neutral and hydrophobic species can be separated with, for example, micellar electrokinetic chromatography (MEKC). In most of the organic solvents (except for formamide and poly (ethylene glycol)) micelles cannot be formed due to very weak hydrophobic (solvophobic) interactions between hydrocarbon chains of the surfactants molecules. Separation of highly hydrophobic substances can be performed using some ion-pairing species as additives. For example, polyaromatic hydrocarbons can be separated employing tetrahexylammonium bromide (THA^+) as an additive to 50 mmol/L ammonium acetate in methanol (Fig. 11.5) [18].

The separation technique used in the above example was called hydrophobic interaction electrokinetic chromatography (HI-EKC). While separating PAHs using HI-EKC with THA^+ the complexes of PAHs molecules with THA^+ are formed, and the number of bound additive molecules depends on hydrophobic interactions and polarizability. Hydrophobic interactions are strong even in methanol, while the influence of polarizability ("electrostatic model") can be explained in such a way that in a medium of low dielectric constant (methanol) delocalized π electrons undergo polarization (induced dipole is formed) in the presence of THA^+ which, in turn, leads to creation of associates of different (depending on hydrophobicity/polarizability of a given PAH) electrophoretic mobility. Both the mentioned mechanisms, with possible domination of the hydrophobic interaction in this particular case, influence the number of THA^+ bound with one PAH molecule (Fig. 11.6) [18].

Other additives that can be used in electrophoretic separation in non-aqueous conditions are: ammonium ions, alkali metal ions, anions ClO_4^-, BF_4^-, NO_3^-, Br^-, Cl^-, $CH_3SO_3^-$, Brij, (+)-18-corona-6-tetraacetic acid, (-)-quinine, 2,4,6-triphenylpyrylium, tetraalkylammonium salts, camphorsulfonic acid, heptanesulfonic acid, tropylium, sodium dodecyl-, tetradecyl-, and hexadecylsulfates [4, 17]. The mentioned species provide the solvophobic, electrostatic (ion–ion, ion–dipole, dipole–dipole), and donor–acceptor interactions. It is believed that in NACE the interactions of analytes with the additives are of much more complex nature when compared to classical CE performed in aqueous conditions. It should be noted, however, that electrostatic and donor–acceptor interactions rather dominate in NACE (due to low dielectric constant of the medium) while hydrophobic interactions dominate in CE.

11.5 Separation of Optically Active Compounds

Separation of optically active compounds using electromigration techniques is quite popular due to high efficiencies obtained and low chemicals consumption. Despite most chiral separations being performed in aqueous or hydro-organic conditions, application of non-aqueous environment may be profitable, which is the result of different properties of organic solvents and water. The separation of enantiomers usually employs the electrostatic interactions of the analyte with the chiral selector [19]. Potential energy of ion–ion (E_{i-i}) and ion–dipole (E_{i-d}) interactions can be expressed by the following Eqs. [4]:

$$E_{i-i} = \frac{z_a z_b e^2}{4\pi\varepsilon_0 \varepsilon r} \qquad (11.1)$$

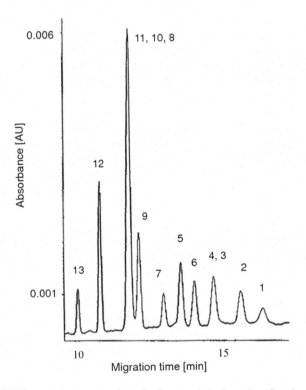

Fig. 11.5 HI-EKC non-aqueous separation of polyaromatic hydrocarbns. Resolved compounds: *13* coronene, *12* benzo[*ghi*]perylene, *11* benzo[*a*]pyrene, *10* benzo[*e*]pyrene, *8* perylene, *9* benzo[*k*]fluoranthrene, *7* chryzene, *5* pyrene, *6* 2,3-benzofluorene, *4* fenanthrene, *3* anthracene, *2* fluorene, *1* naphthalene. Separation medium: 50 mmol/L ammonium acetate–100 mmol/L THA in 100 % methanol. Analyte concentrations ca. 40 μmol/L. Capillary: L_{eff} = 40 cm, E = 532 V/cm, λ = 254 nm. Reprinted with permission from Koch et al. [18]. Copyright 2001 Elsevier

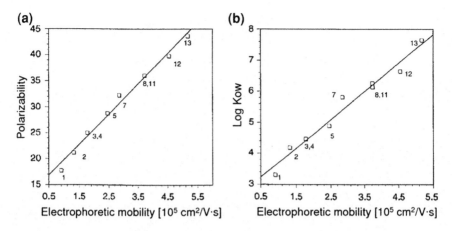

Fig. 11.6 Correlation of electrophoretic mobility of PAHs with polarizability (**a**) and octanol/water partition coefficient (**b**). Analytical conditions the same as in the Fig. 11.4. Reprinted with permission from Koch et al. [18]. Copyright 2001 Elsevier

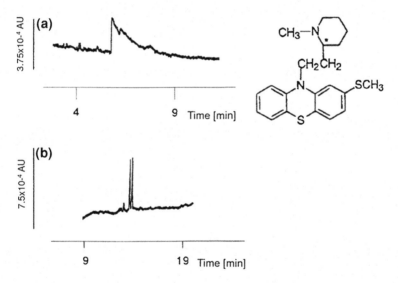

Fig. 11.7 Comparison of separation of thioridazine (antipsychotic drug) enantiomers in aqueous **a** and non-aqueous **b** conditions. Electrolyte: **a** 1.543×10^{-4} w/v % (~ 1 μmol/L) β-CD-$(SO_4-)_4$ in aqueous solution of 50 mmol/L Tris-phosphate (pH 2.50); **b** 1.543×10^{-4} w/v % (~ 1 μmol/L) β-CD-$(SO_4-)_4$ in 150 mmol/L citric acid–100 mmol/L Tris (pH* = 5.1) in formamide. Electric field: **a** 357 V/cm, **b** 595 V/cm. Reprinted with permission from Wang and Khaledi et al. [23]. Copyright 1999 Elsevier

$$E_{i-d} = \frac{z_a e \mu cos\theta}{4\pi\varepsilon_0 \varepsilon r^2} \tag{11.2}$$

where z is charges of the interacting ions a and b, r is distance between the ions, μ is dipole moment, θ is dipole angle.

Because of lower dielectric constant of the organic solvents, such interactions are stronger than in water ($\varepsilon = 78$). It allows to use such chiral selectors that would not provide the interactions strong enough to obtain separation in aqueous conditions. Moreover, some chiral selectors, like modified cyclodextrins, show much better solubility in organic solvents (formamide or N-methylformamide). For example, separations of dansyl derivatized amino acids enantiomers are difficult to be performed in water because of limited solubility of β-cyclodextrin in water. When NMF is employed as a separation medium such analytes are possible to be separated using 80 mmol/L of β-cyclodextrin as and chiral selector [20]. Other selectors that can be successfully used in chiral separations are: heptakis(2,3-diacetyl-6-sulfato)-[β]-cyclodextrin (separation of weakly basic pharmaceuticals in pure methanol) [21], β-CD-(SO_4^-)$_4$ (Fig. 11.6) and QA-β-CD (QA—*quaternary ammonium*) [22, 23], carboxymethyl-β-CD [4], quinine, (+)-18-corona-6-tetraacetic acid [24] or camphorsulfonic acid. Also, some reports have been published on using pure chiral solvents like R-(−)-2-butanol, S-(+)-2-butanol and S-(−)-2-methyl-1-butanol [4] (Fig. 11.7).

11.6 Summary

Non-aqueous electrophoreis and electrochromatography become more and more important in separation sciences and are a profitable complement to classical aqueous and non-aqueous techniques. The most important advantages are: possibility of separation of the water insoluble species and employing of the electrostatic interactions with the additives that would be too weak in water to give a significant separation effect.

References

1. Y. Walbroehl, J.W. Jorgenson, On-column UV absorption detector for open tubular capillary zone electrophoresis. J. Chromatogr. A **315**, 135–143 (1984)
2. Y. Walbroehl, J.W. Jorgenson, Capillary zone electrophoresis of neutral organic molecules by solvophobic association with tetraalkylammonium ion. Anal. Chem. **58**(2), 479–481 (1986)
3. M.L. Riekkola, Recent advances in nonaqueous capillary electrophoresis. Electrophoresis **23**(22–23), 3865–3883 (2002)
4. M.L. Riekkola, M. Jussila, S.P. Porras, I.E. Valko, Non-aqueous capillary electrophoresis. J. Chromatogr. A **892**(1–2), 155–170 (2000)

5. B. Buszewski, Solvents, In: P.J. Worsfold, A. Townshend, C.F. Poole (eds.), Encyclopedia of analytical Science, 2nd edn. (Elsevier, Amsterdam, 2005)
6. S.P. Porras, M.L. Riekkola, E. Kenndler, Capillary zone electrophoresis of basic analytes in methanol as non-aqueous solvent: Mobility and ionisation constant. J. Chromatogr. A **905**(1–2), 259–268 (2001)
7. I.M. Kolthoff, Acid-base equilibriums in dipolar aprotic solvents. Anal. Chem. **46**(13), 1992–2003 (1974)
8. J. Tjørnelund, S.H. Hansen, Non-aqueous capillary electrophoresis of drugs: Properties and application of selected solvents. J. Biochem. Biophys. Methods **38**(2), 139–153 (1999)
9. H. Teng, B.Q. Yuan, T.Y. You, Recent advances in application of nonaqueous capillary electrophoresis. Chin. J. Anal. Chem. **38**(11), 1670–1677 (2010)
10. D. Belder, K. Elke, H. Husmann, Influence of pH*-value of methanolic electrolytes on electroosmotic flow in hydrophilic coated capillaries. J. Chromatogr. A **868**(1), 63–71 (2000)
11. I.E. Valkó, S.P. Porras, M.L. Riekkola, Capillary electrophoresis with wide-bore capillaries and non-aqueous media. J. Chromatogr. A **813**(1), 179–186 (1998)
12. H. Yin, C. Keely-Templin, D. McManigill, Preparative capillary electrophoresis with wide-bore capillaries. J. Chromatogr. A **744**(1–2), 45–54 (1996)
13. M. Jussila, S. Palonen, S.P. Porras, M.L. Riekkola, Compensation of the siphoning effect in nonaqueous capillary electrophoresis by vial lifting. Electrophoresis **21**(3), 586–592 (2000)
14. H. Cottet, M.P. Struijk, J.L.J. Van Dongen, H.A. Claessens, C.A. Cramers, Non-aqueous capillary electrophoresis using non-dissociating solvents: Application to the separation of highly hydrophobic oligomers. J. Chromatogr. A **915**(1–2), 241–251 (2001)
15. E. Drange, E. Lundanes, Determination of long-chained fatty acids using non-aqueous capillary electrophoresis and indirect UV detection. J. Chromatogr. A **771**(1–2), 301–309 (1997)
16. X.H. Xu, R.K. Li, J. Chen, P. Chen, X.Y. Ling, P.F. Rao, Quantification of cholesterol in foods using non-aqueous capillary electrophoresis. J. Chromatogr. B Anal. Technol. Biomed. Life Sci. **768**(2), 369–373 (2002)
17. M.T. Bowser, A.R. Kranack, D.D.Y. Chen, Analyte-additive interactions in nonaqueous capillary electrophoresis: A critical review. Trends Anal. Chem. **17**(7), 424–434 (1998)
18. J.T. Koch, B. Beam, K.S. Phillips, J.F. Wheeler, Hydrophobic interaction electrokinetic chromatography for the separation of polycyclic aromatic hydrocarbons using non-aqueous matrices. J. Chromatogr. A **914**(1–2), 223–231 (2001)
19. I. Bjornsdottir, S. Honoré Hansen, S. Terabe, Chiral separation in non-aqueous media by capillary electrophoresis using the ion-pair principle. J. Chromatogr. A **745**(1–2), 37–44 (1996)
20. I.E. Valkó, H. Sirén, M.L. Riekkola, Chiral separation of dansyl-amino acids in a nonaqueous medium by capillary electrophoresis. J. Chromatogr. A **737**(2), 263–272 (1996)
21. J.B. Vincent, G. Vigh, Nonaqueous capillary electrophoretic separation of enantiomers using the single-isomer heptakis (2, 3-diacetyl-6-sulfato)-[beta]-cyclodextrin as chiral resolving agent. J. Chromatogr. A **816**(2), 233–241 (1998)
22. F. Wang, M.G. Khaledi, Nonaqueous capillary electrophoresis chiral separations with quaternary ammonium [beta]-cyclodextrin. J. Chromatogr. A **817**(1–2), 121–128 (1998)
23. F. Wang, M.G. Khaledi, Non-aqueous capillary electrophoresis chiral separations with sulfated [beta]-cyclodextrin. J. Chromatogr. B Biomed. Sci. Appl. **731**(2), 187–197 (1999)
24. Y. Mori, K. Ueno, T. Umeda, Enantiomeric separations of primary amino compounds by nonaqueous capillary zone electrophoresis with a chiral crown ether. J. Chromatogr. A **757**(1–2), 328–332 (1997)

Chapter 12
Methods of Analyte Concentration in a Capillary

Paweł Kubalczyk and Edward Bald

Abstract Online sample concentration techniques in capillary electrophoresis separations have rapidly grown in popularity over the past few years. During the concentration process, diluted analytes in long injected sample are concentrated into a short zone, then the analytes are separated and detected. A large number of contributions have been published on this subject proposing many names for procedures utilizing the same concentration principles. This chapter brings a unified view on concentration, describes the basic principles utilized, and shows a list of recognized current operational procedures. Several online concentration methods based on velocity gradient techniques are described, in which the electrophoretic velocities of the analyte molecules are manipulated by field amplification, sweeping and isotachophoretic migration, resulting in the online concentration of the analyte.

Capillary electrophoresis (CE) is a modern analytical technique which allows a fast and effective separation of charged particles present in a very small sample. Separation is based on differences in electrophoretic ionic mobility in the electric field, in a narrow capillary [1–12]. CE combines advantages offered by two techniques that constitute the foundation stones upon which modern bio-analysis was built: high performance liquid chromatography and conventional gel electrophoresis (SGE). CE is usually more efficient than any of the two techniques applied separately [13].

When optimizing the electrophoretic conditions in CE, one of many challenges to meet is to lower the limit of detection (LOD) expressed in a number of concentration units. High LOD results from two factors: limited optical path in the most popular CE detection in a capillary and a small volume of a sample which

P. Kubalczyk (✉) · E. Bald
Faculty of Chemistry, University of Łódź, Łódź, Poland
e-mail: pakuba@uni.lodz.pl

B. Buszewski et al. (eds.), *Electromigration Techniques*, Springer Series
in Chemical Physics 105, DOI: 10.1007/978-3-642-35043-6_12,
© Springer-Verlag Berlin Heidelberg 2013

can be injected into the capillary [14]. The optimum volume of the sample applied should not exceed 1–3 % of the capillary volume. The injection of a bigger amount of the sample causes the so-called capillary overload, the result of which is broadening the bands and significantly deteriorated resolution. Additionally, in order to obtain high performance and short analysis time, it is essential to use capillaries of small internal diameters (25–100 μm). Since the capillary cross section is round, the real optical path length is reduced to 80 % of the internal diameter [15].

Low sensitivity is perceived as one of the main limitations of electrophoretic separation methods, particularly when compared to traditional liquid chromatography techniques [16]. The detection limits of the same substances analyzed by means of electromigration techniques are usually 10–100 times higher than those obtained by means of high performance liquid chromatography [17]. In CE, many attempts have been made, and many methods applied, to deal with the problem of low concentration sensitivity. One of those methods is the application of highly sensitive detectors such as an electrochemical detector, chemiluminescent detector, or a detector with laser-induced fluorescence (LIF) [18, 19]. LIF and chemiluminescent detectors enable obtaining more than 1,000 times higher sensitivity; an electrochemical detector, although it does not significantly improve sensitivity, allows the detection of analytes present in a small volume, e.g., femtoliters. Another method is extending the optical path by applying a capillary with a bulb in the detection zone [20, 21], a Z-cell (3 and 10 times higher sensitivity), and a highly sensitive cell [22], the capillary of which in the detection zone is rectangular (10 times higher sensitivity). Another way of LOD lowering is an offline analytes concentration (outside the measuring system) or online analytes concentration (inside the measuring system). Concentration outside the system is commonly used in other techniques, for instance in gas chromatography and high performance liquid chromatography. These are mainly liquid–liquid extraction (LLE), solid phase extraction (SPE), solid phase micro-extraction (SPME), or sorption methods which in optimum conditions allow us to obtain 100 times higher sensitivity.

Nowadays, methods of analytes concentration inside a capillary are highly significant. The term "stacking" [23], is of a broad and ambiguous meaning. It is used by analysts who deal with electromigration methods. On the one hand, it is generally used to describe some analyte concentration method in the electric field; on the other hand, it precisely means a field amplified analyte concentration. In most general reviews, the term "stacking" refers to those of analyte concentration methods which are based on changes in electrophoretic velocities of analytes. It therefore concerns the methods of analyte concentration caused by changes in the strength of field as well as changes in electrophoretic velocity obtained by some other means. In all cases, the key requirements are the presence of an electrophoretic element in the concentration mechanism and the fact that analytes undergo concentration on a sample-background electrolyte boundary by lowering the speed of migration [24].

There exist a few ways to determine the efficiency of an analyte concentration in a sample. In a majority of papers, peak heights obtained when following a conventional procedure are compared with the peak heights obtained as a result of the procedure involving concentration. However, this procedure does not cover all cases, e.g., increased peak heights can be obtained by prolonging the time of a sample injection (1–3 % of a capillary volume). Another means of concentration efficiency determination is the sample zone length to the analyte zone length ratio in a detector (DIBR). It does not, however, apply to the electrokinetic method of introducing a sample in which the sample zone can be narrowed. The most practical way of representing the efficiency is the comparison of LOD in the case of a usual procedure to the one involving concentration.

Currently, in CE, numerous techniques of analyte concentration in a measuring system are applied in order to improve concentration sensitivity, or to reduce LOD limit. In those techniques, analyte concentration occurs at capillary inlet or inside, before separation and detection. The result can be obtained by introducing a large volume of a sample solution followed by the analyte concentration in a narrow band using chromatographic and/or electrophoretic effects. The latter ones are particularly suggested owing to their application simplicity and no need to modify apparatus. The techniques involve (Fig. 12.1) stacking [25–31] and sweeping [32]. The above-mentioned techniques can be applied individually or in combinations [33, 34]. The new methods of sample analysis based on electromigration techniques involving analyte concentration in a measuring system, developed within the last few years, are reviewed in a number of papers [16, 24, 35–40].

12.1 Analytes Concentration in Capillary Zone Electrophoresis

Capillary zone electrophoresis (CZE, Free Solution electrophoresis) is the oldest, simplest, and most commonly applied kind of electrophoresis. When developing this technique, also called free solution electrophoresis, the first attempts to eliminate the main drawback of electromigration techniques, i.e., low concentration sensitivity, were made. Numerous analyte concentration methods in a capillary which are based on the electrophoretic principles were elaborated. The methods were generally named "stacking" [23]. The main concentration methods by stacking mechanisms are discussed below.

12.1.1 Field Amplified Sample Stacking

Analytes concentration by Field Amplified Sample Stacking (FASS) is considered to be one of the simplest concentration techniques in a measuring system. In this

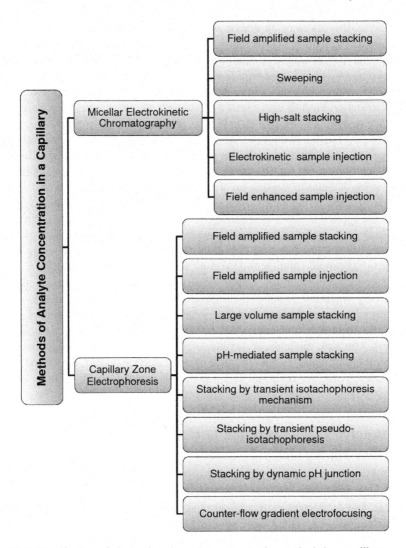

Fig. 12.1 Classification of electrophoretic analyte concentration methods in a capillary

method, a sample is prepared in a buffer characterized by low conductivity and significantly different from the background electrolyte characterized by high conductivity [24]. A convenient way is to dilute the background electrolyte 100 or 1,000 times and dissolve the sample in it. A definite volume of a sample prepared in the way described above is hydrodynamically injected into the capillary filled with a background electrolyte of high conductivity. Then, as a result of applying high voltage of positive polarization, strong electric field appears in the sample zone. Consequently, rapid growth in the ion migration velocity is observed. When ionic analytes reach the boundary between the sample zone and the background electrolyte, electric field strength decreases, the migration velocities are decreased and analytes undergo stacking. If the electroosmotic flow is higher than the speed of the charged analytes, all ions, independently from their charges, move towards the detection window and undergo separation according to the capillary zone electrophoresis mechanism. It enables the determination of anions and cations during one analytical operation. In FASS, a volume of the injected sample must be carefully controlled. When the volume of a sample is excessive, concentration and separation difficulties resulting from a limited capillary length can occur. The maximum amount of the sample should not exceed 5 % of the capillary volume. Increasing the sample volume is followed by band broadening and resolution deterioration due to local EOF velocity in the sample zone higher than in the background electrolyte zone. It is a consequence of the difference in the electric fields and consecutive pressure difference. The pressure difference causes bands broadening and a reduction in resolution [41]. Field amplified analyte concentration allows the method sensitivity to be 10 or even up to 10,000 times higher.

12.1.2 Field Amplified Sample Injection

For FASI concentration, differently from FASS, electrokinetic injection is applied. It has one important advantage: analyte ions are injected to a capillary as a result of two combined phenomena, i.e., electrophoresis and electroosmosis. Dependending on the applied voltage polarization, cations or anions will be injected to the capillary. If electrophoretic mobilities of analytes are in the direction opposite to EOF, smaller amounts of analytes will be injected to a capillary than by means of hydrodynamic injection (FASS). If mobilities are in the same direction as EOF, the situation is reverse, i.e., more analyte ions are injected in comparison to FASS [42]. In the case of FASI, assuming that the same sample volumes are injected (5 % of a capillary volume), sensitivity is usually 100–1,000 times higher. Reassuming, EOF direction and magnitude significantly influence FASI. Dąbek-Złotorzyńska and Piechowski [43] reversed the EOF direction by using dynamically coated capillaries and obtained 400 times higher sensitivity. They did it in order to determine benzoic acid isomeric derivatives in aerosols and car exhaust fumes. It should be emphasized that dynamic coats applied to reverse EOF not only caused shorter separation time but also considerably improved repeatability.

12.1.3 Large Volume Sample Stacking

Concentration by large volume sample stacking (LVSS) used to be called analyte concentration by matrix removal [39]. The technique is comparable to FASS in terms of a buffer system applied. The change in electrodes polarization in order to change electroosmotic flow is the most substantial difference. A sample is dissolved in a buffer of low conductivity or in water. Then it is injected hydrodynamically in an adequate volume into a capillary filled with a background electrolyte of high conductivity. Next, the electrodes polarization is reversed and voltage is applied. As a result, electroosmotic flow towards the capillary outlet is obtained. Negatively charged analytes begin to move towards the end of the capillary until they reach the sample zone—background electrolyte boundary. Anionic analytes stacking appears on the boundary whereas neutral or positively charged particles leave the capillary moving in the opposite direction. It is essential to control current intensity carefully in order to switch polarization and reverse EOF simultaneously. At that moment current intensity should reach 95–99 % of normal value. It will let us avoid a potential loss of anionic analytes. Their separation proceeds according to the capillary zone electrophoresis mechanism. LVSS, in comparison to FASS, enables the introduction of a considerably larger sample volume without remarkable deterioration in separation efficiency. To obtain good repeatability, particular attention should be paid to controlling current intensity as well as the moment of polarization switch to normal. LVSS cannot be applied to determine anions and cations simultaneously. Additionally, analytes should be characterized by low mobility. In order to separate bacteria, Yu and Li [44] applied LVSS with polarization switch to increase CE sensitivity. They obtained 60 times higher sensitivity using a long sample zone (ca. 39 % of the capillary volume). The method enables concentration in a measuring system of intact bacteria cells as well as a quick separation of a bacteria mixture. He and Lee [45] proposed a system similar to LVSS without polarization switch, developed earlier by Chien and Burgi [41]. Still, in order to control matrix removal, they implemented a background electrolyte (BGE) of low pH in connection with a sample zone of high pH. The system was described as LVSS with an EOF pump. The same system was applied for concentrating in measuring systems of methotrexate and its metabolites in human plasma [46]. A sample was hydrodynamically injected (to 74 % of a capillary volume) into a capillary filled with 70 mmol/L of phosphate buffer at pH = 6.0, containing 0.01 % of polyethylene glycol (PEO). Instead of applying pH discontinuity, the authors used PEO to suppress EOF. It allowed them to proceed directly from the matrix removal to separation. The change resulted in roughly 100 times higher sensitivity. Zhang and Yin [47] performed an experiment in which LVSS with negative pressure in multi-T microchips was used to improve the sensitivity of the method used for amino acid derivatives determination with fluorescein isothiocyanide (FITC). 41 and 43 times higher sensitivity values of the method were obtained for valine and alanine, respectively.

LVSS and FASI have a disadvantage. It is a difficulty in applying real samples because a complex matrix causes an increase in a sample conductivity. Therefore, stacking efficiency is reduced. The implementation of adequate extraction methods enabled the application of the techniques for various analytes determination in numerous matrices, e.g., environmental samples [48–51], water extracts of soil [52], or food products [53–55].

12.1.4 pH-Mediated Sample Stacking

Frequently, a sample is a solution of very high conductivity which makes performing an electrophoretic analysis difficult. One of possibilities to obtain a sample zone of low conductivity for its further concentration inside a capillary is pH manipulation. In the case of cations, a sample is electrokinetically injected into a capillary. Simultaneously, anions of the background electrolyte that fills the capillary migrate in the opposite direction, which is towards the capillary inlet. Hence, after a sample injection is accomplished, the sample zone is composed exclusively of the sample cations and background electrolyte anions. At that moment, a strong acid zone is electrokinetically injected into the capillary. H^+ ions from the acid titrate background electrolyte anions in the sample zone the conductivity of which significantly decreases. A further procedure is identical to the one applied in the concentration by FASS or LVSS. It is also possible to determine anions. Then, instead of a strong acid, a strong base is electrokinetically injected to the capillary. Thus, background electrolyte cations (usually NH_4^+) are titrated with OH^- ions creating a zone of low conductivity [37]. The technique described above was implemented for the first time by Hadwiger et al. [56].

12.1.5 Stacking by Transient Isotachophoresis Mechanism

In classical isotachophoresis (ITP), a sample zone is injected between two electrolytes: leading electrolytes and terminating. Mobility of the first electrolyte is higher, of the latter one it is lower than mobility of analytes in a sample. In practice, first, a capillary is filled with the leading electrolyte and then, a sample is injected. The capillary inlet is immersed in a vial filled with the terminating electrolyte, while the outlet in the leading electrolyte. The analyte and the corresponding leading and terminating ions must be of the same sign. They also have to migrate in the same direction. After applying high voltage, the analyte is concentrated to reach the concentration value of the leading ion, which is a unique ITP feature [37]. In the case of concentration by stacking according to transient isotachophoresis (tITP) mechanism, high mobility ions are injected to a sample and background electrolyte ions function as terminating ions. As the electric field is generated, analytes create a narrow zone typical of ITP. Then, the leading electrolyte concentration decreases as a result of ions diffusion. Therefore, the electric field is the same within the whole capillary and isotachophoresis fluently changes into capillary zone electrophoresis [15, 57]. A tITP advantage is a possibility to inject a larger volume of a sample which does not have to be of low conductivity.

It is also possible to combine tITP with electrokinetic sample injection, which was first suggested by Hirokawa et al. [58]. They called the technique electrokinetic supercharging (EKS). Botello et al. [59] implemented the technique to capillary electrophoresis of non-steroidal anti-inflammatory drugs. They obtained 2,000 times higher sensitivity in comparison to the regular procedure involving hydrodynamic sample injection.

12.1.6 Stacking by Transient Pseudo Isotachophoresis

Concentration by pseudo-tITP is based on injecting a large sample zone mixed with acetonitrile or other organic solvent in the ratio 1:2 into a capillary. This kind of concentration was introduced by Shihabi [60, 61] and Shihabi et al. [62] to the electrophoretic analysis of samples containing high salt concentration. The technique allows the injection of a much larger sample zone (more than 30 % of the capillary volume) in comparison to the conventional CZE (1–3 % of the capillary volume). A probable stacking mechanism (Fig. 12.2) was called transient pseudo-isotachophoresis.

ITP exhibits many characteristic features but the most important one is that all analytes, having reached equilibrium, move at the same speed which is imposed by the leading ion mobility and its concentration. The terminating ion, which determines the high electrical field, is crucial for keeping the analytes in the band and prevents its diffusion. In pseudo-tITP, ACN functions as a terminating pseudo-ion causing high field intensity required to accelerate analytes migration, not being an ion itself. Pseudo-tITP is successfully applied as an analytes concentration technique for determining thiols (Fig. 12.3), peptides, and other compounds [35, 37, 63–68] in real samples of high salinity.

12.1.7 Stacking by Dynamic pH Junction

The concept of such an analytes concentration technique in a sample was suggested by Aebersold and Morrison [69]. The technique takes advantage of pH discontinuity between the background electrolyte and the sample which results in a significant analytes mobility change causing the stacking effect [16, 70]. Aebersold and Morrison obtained only 5 times higher sensitivity while Britz-McKibbin et al. [71] and Britz-McKibbin and Chen [72] later reported 100 times higher amplification, naming the method "dynamic pH junction". It can be successfully applied to a selective weak electrolytes, the charge and mobility of which are the function of pH. In order to determine anions, a long sample zone is hydrodynamically injected to a capillary at pH lower than that for the background electrolyte. After the vial containing a sample is exchanged with the one containing the background electrolyte, and high voltage is applied, OH^- ions permeate from the alkali BGE to the sample zone through the boundary. The higher pH zone causes the non-dissociated weak acids acceleration in the sample zone. The acids undergo stacking in the narrow band. The resulting pH boundary between the background electrolyte and

Initial conditions

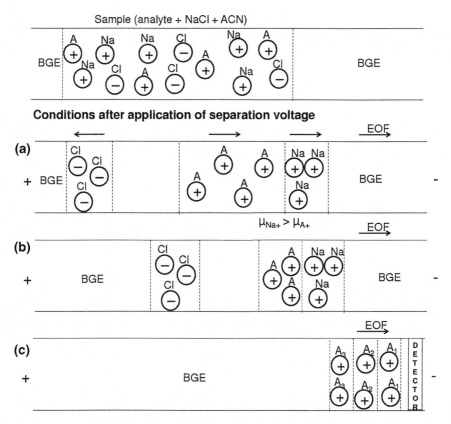

Fig. 12.2 Concentration by stacking by transient pseudo-isotachophoresis. A sample with high salt concentration and an organic modifier (ACN; 2:1 v/v) is hydrodynamically injected to a capillary filled with a background electrolyte (BGE). **a** After applying voltage, salt ions move rapidly in ACN environment, and then decelerate on the sample zone/BGE boundary. Two zones of different field strength values are created, the low field strength one and the high field strength one in ACN. **b** Analytes, since they are in the high strength zone (ACN), are significantly accelerated, and then undergo stacking as they reach the low field strength zone. **c** After stacking, the analytes, permeate to BGE, and then undergo separation according to CZE principles

the sample zone is present only for a short period of time till OH⁻ ions reach the end of the sample zone. The analytes which have undergone stacking are further separated according to CZE. In this technique, the sample conductivity is of little importance and can be lower, equal, or higher than the background electrolyte conductivity. The same theoretical fundamentals are applied for cations determination in which the combination of the reversed polarization and reversed electroosmotic flow is used. The boundary between a sample and a background electrolyte moves together with EOF towards the anode; positively charged analytes, however, migrate backwards, to the cathode. The pH value increases at the

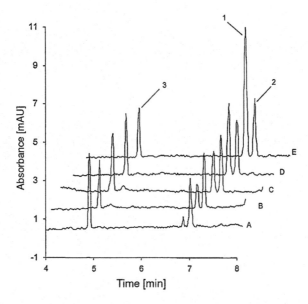

Fig. 12.3 Electropherograms of human blood plasma after reducing with tris 2-carboxyethyl-phosphine and derivatization with 2-chloro-1-methylquinolinium tetrafluoroborate (CMQT), obtained with the application of pseudo-tITP stacking. **a** Plasma with homocysteine at concentration 6.29 μmol/L. **b–e** Plasma spiked with homocysteine to its final concentration of 16.29; 26.29; 46.29; 86.29 μmol/L plasma, respectively. Peaks on the electropherogram: *1.* Homocysteine-CMQT; *2.* Cysteinylglycine-CMQT; *3.* Excess of derivatizing reagent CMQT. Reprinted with permission from Kubalczyk and Bald [63]. Copyright 2006 Springer

end of the sample zone because OH⁻ ions permeate into it from the background electrolyte. The cation analytes partially lose the charge, their mobility is lower, and they undergo stacking on the background electrolyte—sample zone boundary. After stacking, ions are carried by the electroosmotic flow towards the anode. When EOF is suppressed, cations undergo stacking on the sample-background electrolyte boundary, because after applying positive voltage, OH⁻ ions from the background electrolyte permeate into the sample zone through the boundary and raise pH in the sample zone. After stacking is accomplished, cations migrate at lower velocity which results from a higher pH value as well as from partial deprotonating [37]. Imami et al. [73] presented a peptides analysis method by using a simple formic acid/ammonium formate electrolyte system. The biggest advantage of the method is the possibility to analyze a large number of peptides at a wide pI range simultaneously. Under the optimum conditions of cytochrome c analysis, 30 % of the capillary is filled with a sample. 550–1,000 times higher sensitivity was therefore obtained. Jaafar et al. [74] developed a method of arsenic compounds determination in soil and plant litter samples. The method was characterized by 100–800 times higher sensitivity and a very low LOD at 0.34–1.93 ppb, after the SPE extraction. Kazarian et al. [75] implemented the stacking by dynamic pH junction technique to analyze saccharides such as glucose, lactose, or hydrated maltotriose. They employed strong 5-aminofluorescein absorption and acidic environment of the derivatizing mixture; they filled 20–100 % of the capillary volume with it. A very high efficiency of 150,000 plates/m and the LOD at nanomoles level was obtained using an absorptiometric detector (UV–Vis) based on light emitting diodes (LED).

12.1.8 Concentration by Counter-Flow Gradient Electrofocusing

The technique relates to methods that allow every component of a sample to be focused in a strictly defined place of the separation space. In many respects, it is related to isoelectro-focusing (IEF) in which every component is focused. Within this decade, many new methods which enable focusing analytes in places corresponding to their electrophoretic mobility have been developed, e.g., electric field gradient focusing (EFGF) or temperature gradient focusing (TGF). Some attention, however, should be paid to the fact that concentration by counter-flow gradient electrofocusing methods are exceptionally flexible and can be applied as analytical methods themselves or as an electrophoretic equivalent of SPE in which analytes are selectively retained and/or released in a definite and controlled way [24].

12.2 Sample Concentration in Micellar Electrokinetic Chromatography

Micellar electrokinetic chromatography (MEKC) is a type of CE applied mainly for the separation of neutral particles which all migrate according to electroosmotic flow. In order to enable their separation, it is essential to apply a proper additive to the background electrolyte. Examples of such additives are surfactants, the molecules which at certain concentration values create aggregates called micelles. Micelles are usually of spherical shape. Hydrophobic tails of surfactant monomers are directed inside a micelle; the charged (hydrophilic) heads, however, are directed outside, towards the buffer. Micelles can be positively or negatively charged which results in their migration according to EOF, or in the opposite direction. At the neutral or alkaline pH, EOF is faster than the negatively charged micelles velocity. Therefore, the migration occurs in the same direction as EOF. The higher the analyte affinity towards the micelle, the longer its migration time. If a substance does not interact with micelles, it migrates according to EOF.

In MEKC, similar to CZE, a range of methods enabling the concentration of analytes in a capillary are applied. Since the separated analytes are not charged, their concentration mechanism must be based on some other rules. Analytes concentration in MEKC involves the following techniques: field amplified sample stacking (FASS), sweeping, stacking by high salt concentration, electrokinetic sample injection, stacking by field enhanced sample injection (FESI).

12.2.1 Field Amplified Sample Stacking

In the field amplified sample stacking (FASS) technique in MEKC, the same principles are applied as in CZE. There is one restriction, however, that there are micelles in a sample and BGE, which interact with neutral analytes enabling their migration in the electric field. The procedure involves the preparation of a sample in a diluted micellar solution so that its conductivity should be lower than that of a background electrolyte, and thus, the electric field should be higher in the sample zone [76]. With normal polarization applied and at high EOF very fast transport of analytes with micelles from the sample zone towards the negative electrode take place. After permeating into the BGE, analytes rapidly decelerate (low field) and undergo stacking into a narrow band (Fig. 12.4).

12.2.2 Sweeping

Analytes concentration by sweeping technique in MEKC was first described by Terabe in 1998. He compared the method to the way in which a broom is used in order to sweep rice grains from the floor [77]. In this phenomenon, micelles, as they move after applying voltage through the sample zone, sweep and concentrate analytes into a very narrow zone. The capillary is initially conditioned with a BGE with low pH. Next, the sample without micelles is hydrodynamically injected into the capillary as a long

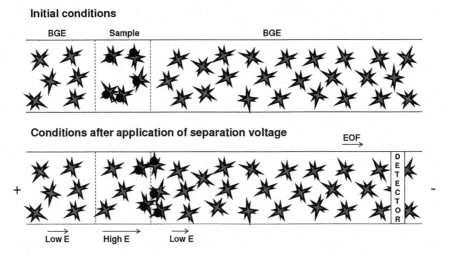

Fig. 12.4 Concentration by field amplified sample stacking in MEKC. A low conductivity sample, prepared in a micellar solution is hydrodynamically injected into a capillary filled with the high conductivity micellar BGE. After applying voltage, analytes and micelles are accelerated towards the cathode, and when having reached the sample zone/BGE boundary, they rapidly decelerate (low field strength) undergoing stacking into a narrow band, according to [38]

Initial conditions

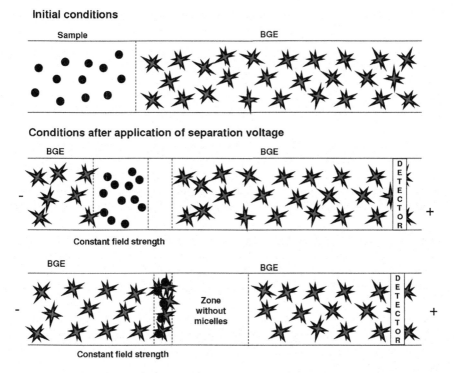

Fig. 12.5 Concentration by sweeping. A long sample zone without micelles is hydrodynamically injected into a capillary filled with the BGE at a low pH value with negatively charged micelles. In a homogeneous electric field with the negative polarization, micelles permeate into a sample zone and sweep analytes into a narrow zone. Then, the analytes are separated according to the MEKC mechanism, according to [14]

zone. The sample conductivity can be lower, the same or higher than the BGE conductivity. Even if authors use different nomenclature for the technique, when a sample is of higher conductivity than the BGE, the term "sweeping" can be used due to the same analytes concentration mechanism in all the three cases [24]. In a homogeneous electric field, with the negative polarization, at the retained EOF, micelles permeate into a sample zone sweeping the analytes into a narrow zone (Fig. 12.5). Consequently, analytes interact with micelles and are separated according to MEKC [14]. The method can be employed to both ions and uncharged molecules. The concentration efficiency depends on the analytes affinity in relation to micelles. Namely, the higher the affinity, the higher the concentration level. Terabe [38] suggested the following equation which relates the concentration level to the micelles affinity:

$$l_{swept} = l_{injected}\left(\frac{1}{1+k}\right)$$
(12.1)

where l_{swept} is length of swept analyte zone; $l_{injected}$ length of injected sample zone; k is retention coefficient for given analyte.

Currently, sweeping is a popular technique of neutral analytes concentration. Owing to the method, the detector signal can be amplified by 5,000 times. The sweeping method has an important advantage: in order to improve the sweeping effect [78], a sample matrix can be manipulated (pH change, organic solvents addition) to increase analytes solubility or optimize the content and pH of the buffer. The method was successfully implemented to the determination of steroids, alkyl phenyl ketones and naphthalene derivatives [77], phenols [79], herbicides [80], pesticides [81], alkaloids [82], etc.

A certain variant of the above-mentioned concentration technique is also noteworthy: concentration by electrokinetic sample injection with the use of a cation-selective exhausting and sweeping technique. The capillary is conditioned with a low pH buffer (without micelles), and then, a long buffer zone of high conductivity and a short water zone are injected. In turn, with the use of positive polarization, cations from the sample previously prepared in water or a low conductivity solution are injected electrokinetically into a capillary for a long time (10 min). In a capillary, a very long cationic analytes zone at a concentration higher than in the sample is created. Immediately, after injecting cations, the ends of the capillary are inserted into the vials containing low pH buffers with anionic micelles. After applying voltage of negative polarization to the capillary, anionic micelles enter the capillary and sweep cationic analytes into narrow zones. Finally, separation occurs according to the MEKC mechanism. The technique enables the improvement of the detector signal in a hydrophobic cationic compounds analysis by ca. million times [33].

12.2.3 Stacking with High Salt Concentration

This one is a technique, the concentration mechanism of which is based on micelles stacking on the cathode side of the sample zone [78]. Analytes, migrating with EOF through the stacked micelles zone, undergo strong local retention and stacking into a narrow zone (Fig. 12.6). The name of the technique is derived from the fact that in order to enable strong micelles concentration, the sample zone conductivity must be a few times higher than the BGE conductivity; in other words, the ionic strength of a sample matrix must be 1.5–2.5 times higher than the ionic strength of a background electrolyte.

12.2.4 Electrokinetic Sample Injection

In the technique, analytes from the sample are introduced into a capillary with EOF, interacting with micelles included in the background electrolyte (Fig. 12.7). In the paper by Palmer et al. [83], analytes undergo concentration on the sample

Initial conditions

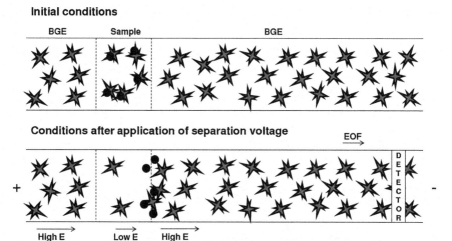

Fig. 12.6 Concentration by stacking with high salt concentration. A high conductivity sample is injected into a capillary filled with a low conductivity micellar BGE. Analytes migrating according to EOF through the stacked micelles zone are strongly retained and undergo stacking into a narrow zone themselves, according to [38]

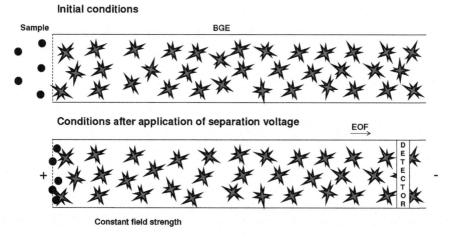

Fig. 12.7 Concentration by electrokinetic sample injection. Sample analytes are injected electrokinetically into a capillary with EOF and by interacting with micelles contained in the background electrolyte create a narrow zone, according to [38]

zone-BGE boundary. If the velocity difference between EOF and micelles is big enough, it is possible to introduce a large amount of an analyte to a capillary by injecting the sample volume larger than the capillary volume.

12.2.5 Field Enhanced Sample Injection

In field enhanced sample injection (FESI), into capillary filled with neutral buffer, water plug is hydrodynamically introduced. Next, a sample prepared in neutral pH low conductivity micellar solution is injected under negative polarization. A vial containing a sample is replaced by a vial with a buffer when the current intensity reaches 97–99 % of the value measured after the capillary conditioning. The advantage of FESI in MEKC is the fact that samples are prepared in micellar solutions which improves the solubility of hydrophobic compounds [15].

12.3 Summary

The main techniques of analytes concentration in the capillary employed in CZE and MEKC based on the electrophoretic mechanisms were presented. Their numerous variants and combinations are widely described in the literature which is extensive [16, 24, 36]. What distinguishes the techniques is the fact that they are always used in electrophoretic measuring systems which require no modification of commercially available measuring instruments. Concentrating according to these techniques (stacking and sweeping) is based on different analytes velocities in different capillary fragments, which result from (1) a change of an electric field strength or (2) a chemical interaction of an analyte with a separating electrolyte or other substances injected to a capillary (e.g., stacking by pH junction).

A change in an analyte velocity can be physically induced. It occurs when the analyte cannot continue moving in the separating electrolyte at the same speed within the whole capillary due to physical reasons. This concentration technique was applied in peptides and DNA containing samples [84] giving good concentration results at the inlet of the capillary. The concentration methods based on physical limitations in the analytes mobility were applied mainly in the microchip analysis [24, 85–91]. These techniques, as well as chromatographic concentration methods, require certain intervention in the measuring system construction. Chromatographic concentration methods in the CE measuring system and outside it are only mentioned in this study, and their review is the topic of numerous papers [16, 24, 92]. Some information is also included in other parts of the book.

References

1. J.W. Jorgenson, K.D. Lukacs, Zone electrophoresis in open-tubular glass capillaries. Anal. Chem. **53**, 1298–1302 (1981)
2. J.W. Jorgenson, K.D. Lukacs, High-resolution separations based on electrophoresis and electroosmosis. J. Chromatogr. **218**, 209–216 (1981)
3. R.A. Wallingford, A.G. Ewing, Capillary electrophoresis. Adv. Chromatogr. **29**, 1–76 (1989)

4. A.G. Ewing, R.A. Wallingford, T.M. Olefirowicz, Capillary electrophoresis. Anal. Chem. **61**, 292A–303A (1989)
5. B.L. Karger, A.S. Cohen, A. Guttman, High-performance capillary electrophoresis in the biological sciences. J. Chromatogr. B **492**, 585–614 (1989)
6. M.J. Gordon, X. Huang, S.L. Pentoney, R.N. Zare, Capillary electrophoresis. Science **241**, 224–228 (1988)
7. E.S. Yeung, Indirect detection methods: looking for what is not there. Acc. Chem. Res. **22**, 125–130 (1989)
8. P.D. Grossman, J.C. Colburn, H.H. Lauer, R.G. Nielsen, R. Riggin, G.S. Sittampalam, E.C. Rickard, Application of free-solution capillary electrophoresis to the analytical scale separation of proteins and peptides. Anal. Chem. **61**, 1186–1194 (1989)
9. J. Snopek, I. Jelinek, E. Smolkova-Keulemansova, Micellar, Inclusion and metal–complex enantioselective pseudophases in high-performance electromigration methods. J. Chromatogr. A **452**, 571–590 (1988)
10. N.A. Guzman, L. Hernandez, B.G. Hoebel, Capillary electrophoresis: a new era in micro separations. BioPharm. **2**, 22–37 (1989)
11. J.W. Jorgenson, K.D. Lukacs, Free-zone electrophoresis in glass capillaries. Clin. Chem. **27**, 1551–1553 (1981)
12. J.W. Jorgenson, K.D. Lukacs, Microcolumn separations. J. Chromatogr. Libr. **30**, 121 (1985)
13. P. Mucha, P. Rekowski, A. Szyk, G. Kupryszewski, J. Barciszewski, Capillary electrophoresis: a new tool for biomolecule separation (Elektroforeza kapilarna: nowe narzedzie analizy biomolekuł). Postępy Biochem ii **43**, 208–216 (1997)
14. A.T. Aranas, A.M. Guidote Jr, J.P. Quirino, Sweeping and new on-line sample preconcentration techniques in capillary electrophoresis. Anal. Bioanal. Chem. **394**, 175–185 (2009)
15. A. Mrass, E. Bald, Metody zatężania analitów w kapilarze w wysokosprawnej elektroforezie kapilarnej. Wiad. Chem. **55**, 933–953 (2001)
16. M.C. Breadmore, Recent advances in enhancing the sensitivity of electrophoresis and electrochromatography in capillaries and microchips. Electrophoresis **28**, 254–281 (2007)
17. Z. Witkiewicz (ed.), *Podstawy Chromatografii* (WNT, Warsaw, 2005)
18. D.B. Craig, J.C.Y. Wong, N.J. Dovichi, Detection of Attomolar concentrations of alkaline phosphatase by capillary electrophoresis using laser-induced fluorescence detection. Anal. Chem. **68**, 697–700 (1996)
19. C.J. Smith, J. Grainger, D.G. Patterson Jr, Separation of polycyclic aromatic hydrocarbon metabolites by γ-cyclodextrin-modified micellar electrokinetic chromatography with laser-induced fluorescence detection. J. Chromatogr. A **803**, 241–247 (1998)
20. N.M. Djordjevic, M. Widder, R. Kuhn, Signal enhancement in capillary electrophoresis by using a sleeve cell arrangement for optical detection. J. High Res. Chromatogr. **20**, 189–192 (1997)
21. G. Ross, P. Kaltenbach, D. Heilger, High sensitivity is key to CE detection boost: with a new detection cell, sensitivity and linearity for CE are comparable to that of HPLC. Today's Chem. Work **6**, 31–36 (1997)
22. Technical Bulletin/Hewlett Packard (1997) 12–5965–5984E
23. Z. Witkiewicz, J. Hetper, *Słownik Chromatografii i Elektroforezy* (PWN, Warszawa, 2004)
24. M.C. Breadmore, J.R.E. Thabano, M. Davod, A.A. Kazarian, J.P. Quirino, R.M. Guijt, Recent advances in enhancing the sensitivity of electrophoresis and electrochromatography in capillaries and microchips (2006–2008). Electrophoresis **30**, 230–248 (2009)
25. R.L. Chien, D.S. Burgi, On-column sample concentration using field amplification in CZE. Anal. Chem. **64**, 489A–496A (1992)
26. J.P. Quirino, S. Terabe, Sample stacking of cationic and anionic analytes in capillary electrophoresis. J. Chromatogr. A **902**, 119–135 (2000)
27. P. Gebauer, W. Thormann, P. Bocek, Sample self-stacking in zone electrophoresis theoretical description of the zone electrophoretic separation of minor compounds in the presence of

bulk amounts of a sample component with high mobility and like charge. J. Chromatogr. **608**, 47–57 (1992)

28. Z.K. Shihabi, Peptide stacking by acetonitrile-salt mixtures for capillary zone electrophoresis. J. Chromatogr. A **744**, 231–240 (1996)
29. Y.P. Zhao, C.E. Lunte, pH-mediated field amplification on-column preconcentration of anions in physiological samples for capillary electrophoresis. Anal. Chem. **71**, 3985–3991 (1999)
30. P. Britz-McKibbin, G.M. Bebault, D.D.Y. Chen, Velocity–difference induced focusing of nucleotides in capillary electrophoresis with a dynamic pH junction. Anal. Chem. **72**, 1729–1735 (2000)
31. J.P. Quirino, M.T. Dulay, R.N. Zare, On-line preconcentration in capillary electrochromatography using a porous monolith together with solvent gradient and sample stacking. Anal. Chem. **73**, 5557–5563 (2001)
32. J.P. Quirino, J.B. Kim, S. Terabe, Sweeping: concentration mechanism and applications to high-sensitivity analysis in capillary electrophoresis. J. Chromatogr. A **965**, 357–373 (2002)
33. J.P. Quirino, S. Terabe, Approaching a million-fold sensitivity increase in capillary electrophoresis with direct ultraviolet detection: cation-selective exhaustive injection and sweeping. Anal. Chem. **72**, 1023–1030 (2000)
34. P. Britz-McKibbin, K. Otsuka, S. Terabe, On-line focusing of flavin derivatives using dynamic ph junction-sweeping capillary electrophoresis with laser-induced fluorescence detection. Anal. Chem. **74**, 3736–3743 (2002)
35. A.R. Timerbaev, T. Hirokawa, Recent advances of transient isotachophoresis-capillary electrophoresis in the analysis of small ions from high-conductivity matrices. Electrophoresis **27**, 323–340 (2006)
36. Z. Mala, A. Slampova, P. Gebauer, P. Bocek, Contemporary sample stacking in CE. Electrophoresis **30**, 215–229 (2009)
37. Z. Mala, L. Krivankova, P. Gebauer, P. Bocek, Contemporary sample stacking in CE: a sophisticated tool based on simple principles. Electrophoresis **28**, 243–253 (2007)
38. D.S. Burgi, B.C. Giordano, in *Online sample preconcentration for capillary electrophoresis*, ed. by J.P. Landers. Handbook of capillary and microchip electrophoresis and associated microtechniques, vol 3, (CRC Press, Taylor&Francis Group, Boca Raton, New York 2008)
39. S.L. Simpson Jr, J.P. Quirino, S. Terabe, On-line sample preconcentration in capillary electrophoresis fundamentals and applications. J. Chromatogr. A **1184**, 504–541 (2008)
40. K. Sueyoshi, F. Kitagawa, K. Otsuka, Recent progress of online sample preconcentration techniques in microchip electrophoresis. J. Sep. Sci. **31**, 2650–2666 (2008)
41. R.L. Chien, D.S. Burgi, Sample stacking of an extremely large injection volume in high-performance capillary electrophoresis. Anal. Chem. **64**, 1046–1050 (1992)
42. R.L. Chien, D.S. Burgi, Field amplified sample injection in high-performance capillary electrophoresis. J. Chromatogr. **559**, 141–152 (1991)
43. E. Dąbek-Złotorzynska, M. Piechowski, Application of CE with novel dynamic coatings and field-amplified sample injection to the sensitive determination of isomeric benzoic acids in atmospheric aerosols and vehicular emission. Electrophoresis **28**, 3526–3534 (2007)
44. L. Yu, S.F.Y. Li, Large-volume sample stacking with polarity switching for the analysis of bacteria by capillary electrophoresis with laser-induced fluorescence detection. J. Chromatogr. A **1161**, 308–313 (2007)
45. Y. He, H.K. Lee, Large-volume sample stacking in acidic buffer for analysis of small organic and inorganic anions by capillary electrophoresis. Anal. Chem. **71**, 995–1001 (1999)
46. C.Y. Kuo, S.S. Chiou, S.M. Wu, Solid-phase extraction and large-volume sample stacking with an electroosmotic flow pump in capillary electrophoresis for determination of methotrexate and its metabolites in human plasma. Electrophoresis **27**, 2905–2909 (2006)
47. L. Zhang, X.F. Yin, Field amplified sample stacking coupled with chip-based capillary electrophoresis using negative pressure sample injection technique. J. Chromatogr. A **1137**, 243–248 (2006)

48. Z.L. Chen, G. Owens, R. Naidu, Confirmation of vanadium complex formation using electrospray mass spectrometry and determination of vanadium speciation by sample stacking capillary electrophoresis. Anal. Chim. Acta **585**, 32–37 (2007)
49. Z.L. Chen, M. Megharaj, R. Naidu, Confirmation of iron complex formation using electrospray ionization mass spectrometry (ESI-MS) and sample stacking for analysis of iron polycarboxylate speciation by capillary electrophoresis. Microchem. J. **86**, 94–101 (2007)
50. C. Quesada-Molina, A.M. Garcia-Campana, L. del Olmo-Iruela, M. del Olmo, Large volume sample stacking in capillary zone electrophoresis for the monitoring of the degradation products of metribuzin in environmental samples. J. Chromatogr. A **1164**, 320–328 (2007)
51. M.L. Bailon-Perez, A.M. Garcia-Campana, C. Cruces-Blanco, M.D. Iruela, Trace determination of β-lactam antibiotics in environmental aqueous samples using off-line and on-line preconcentration in capillary electrophoresis. J. Chromatogr. A **1185**, 273–280 (2008)
52. K. Kutschera, A.C. Schmidt, S. Kohler, M. Otto, CZE for the speciation of arsenic in aqueous soil extracts. Electrophoresis **28**, 3466–3476 (2007)
53. M.I. Bailon-Perez, A.M. Garcia-Campana, C. Cruces-Blanco, M. del Olmo Iruela, Large-volume sample stacking for the analysis of seven β-lactam antibiotics in milk samples of different origins by CZE. Electrophoresis **28**, 4082–4090 (2007)
54. J.J. Soto-Chinchilla, A.M. Garcia-Campana, L. Gamiz-Garcia, C. Cruces-Blanco, Application of capillary zone electrophoresis with large-volume sample stacking to the sensitive determination of sulfonamides in meat and ground water. Electrophoresis **27**, 4060–4068 (2006)
55. E. Bermudo, O. Nunez, L. Puignou, M.T. Galceran, Analysis of acrylamide in food products by in-line preconcentration capillary zone electrophoresis. J. Chromatogr. A **1129**, 129–134 (2006)
56. M.E. Hadwiger, S.R. Torchia, S. Park, M.E. Biggin, C.E. Lunte, Optimization of the separation and detection of the enantiomers of isoproterenol in microdialysis samples by cyclodextrin-modified capillary electrophoresis using electrochemical detection. J. Chromatogr. B **681**, 241–249 (1996)
57. F. Foret, E. Szoko, B.L. Karger, On-column transient and coupled column isotachophoretic preconcentration of protein samples in capillary zone electrophoresis. J. Chromatogr. **608**, 3–12 (1992)
58. T. Hirokawa, H. Okamoto, B. Gas, High-sensitive capillary zone electrophoresis analysis by electrokinetic injection with transient isotachophoretic preconcentration: electrokinetic supercharging. Electrophoresis **24**, 498–504 (2003)
59. I. Botello, F. Borrull, C. Aguilar, M. Calull, Electrokinetic supercharging focusing in capillary zone electrophoresis of weakly ionizable analytes in environmental and biological samples. Electrophoresis **31**, 2964–2973 (2010)
60. Z.K. Shihabi, Stacking and discontinuous buffers in capillary zone electrophoresis. Electrophoresis **21**, 2872–2878 (2000)
61. Z.K. Shihabi, Transient pseudo-isotachophoresis for sample concentration in capillary electrophoresis. Electrophoresis **23**, 1612–1617 (2002)
62. Z.K. Shihabi, M.E. Hinsdale, C.P. Cheng, Analysis of glutathione by capillary electrophoresis based on sample stacking. Electrophoresis **22**, 2351–2354 (2001)
63. P. Kubalczyk, E. Bald, Transient pseudo-isotachophoretic stacking in analysis of plasma for homocysteine by capillary zone electrophoresis. Anal. Bioanal. Chem. **384**, 1181–1185 (2006)
64. P. Kubalczyk, E. Bald, Method for determination of total cysteamine in human plasma by high performance capillary electrophoresis with acetonitrile stacking. Electrophoresis **29**, 3636–3640 (2008)
65. P. Kubalczyk, E. Bald, Analysis of orange juice for total cysteine and glutathione content by CZE with UV-absorption detection. Electrophoresis **30**, 2280–2283 (2009)
66. Y. Kong, N. Zheng, Z. Zhang, R. Gao, Optimization stacking by transient pseudo-isotachophoresis for capillary electrophoresis: example analysis of plasma glutathione. J. Chromatogr. B **795**, 9–15 (2003)

67. Y. Chen, L. Xu, L. Zhang, G. Chen, Separation and determination of peptide hormones by capillary electrophoresis with laser-induced fluorescence coupled with transient pseudo-isotachophoresis preconcentration. Anal. Biochem. **380**, 297–302 (2008)

68. Z.X. Zhang, X.W. Zhang, J.J. Wang, S.S. Zhang, Sequential preconcentration by coupling of field amplified sample injection with pseudo isotachophoresis-acid stacking for analysis of alkaloids in capillary electrophoresis. Anal. Bioanal. Chem. **390**, 1645–1652 (2008)

69. R. Aebersold, H.D. Morrison, Analysis of dilute peptide samples by capillary zone electrophoresis. J. Chromatogr. A **516**, 79–88 (1990)

70. J. Horakova, J. Petr, V. Maier, J. Znaleziona, A. Stanová, J. Marák, D. Kaniansky, J. Sevcík, Combination of large volume sample stacking and dynamic pH junction for on-line preconcentration of weak electrolytes by capillary electrophoresis in comparison with isotachophoretic techniques. J. Chromatogr. A **1155**, 193–198 (2007)

71. P. Britz-Mckibbin, A.R. Kranack, A. Paprica, D.D.Y. Chen, Quantitative assay for epinephrine in dental anesthetic solutions by capillary electrophoresis. Analyst **123**, 1461–1463 (1998)

72. P. Britz-McKibbin, D.D.Y. Chen, selective focusing of catecholamines and weakly acidic compounds by capillary electrophoresis using a dynamic pH junction. Anal. Chem. **72**, 1242–1252 (2000)

73. K. Imami, M.R.N. Monton, Y. Ishihama, S. Terabe, Simple on-line sample preconcentration technique for peptides based on dynamic pH junction in capillary electrophoresis-mass spectrometry. J. Chromatogr. A **1148**, 250–255 (2007)

74. J. Jaafar, Z. Irwan, R. Ahamad, S. Terabe, T. Ikegami, N. Tanaka, Online preconcentration of arsenic compounds by dynamic pH junction-capillary electrophoresis. J. Sep. Sci. **30**, 391–398 (2007)

75. A.A. Kazarian, E.F. Hilder, M.C. Breadmore, Utilisation of pH stacking in conjunction with a highly absorbing chromophore, 5-aminofluorescein, to improve the sensitivity of capillary electrophoresis for carbohydrate analysis. J. Chromatogr. A **1200**, 84–91 (2008)

76. Z. Liu, P. Sam, S.R. Sirimanne, P.C. McClure, J. Grainger, D.G. Patterson, Field-amplified sample stacking in micellar electrokinetic chromatography for on-column sample concentration of neutral molecules. J. Chromatogr. A **673**, 125–132 (1994)

77. J.P. Quirino, S. Terabe, Exceeding 5000-fold concentration of dilute analytes in micellar electrokinetic chromatography. Science **282**, 465–468 (1998)

78. J. Palmer, N.J. Munro, J.P. Landers, A universal concept for stacking neutral analytes in micellar capillary electrophoresis. Anal. Chem. **71**, 1679–1687 (1999)

79. Y. Hongyuan, Y. Gengliang, Q. Fengxia, L. Haiyan, C. Li, Determination of phenol pollutants in industrial waste water by MEKC and on-line sweeping technique. Chem J Internet **6**, 26 (2004)

80. O. Nunez, J.B. Kim, E. Moyano, M.T. Galceran, S. Terabe, Analysis of the herbicides paraquat, diquat and difenzoquat in drinking water by micellar electrokinetic chromatography using sweeping and cation selective exhaustive injection. J. Chromatogr. A **961**, 65–75 (2002)

81. C.L. Da Silva, E.C. de Lima, M.F.M. Tavares, Investigation of preconcentration strategies for the trace analysis of multi-residue pesticides in real samples by capillary electrophoresis. J. Chromatogr. A **1014**, 109–116 (2003)

82. S.W. Sun, H.M. Tseng, Sensitivity improvement on detection of Coptidis alkaloids by sweeping in capillary electrophoresis. J. Pharma. Biomed. Anal. **37**, 39–45 (2005)

83. J. Palmer, D.S. Burgi, J.P. Landers, Electrokinetic stacking injection of neutral analytes under continuous conductivity conditions. Anal. Chem. **74**, 632–638 (2002)

84. C.J. Yu, H.C. Chang, W.L. Tseng, On-line concentration of proteins by SDS-CGE with LIF detection. Electrophoresis **29**, 483–490 (2008)

85. J. de Jong, R.G.H. Lammertink, M. Wessling, Membranes and microfluidics: a review. Lab Chip **6**, 1125–1139 (2006)

86. A. Holtzel, U. Tallarek, Ionic conductance of nanopores in microscale analysis systems: where microfluidics meets nanofluidics. J. Sep. Sci. **30**, 1398–1419 (2007)

87. M.L. Kovarik, S.C. Jacobson, Integrated nanopore/microchannel devices for ac electrokinetic trapping of particles. Anal. Chem. **80**, 657–664 (2008)
88. Y.H. Kim, I. Yang, S.R. Park, Well-less capillary array electrophoresis chip using hydrophilic sample bridges. Anal. Chem. **79**, 9205–9210 (2007)
89. R. Dhopeshwarkar, R.M. Crooks, D. Hlushkou, U. Tallarek, Transient effects on microchannel electrokinetic filtering with an ion-permselective membrane. Anal. Chem. **80**, 1039–1048 (2008)
90. J.H. Lee, S. Chung, S.J. Kim, J.Y. Han, Poly(dimethylsiloxane)-based protein preconcentration using a nanogap generated by junction gap breakdown. Anal. Chem. **79**, 6868–6873 (2007)
91. Z.C. Long, D.Y. Liu, N.N. Ye, J.H. Qin, B.C. Lin, Integration of nanoporous membranes for sample filtration/preconcentration in microchip electrophoresis. Electrophoresis **27**, 4927–4934 (2006)
92. P. Puig, F. Borrull, M. Calull, C. Aguilar, Sorbent preconcentration procedures coupled to capillary electrophoresis for environmental and biological applications. Anal. Chem. Acta. **616**, 1–18 (2008)

Chapter 13
Stereoisomers Separation

Piotr Wieczorek

Abstract The use of capillary electrophoresis for enantiomer separation and optical purity determination is presented. The contents start with basic information about the nature of stereoizomers and the mechanism of enantioseparation using capillary electrophoresis techniques. The molecules to be separated show identical chemical structure and electrochemical behavior. Therefore, the chiral recognition of enantiomers is possible only by bonding to chiral selector and the separation based on very small differences in complexation energies of diastereomer complexes formed. This method is useful for this purpose due to the fact that different compounds can be used as chiral selectors. The mostly used chiral selectors like cyclodextrins, crown ethers, chiral surfactants, macrocyclic antibiotics, transition metal complexes, natural, and synthetic polymers and their application for this purpose is also discussed. Finally, examples of practical applications of electromigration techniques for enantiomers separation and determination are presented.

The branch of chemistry called stereochemistry deals with examination of three-dimensional structure of molecules and atoms arrangement in the space. The compounds of an identical atomic composition but differing in structure are called isomers. Two basic types of isomerism are distinguished: structural isomerism (constitutional) where atoms are connected with each other in various manners and sequences, and stereoisomerism. In the second case, the atoms in two objects are connected in an identical manner and in the same sequence, but they differ with spatial arrangement. Stereoisomerism is divided into two kinds: geometric and optical isomerism. Geometric isomers are such compounds, which differ with arrangement of substituents by two carbon atoms present in a ring or connected with double bond. Such compounds cannot be transformed into each other in normal conditions by a

P. Wieczorek (✉)
Faculty of Chemistry, University of Opole, Opole, Poland
e-mail: Piotr.Wieczorek@uni.opole.pl

B. Buszewski et al. (eds.), *Electromigration Techniques*, Springer Series
in Chemical Physics 105, DOI: 10.1007/978-3-642-35043-6_13,
© Springer-Verlag Berlin Heidelberg 2013

Fig. 13.1 Model of chiral compound, according to Ref. [1]

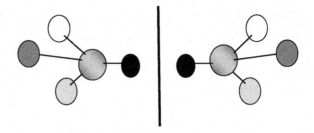

rotation around the bond, and are characterized by different physicochemical properties, such as melting and boiling points, solubility, density, vapour pressure, etc. Whereas chiral molecules having asymmetry center or asymmetric carbon atom give optical isomers. The substituents at asymmetric carbon atoms may have one of two configurations, which are related as an object and its mirror image and are referred as enantiomers. The concept of chirality originates from the Greek word "*kheir*", which means "*hand*" and describes the objects differing with their mirror image, like left and right hand (Fig. 13.1). Such molecules have identical physical and chemical properties, except the direction of polarized light rotating, which is called optical activity. One of them rotates the light surface left and is referred as "*L*" or (−), and the second "*D*" or (+). Meanwhile according to Cahn–Ingold–Prelog priority rules, taking into consideration the so-called substituents priority, one is called "*S*" and the second "*R*".

Since such isomers differ only with direction, and not the size of an angle of polarized light rotation, equimolar mixtures of enantiomers, referred as racemates, are inactive optically. Due to the lack of differences in physicochemical properties, the racemic mixtures are difficult to separate. Molecules containing two or more stereogenic centers may occur in pairs of enantiomers or diastereoisomers. Diastereoisomers have an identical composition, but are not mirror images, and therefore differ in physical properties and may be separated [1].

Numerous chemical substances, including natural compounds, food additives, pesticides, and first of all medicines, are optically active substances which isomers differ in biological properties. For example, *S* isomer of asparagine has bitter taste, and *R* isomer is sweet, while in the case of limonene, its *L* isomer has lemon aroma, and *D* isomer orange aroma (Fig. 13.2).

The fact of presence of differences in biological properties of particular enantiomers is of significant meaning in case of medicines and food additives. In numerous cases, one of the isomers is a medicine, and the second may have toxic activity, or in the best case may not demonstrate biological effect. Examples of such medicines may be hypotonic drug—thalidomide, which second isomer exhibits teratogenic activity, or ethambutol used in tuberculosis treatment, which second enantiomer causes blindness (Fig. 13.3).

For that reason, one of the most important problems in modern chemistry is elaboration of methods for obtaining pure enantiomers. Most often they are obtained directly during synthesis in biotransformation reactions, stereoselective synthesis, or by synthesis of racemic mixtures and elaboration of effective method of their separation.

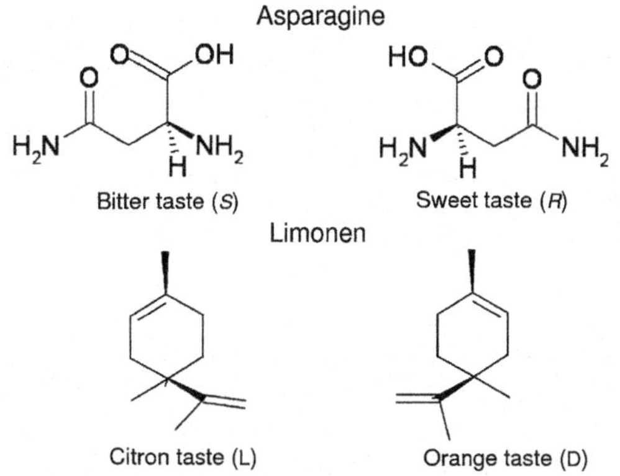

Fig. 13.2 Asparagine and
limonene isomers

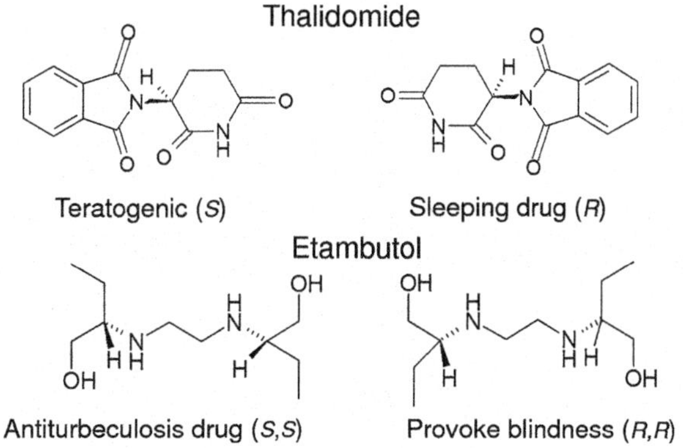

Asparagine

Bitter taste (*S*)   Sweet taste (*R*)

Limonen

Citron taste (L)   Orange taste (D)

Fig. 13.3 Thalidomide and
etambutol isomers

Thalidomide

Teratogenic (*S*)   Sleeping drug (*R*)

Etambutol

Antiturbeculosis drug (*S,S*)   Provoke blindness (*R,R*)

13.1 Mechanism of Chiral Separation

Enantiomers may be generally separated in two ways: using microorganisms or
enzymes, which as a result of suitable reactions may change or decompose only
one of enantiomers, or transforming them into the mixtures of diastereomers of
various properties. For that purpose, the following methods are used: crystallization, extraction, chromatographic methods, membrane techniques, and—more and
more often—electromigration methods. Moreover, an assessment of optical purity
of particular enantiomers obtained both as a result of chemical reactions, and by
racemic mixtures separation is often essential [2]. For that purpose, the following
methods are used: NMR, polarimetry, immunological methods, chiral sensors,
isotopes dilution, chromatographic techniques (TLC, GC, and HPLC), and also
capillary electrophoresis. From the known techniques of enantiomers separation,
the following are used the most often: capillary zone electrophoresis (CZE),
micellar electrokinetic chromatography (MEKC), capillary electrochromatography
using chiral stationary phase (CEC) and also gel electrophoresis (GCE) [3, 4].

In separation with an application of capillary electrophoresis (CZE, MEKC) the differences in mobility of analyzed substances present in the mixture are used. Since enantiomers possess identical ionic radii and the same charge density, and thus the same apparent mobility, their separation in achiral environment of carrying buffer is impossible.

$$\mu_a = \mu_s = \mu_R, \tag{13.1}$$

where μ_a is apparent mobility of free enantiomers; μ_S is apparent mobility of S enantiomer; μ_R is apparent mobility of R enantiomer.

Thus, in order to make possible the separation of the pair of enantiomers, it is necessary to introduce chiral discrimination factor (chiral selector) to the basic electrolyte. Such selectors, forming transitory diastereomeric complexes with examined isomers, enable their separation [5, 6].

$$E_R + CS \overset{K_1}{\leftrightarrow} E_R CS \tag{13.2}$$

$$E_S + CS \overset{K_2}{\leftrightarrow} E_S CS \tag{13.3}$$

where E_R, E_S are enantiomers, CS is chiral selector, K_1, K_2 are complexing constants, $E_R CS$, $E_S CS$ are enantiomer–selector complex.

Complexing constants for these reactions may be defined as follows:

$$K_1 = \frac{[E_R CS]}{[E_R][CS]} \quad K_2 = \frac{[E_S CS]}{[E_S][CS]}. \tag{13.4}$$

The condition of separation is the presence of differences in the values of complexing constants, which is the reason for the occurrence of the difference in electrophoretic activity of their complexes according to the formula [7, 8]:

$$\Delta\mu = \frac{(\mu^f - \mu^c)\Delta K[CS]}{(1 + K_R[CS])(1 + K_S[CS])}, \tag{13.5}$$

where μ^f is free compound mobility; μ^c is complexed compound mobility.

The effectiveness of enantiomers selection is in turn determined usually using selectivity (α) and/or separability (R_S):

$$\alpha = \mu_{ERCS}/\mu_{ESCS} \tag{13.6}$$

where μ_{ERCS}, μ_{ESCS} are electrophoretic mobility of analyzed enantiomers complexes

$$R_S = \frac{1}{4}\sqrt{N}\left(\frac{\Delta\mu}{\mu}\right), \tag{13.7}$$

where N is number of theoretical plates; μ is mean electrophoretic mobility of analyzed substances [7].

13.2 Chiral Selectors

Chiral selectors are optically active compounds able to form transitory diastereomeric complexes with separated substances, using hydrogen bonds, hydrophobic, electrostatic interactions (dipol–dipol), π–π, and van der Waals forces [9]. The stability of such complexes depends on spatial matching and kind and strength of interactions between the components. Good chiral selectors should be thus characterized by high optical purity, ability to formation of transitory diastereomeric complexes with separated enantiomers. Moreover, they should be stable chemically and soluble in buffers used as electrolytes, should not influence the detector operation and finally have affordable price. For that reason, as chiral discriminators in capillary electrophoresis, usually cyclodextrins and their derivatives, chiral crown ethers and surfactants, macrocyclic antibiotics, chiral metal complexes, and also proteins, polymers and polysaccharides are used, those obtained from natural sources and via chemical synthesis [10, 11].

13.3 Cyclodextrins

In recent years cyclodextrins are chiral selectors used last years the most often in enantiomers separation using electrophoresis. Naturally occurring α, β, and γ cyclodextrins are cyclic oligosaccharides formed of 6, 7, and 8 α-D-glucopyranoside units, respectively [10] (Fig. 13.4).

Cyclodextrins have a structure of cut conus inside which the hydrophobic cavity is formed, whose width and depth determine the possibilities of hydrophobic fragment of a given compound binding. Hydroxyl groups enabling

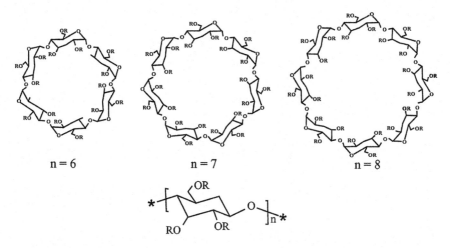

$$n = 6 \qquad n = 7 \qquad n = 8$$

Fig. 13.4 Cyclodextrins structures, according to Ref. [12]

Fig. 13.5 Scheme of
substance bonding by
cyclodextrin

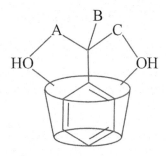

formation of hydrogen bonds are observed outside of the molecules of cyclo-dextrins. Thanks to such a structure, the stereoselective binding of other molecules or ions in enantiomer-selector complex may occur in cyclodextrins molecules. The condition of complex formation is that the molecule is entirely or only partially located in cyclodextrin molecule. This way the complex is stabilized by hydro-phobic bonds formed in cyclodextrin cavity and hydrogen bonds between sub-stituents by asymmetric carbon atom of analyzed compound, and hydroxyl groups of cyclodextrins [9]. For better understanding of that issue, the way of substance binding by cyclodextrins molecules is presented in Fig. 13.5.

The size of cavity, and also cyclodextrins solubility may be changed by chemical modification of these molecules. Various cyclodextrins derivatives, including their anionic and cationic analogues, are used in capillary electropho-resis as chiral selectors [13]. Among cyclodextrins derivatives, their acetyl-, hydroxypropyl-, methyl-, carboxymethyl sulfonic, and phosphonic analogues are used the most often [10]. However, charged, modified cyclodextrins, besides some advantages have also numerous disadvantages. They cause an increase in current flow by the system often require polarity of electrodes inversion so that the formed diastereoisomeric complexes migrate to anode and cathode. Moreover, their increased mobility may cause an excessive electrodispersion causing peaks deformation [10].

Cyclodextrins and their derivatives due to the possibility of formation of complexes with many various compounds, good solubility in water solutions, stability, and low price are the most often used chiral selectors in capillary elec-trophoresis. They find an application in drugs analysis (e.g., antimalarial, anti-asthmatic, and antidepressants), amino acids and their derivatives, peptides, alkaloids, alcohols, and also pesticides [14].

13.4 Chiral Crown Ethers

Crown ethers are synthetic macrocyclic polyethers similar in shape to the crown [15]. They are able to form inclusion complexes enantiomer-selector with various compounds. Such complexes, due to the presence of donor atoms (oxygen atoms) in crown ethers molecules are usually stabilized by electrostatic interactions and

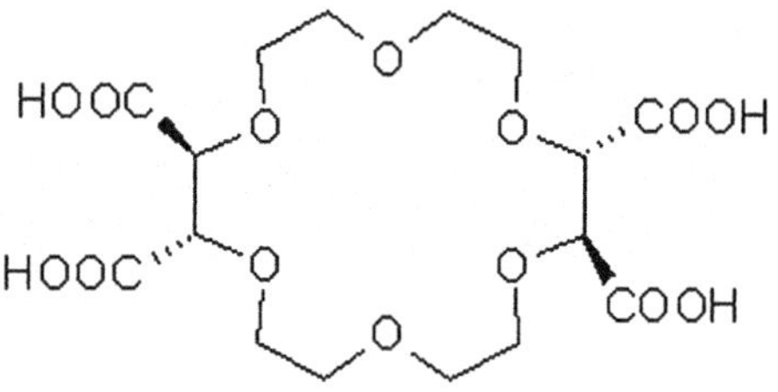

Fig. 13.6 18-crown-tetracarboxylic acid structure, according to Ref. [8]

hydrogen bonds. For that reason, chiral crown ethers have been used for many years as selectors for separation of enantiomers of amines, amino acids and their derivatives, amino alcohols, etc. However, due to poor solubility in water, tetra-carboxylic, e.g., 18-crown-tetracarboxylic acid (Fig. 13.6), and tetraoxa diazo derivatives of crown ethers are used more often as chiral selectors in capillary electrophoresis [16]. In case of other derivatives, it is possible to use an addition of organic solvent for the basic electrolyte, which increased solubility both of selectors used, and in some cases also substances separated [17]. The effect of an application of organic solvents and their mixtures as additives for mobile phase is also the change in constants of binding of diastereometric guest–host complexes, which in numerous cases improves separation selectivity [18].

Crown ethers and their derivatives were used among others to separation of enantiomers of amino acids, amines, amino alcohols, peptides [15, 16], or inhibitors of HIV virus proteases [19]. Moreover, in enantioselective separation of amino acids methyl esters using capillary electrophoresis, better separation was obtained using the mixture of two selectors, tetraoxa diazo derivatives of crown ethers, and methyl-β-cyclodextrin [20].

13.5 Chiral Surfactants

Separation of optical isomers using MEKC method is possible in two ways. One of them is an application of achiral surfactants in a concentration above CMC and an addition of suitable chiral selectors to the system, and the second is an application of chiral surfactants. Chiral surfactants are relatively large groups of chiral selectors and are divided into natural (bile acids salts, digitonins, saponins) and synthetic (sugars derivatives, amino acids). As micellar chiral selectors, both polymers obtained from L-amino acids derivatives and peptides are used [21]. These polymers are also used as selective surfactants in micellar electrokinetic chromatography (MEKC) [22], whereas among natural surfactants, the most often used chiral detergents are bile acids salts formed from four rings and side chains containing usually hydroxyl groups (Fig. 13.7). They are successfully used for separation of amino acids dansyl derivatives enantiomers, dinaphtyl derivatives, analogues of diltiazem, and trimethohydroquinone [18].

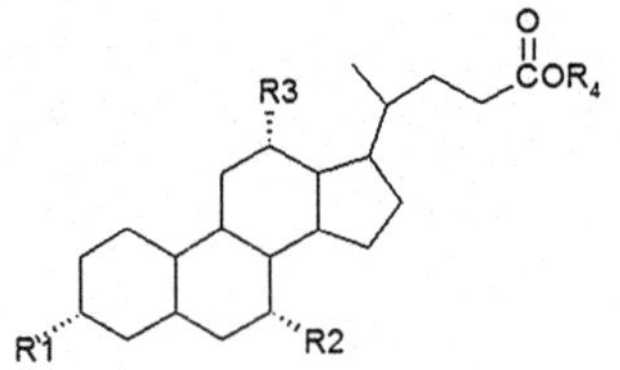

Fig. 13.7 Bile acids salts
applied as chiral surfactants,
according to Ref. [11]

Bile salt	R_1	R_2	R_3	R_4
sodium chlorate	OH	OH	OH	ONa
sodium taurochlorate	OH	OH	OH	$NHCH_2CH_2SO_3Na$
sodium deoxychlorate	OH	H	OH	ONa
sodium deoxytaurochlorate	OH	H	OH	$NHCH_2CH_2SO_3Na$

13.6 Macrocyclic Antibiotics

Macrocyclic antibiotics are a diversified group of chiral selectors often used for
separation of optical isomers of numerous groups of compounds, both polar (acids
and bases) and neutral. This results from the presence of many chiral centers and
functional groups able to interact with analyzed substance. Six types of antibiotics
are used in capillary electrophoresis as chiral selectors. They are: polypeptides,
glycopeptides, ansamycins, macrolides, aminoglycosides, and lincosamides [23–
25]. The best and most often used in CE chiral selectors among macrocyclic
antibiotics are: vancomycin (Fig. 13.8), teicoplanin and ristocetin A.

Due to wide spectrum of functional groups able to interact with separated macrocyclic compounds, the antibiotics are used in enantiomers separation, e.g., anti-
inflammatory drugs which do not contain steroid arrangement, anticarcinogenic

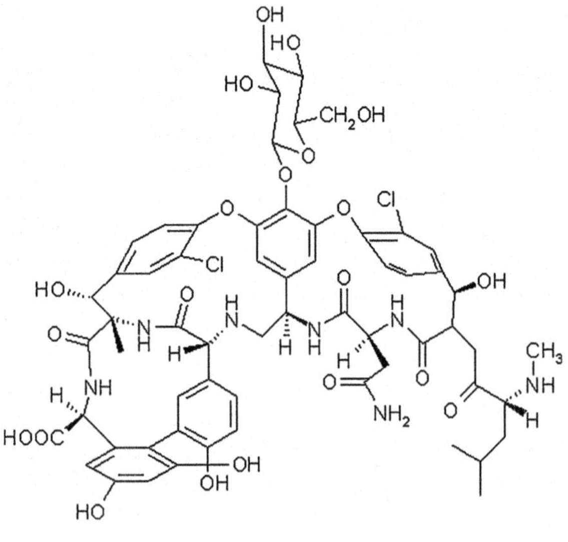

Fig. 13.8 Vankomycin
structure, according
to Ref. [8]

drugs, and amino acids derivatives [25]. However, the factors limiting their application are limited stability and strong absorption of UV above $\lambda = 250$ nm [15].

13.7 Chiral Metals Complexes

Another kind of chiral selectors used for separation of enantiomers are so-called chiral metal complexes. Separation of enantiomers with ligands replacement involves in that case the formation of chelating complex created from central cation and at least two bifunctional ligands (usually L-amino acids) added as selectors for carrying electrolyte [10]. The most commonly used metal ions are: cations of copper (Cu^{2+}), nickel (Ni^{2+}) and zinc (Zn^{2+}), which results from the fact that they form relatively stable complexes. Examples of chemical structures formed as a result of Cu II complexes formation with aspartame are presented in Fig. 13.9.

Chiral metal complexes are used first of all in amino acids separation, and also peptides, hydroxy acids, diamines, and amino alcohols [10].

13.8 Natural and Synthetic Polymers

Also, natural and synthetic polymers are used as chiral selectors in capillary electrophoresis, in capillary zone electrophoresis (CZE), capillary electrochromatography (CEC), and in electrokineticchromatography (EKC). These selectors similar to low-molecular selectors, may be components of buffer or immobilized on capillary walls or in a gel. Proteins are natural polymers composed of numerous optically active amino acids of positive or negative charge. Due to complex structure and presence of various functional groups, electrostatic, hydrophobic interactions, hydrogen bonds, etc., are responsible for stereoselective interaction of protein with analyte, depending on its structure [26]. Using proteins as selectors for enantiomers separation, the strong influence of pH of basic electrolyte and its Ionic strength on ability of enantiomers complexing and stability of complexes formed should be taken into consideration [15]. Proteins used most often in medicines

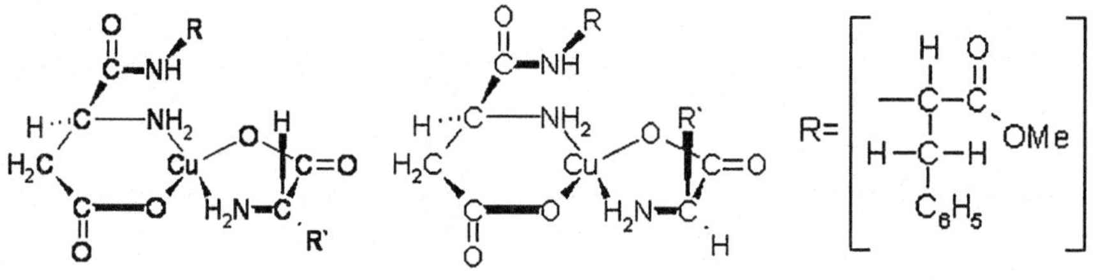

Fig. 13.9 Structure of Cu (II) and aspartam complex, according to Ref. [8]

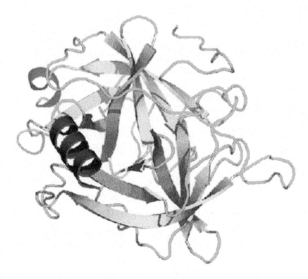

Fig. 13.10 Chymotrypsin [27]

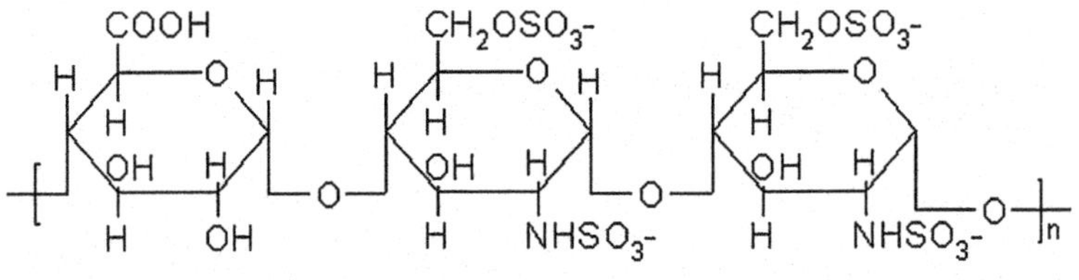

Fig. 13.11 Heparine

analysis are, among others, bovine, rats and porcine albumin, and also riboflavin, trypsin and chymotrypsin [26] (Fig. 13.10).

The second group of naturally occurring polymers, which due to the asymmetric structure commonly used for enantiomers separations, are polysaccharides. Similar to the case of macrocyclic antibiotics and proteins, the mechanism of enantiomers separation involves numerous interactions (among other hydrogen, dipole–dipole), thus giving the possibility of analysis of compounds of differentiated structure [28]. Both neutral and ionic polysugars are used as selectors in capillary electrophoresis. The most often used polysaccharides are cellulose, amylose, starch, dextranes, pectins, and heparin (Fig. 13.11). These compounds found an application inter alia in chiral analysis of diltiazem, laudanosine, laudanosoline, propranolol, nornicotine, or doxylamine [28].

From many known synthetic polymers, polyacrylamides, polyacrylates, and molecularly imprinted polymers (MIP) are used the most often in enantiomers separation. These polymers found an application, among others, in chiral analysis of amino acids and their derivatives, amines, carboxylic acids, peptides, and also alkaloids [15].

Table 13.1 Enantiomer separation and determination examples

Type of analytes	Sample	Chiral selector	Sample preparation	Detection method	Ref.
α-amino acids	Water solutions	α-CD	Direct injection	CE-UV	[29]
	Water solutions	18-crown-6- tetrakarboksy acid	Direct injection	CE-UV	[30]
	Orange juices	β-CD	Direct injection	CE-MS	[31]
	Transgenic soja	3-monodeoksy-3-monoamino-β-CD	Extraction	CE-TOF-MS	[32]
	Rice	Zn (II)-L-arginine complex	Extraction	CE-UV	[33]
	Milk, yogurts	Wankomycine	Extraction	CE-UV	[34]
Dansyl-AA	Water solutions	Beef albumine/dextran	Direct injection	MEKC-UV	[35]
Non-steroid antiflamaric drugs	Water solutions	SDS-wankomycine	Direct injection	MEKC-UV	[36]
	Water solutions	2-O-(2-aminomethyl-imino-propyl)-β-O-hydroksypropyl-CD	Direct injection	CE-UV	[37]
	Human blood serum, urine	2,3,6-tri-O-methyl- β-CD;	SPE	CE-UV	[38]
Metamfetamine	Urine	2,3-di-acetyic-6-sulfo- β-CD	Direct injection	CE-MS	[39]
Salbutamol	Urine	2,3-di-acetic-6-sulfo- β-CD	SPE	CE-UV	[40]
Lactic acid	Plazma	Hydroksypropyl- β-CD	Extraction	CE-UV	[41]
cis-β-laktams	Water solutions	Methyl and sulfo β-CD mixture	Direct injection	CE-UV	[42]
β-blokers (pindolol, atenolol)	Water solutions	Rifamycine B	Direct injection	Indirect CE-UV	[43]

(continued)

Table 13.1 (continued)

Type of analytes	Sample	Chiral selector	Sample preparation	Detection method	Ref.
Anti malaria drugs (primaquin)	Water solutions	α, β, γ-CD and their derivatives	Direct injection	CE-UV	[44]
	Rats liver	Maltodekstrine	Direct injection	CE-UV	[45]
Katechins	Black tea	6-O-α-D-glukosylo-β-CD/SDS	Direct injection	MEKC-UV	[46]
Flawonoids	Citrus fruits	α, β, γ-CD and their derivatives	SPE	CE-UV	[47]
Tiobenkarb	Rats liver	γ-CD/SDS	Extraction	MEKC-UV	[48]
Fenoxypropionic acids	Surface water solutions	γ-CD; wankomycine	SPE	CE-UV	[49]
Aminoacids phosphonic analoques	Water solutions	α-CD	Direct injection	CE-UV	[50]
Aromatic aminophosphonic acids	Water solutions	α-CD, β-CD and their hydroxypropyl derivatives	Direct injection	CE-UV	[51]
Aminoalkane phosphonic acids	Water solutions	α-CD, β-CD, 2,3-O-dimetylo-β-CD	Direct injection	Indirect CE-UV	[52]
HIV protease inhibitors	Water solutions	Benzyloxcarbonyl-methyl derivative of 16-crown-4/SDS	Direct injection	MEEKC-UV	[19]

13.9 Application of Electromigration Techniques on Stereoisomers Analysis

Numerous natural and synthetic compounds exhibiting biological activity are optically active compounds, which particular enantiomers differ in biological properties. Also basic biopolymers, i.e., proteins, polysaccharides, and nucleic acids are formed from one kind of enantiomers. Proteins and peptides are formed from L-amino acids and polysaccharides from D-sugars. Therefore, effective methods of separation and determination of optical purity of optical isomers have been continuously searched for. Due to numerous advantages, electromigration techniques, and especially capillary electrophoresis, have been used recently more and more often for that purpose. A variety of chiral selectors usable in electromigration methods, allows to apply the electrophoresis for determination of optical isomers of various compounds both high-molecular and those of a smaller molecular mass, including medicines, pesticides and food products components. Different types of electromigration methods are thus used for an analysis of stereoisomers in samples of water, food, and in physiological fluids. Examples of an application of electromigration techniques in stereoisomers separation and determination are presented in Table 13.1.

13.10 Summary

Capillary electrophoresis is a useful analytical technique for separation of enantiomers of organic compounds and also for optical purity investigation. Due to an application of water solutions, buffers being basic electrolyte, it is especially often used for determination of optical purity of compounds exhibiting biological activity, which are naturally well soluble in water. First of all, this results from the possibility of an application of various chiral selectors, both low- and high-molecular, which enables the analysis of numerous classes of compounds. An application of that method allows also to determine the optical purity of compounds of similar structures and present on similar concentrations level. Electromigration techniques are used in control of medicines' quality and levels of enantiomers in medicines and their metabolites in body fluids. They are also used for determination of chiral substances of natural origin and residues of pesticides in food products and environmental samples, and also in control of the processes of enantiomers production. Due to the numerous advantages of the CE technique, such as high separation degree, short time of analysis, and small sample volume, the method is also commonly used in routine medical analysis.

References

1. J. McMurry, *Chemia Organiczna* (Wydawnictwo Naukowe PWN, Warsaw, 2000)
2. P. Dżygiel, P.P. Wieczorek, Zastosowanie ekstrakcji i technik membranowych do rozdziału stereoizomerów aminokwasów i ich pochodnych. Wiadomości Chemiczne **58**, 943–962 (2004)
3. W. Szczepaniak, Metody instrumentalne w analizie chemicznej. Wydawnictwo Naukowe PWN Warszawa (2002)
4. C. Simo, V. Garcia-Canas, A. Cifuentes, Chiral CE–MS. Electrophoresis **31**, 1442–1456 (2010)
5. F. Wang, M.G. Khaledi, Enantiomeric separations by nonaqueous capillary electrophoresis. J. Chromatogr. A **875**, 277–293 (2000)
6. N.M. Maier, P. Franco, W. Lindner, Separation of enantiomers: Needs, challenges, perspectives. J. Chromatogr. A **906**, 3–33 (2001)
7. B. Chankvetadze, *Capillary Elctrophoresis in Chiral Analysis* (Willey, Chichester, 1997)
8. B. Chankvetadze, Enantioseparations by using capillary electrophoretic techniques the story of 20 and a few more years. J. Chromatogr. A **1168**, 45–70 (2007)
9. S. Fanali, Enantioselective determination by capillary electrophoresis with cyclodextrins as chiral selectors. J. Chromatogr. A **875**, 89–122 (2000)
10. M. Blanco, I. Valverde, Choice of chiral selector for enantioseparation by capillary electrophoresis. Trends Anal. Chem. **22**, 428–439 (2003)
11. G. Gübitz, M.G. Schmid, Chiral separation by capillary electromigration techniques. J. Chromatogr. A **1204**, 140–156 (2008)
12. www.uni-duesseldorf.de
13. V. Cucinotta, A. Contino, A. Giuffrida, G. Maccarrone, M. Messina, Application of charged single isomer derivatives of cyclodextrins in capillary electrophoresis for chiral analysis. J. Chromatogr. A **1217**, 953–967 (2010)
14. E. Rudzińska, *Na pograniczu chemii i biologii IX* (Wyd. UAM, Poznań, 2003), pp. 199–205
15. I. Ali, A.K. Kumerer, H.Y. Aboul-Enein, Mechanistic principles in chiral separations using liquid chromatography and capillary electrophoresis. Chromatographia **63**, 295–307 (2006)
16. T. Ivanyi, K. Pal, I. Lazar, D. Luc Massart, Y. Vander Heyden, Application of tetraoxadiaza-crown ether derivatives as chiral selector modifiers in capillary electrophoresis. J. Chromatogr. A **1028**, 325–332 (2004)
17. S. Fanali, Controlling enantioselectivity in chiral capillary electrophoresis with inclusion-complexation. J. Chromatogr. A **792**, 227–267 (1997)
18. M.L. Riekkola, S.K. Wiedmer, I.E. Valko, H. Siren, Selectivity in capillary electrophoresis in the presence of micelles, chiral selectors and non-aqueous media. J. Chromatogr. A **792**, 13–35 (1997)
19. S. Leonard, A. Van Schepdael, T. Ivanyi, I. Lazar, J. Rosier, M. Vanstockem, H. Vermeersch, J. Hoogmartens, Development of a capillary electrophoretic method for the separation of diastereoisomers of a new human immunodeficiency virus protease inhibitor. Electrophoresis **26**, 627–632 (2005)
20. J. Elek, D. Mangelings, T. Ivanyi, I. Lazar, Y. Vander Heyden, Enantioselective capillary electrophoretic separation of tryptophane—and tyrosine-methylesters in a dual system with a tetra-oxadiaza-crown-ether derivative and a cyclodextrin. J. Pharm. Biomed. Anal. **38**, 601–608 (2005)
21. L. Bluhm, J. Huang, T.Y. Li, Recent advances in peptide chiral selectors for electrophoresis and liquid chromatography. Anal. Bioanal. Chem. **382**, 592–598 (2005)
22. R. Iqbal, S.A.A. Rizvi, C. Akbay, S.A. Shamsi, Chiral separations in microemulsion electrokinetic chromatography: Use of micelle polymers and microemulsion polymers. J. Chromatogr. A **1043**, 291–302 (2004)
23. T.J. Ward, T.M. Oswald, Enantioselectivity in capillary electrophoresis using the macrocyclic antibiotics. J. Chromatogr. A **792**, 309–325 (1997)

24. T.J. Ward, A.B. Farris III, Chiral separations using the macrocyclic antibiotics: A review. J. Chromatogr. A **906**, 73–89 (2001)

25. A.F. Prokhorova, E.N. Shapovalova, O.A. Shpigun, Chiral analysis of pharmaceuticals by capillary electrophoresis using antibiotics as chiral selectors. J. Pharm. Biomem. Anal. **53**, 1170–1179 (2010)

26. D.K. Lloyd, A.F. Aubry, E. De Lorenzi, Selectivity in capillary electrophoresis: The use of proteins. J. Chromatogr. A **792**, 349–369 (1997)

27. Chemistry.umeche.maine.edu

28. H. Nishi, Enantioselectivity in chiral capillary electrophoresis with polysaccharides. J. Chromatogr. A **792**, 327–347 (1997)

29. P. Dżygiel, P. Wieczorek, J.Å. Jönsson, Enantiomeric separation of amino acids by capillary electrophoresis with α-cyclodextrin. J. Chromatogr. A **793**, 414–418 (1998)

30. R. Kuhn, Enantiomeric separation by capillary electrophoresis using a crown ether as chiral selector. Electrophoresis **20**, 2605–2613 (1999)

31. C. Simo, A. Rizzi, C. Barbas, A. Cifuentes, Chiral capillary electrophoresis-mass spectrometry of amino acids in foods. Electrophoresis **26**, 1432–1441 (2005)

32. A. Giuffrida, C. Leon, V. Gracia-Canas, V. Cucinota, A. Cifuentes, Modified cyclodextrins for fast and sensitive chiral-capillary electrophoresis-mass spectrometry. Electrophoresis **30**, 1734–1742 (2009)

33. L. Qi, M. Liu, Z. Guo, M. Xie, C. Qiu, Y. Chen, Assay of aromatic amino acid enantiomers in rice-brewed suspensions by chiral ligand-exchange CE. Electrophoresis **28**, 4150–4155 (2007)

34. W. Pormsila, X.Y. Gong, P.C. Hauser, Determination of the enantiomers of α-hydroxy-and α-amino acids in capillary electrophoresis with contactless conductivity detection. Electrophoresis **31**, 2044–2048 (2010)

35. P. Sun, N. Wu, G. Barker, R.A. Hartwick, Chiral separations using dextran and bovine serum albumin as run buffer additives in affinity capillary electrophoresis. J. Chromatogr. **648**, 475–480 (1993)

36. K.L. Rundlett, D.W. Armstrong, Effect of micelles and mixed micelles on efficiency and selectivity of antibiotic-based capillary electrophoretic enantioseparations. Anal. Chem. **67**, 2088–2095 (1995)

37. X. Lin, C. Zhu, A. Hao, Enantiomeric separations of some acidic compounds with cationic cyclodextrin by capillary electrophoresis. Anal. Chim. Acta **517**, 95–101 (2004)

38. F.K. Główka, M. Karaźniewicz, High performance capillary electrophoresis method for determination of ibuprofen enantiomers in human serum and urine. Anal. Chim. Acta **540**, 95–102 (2005)

39. R. Iio, S. Chinaka, N. Takayama, K. Hayakawa, Simultaneous chiral analysis of methamphetamine and related compounds by capillary electrophoresis/mass spectrometry using anionic cyclodextrin. Anal. Sci. **21**, 15–19 (2005)

40. A.C. Servais, P. Chiap, Ph Hubert, J. Crommen, M. Fillet, Determination of salbutamol enantiomers in human urine using heptakis(2,3-di-O-acetyl-6-O-sulfo)-β-cyclodextrin in nonaqueous capillary electrophoresis. Electrophoresis **25**, 1632–1640 (2004)

41. L. Tan, Y. Wang, X.Q. Liu, H.X. Ju, J.H. Li, Simultaneous determination of L- and D-lactic acid in plasma by capillary electrophoresis. J. Chromatogr. B **814**, 393–398 (2005)

42. K. Nemeth, E. Varga, R. Ivanyi, J. Szeman, J. Visy, L. Jicsinszky, L. Szente, E. Forro, F. Fulop, A. Peter, M. Simonyi, Separation of cis-β-lactam enantiomers by capillary electrophoresis using cyclodextrin derivatives. J. Pharm. Biomed. Anal. **53**, 382–388 (2010)

43. D.W. Armstrong, K. Rundlett, G.L. Reid III, Use of a macrocyclic antibiotic, rifamycin B, and indirect detection for the resolution of racemic amino alcohols by CE. Anal. Chem. **66**, 1690–1695 (1994)

44. K. Nemeth, G. Tarkanyi, E. Varga, T. Imre, R. Mizsei, R. Ivanyi, J. Visy, J. Szeman, L. Jicsinszky, L. Szente, M. Simonyi, Enantiomeric separation of antimalarial drugs by capillary electrophoresis using neutral and negatively charged cyclodextrins. J. Pharm. Biomed. Anal. **54**, 475–481 (2011)

45. R. Bortocan, P.S. Bonato, Enantioselective analysis of primaquine and its metabolite carboxyprimaquine by capillary electrophoresis. Electrophoresis **25**, 2848–2853 (2004)
46. S. Kodama, A. Yamamoto, A. Matsunaga, H. Yanai, Direct enantioseparation of catechin and epicatechin in tea drinks by 6-O-α-D-glucosyl-β-cyclodextrin-modified micellar electrokinetic chromatography. Electrophoresis **25**, 2892–2898 (2004)
47. N. Gel-Moreto, R. Streich, R. Galensa, Chiral separation of diastereomeric flavanone–7–O–glycosides in citrus by capillary electrophoresis. Electrophoresis **24**, 2716–2722 (2004)
48. S. Kodama, A. Yamamoto, A. Matsunaga, K. Okamura, R. Kizu, K. Hayakawa, Enantioselective analysis of thiobencarb sulfoxide produced by metabolism of thiobencarb by hydroxypropyl-γ-cyclodextrin modified micellar electrokinetic chromatography. J. Sep. Sci. **25**, 1055–1062 (2002)
49. C.M. Polcaro, C. Mara, C. Desiderio, S. Fanali, Stereoselective analysis of acid herbicides in natural waters by capillary electrophoresis. Electrophoresis **20**, 2420–2424 (1999)
50. P. Dżygiel, E. Rudzińska, P. Wieczorek, P. Kafarski, Determination of optical purity of phosphonic acid analogues of aromatic amino acids by capillary electrophoresis with α-cyclodextrin. J. Chromatogr. A **895**, 301–307 (2000)
51. E. Rudzińska, P. Dżygiel, P. Wieczorek, P. Kafarski, Separation of aromatic aminophosphonic acid enantiomers by capillary electrophoresis with the application of cyclodextrins. J. Chromatogr. A **979**, 115–122 (2002)
52. E. Rudzińska, P. Wieczorek, P. Kafarski, Separation of aminoalkanephosphonic acid enantiomers by indirect UV detection capillary electrophoresis with application of cyclodextrins. Electrophoresis **24**, 2693–2697 (2003)

Chapter 14
"Lab-on-a-Chip" Dedicated for Cell Engineering

Elżbieta Jastrzębska, Aleksandra Rakowska and Zbigniew Brzózka

Abstract In this chapter various microsystems (called Lab-on-a-Chip) dedicated to cell engineering are described. We present methods for microsystems fabrication and also essential parameters, which have influence on cell culture in microscale. This area of research is significant, because microsystems nowadays play an important role in cell studies and drug development. It can be caused by possibility to perform *high-throughput* screening in microscale. In addition, the size scale of microfluidic devices is especially suitable for biological applications at the cellular level, the scale of microchannels corresponds well with the native cellular microenvironment. The volume of cells to extracellular fluid in microsystems is similar to in vivo conditions, the next advantage of microchips. Moreover, a special architecture of microdevices gives the possibility for control of growth factors, reagents, and oxygen inside the system.

14.1 Miniaturization

The development of new technologies enables miniaturization of devices used in research. It also allows to reduce the amount of materials and minimize the quantity of reagents required for determinations. Microbioanalytics is the field, in which for example, process control in biotechnology or *online* monitoring of environmental pollution is performed. New miniaturized analytical systems allowing for fast, cheap, and easy way to carry out such determinations and chemical analysis applicable to the areas of laboratory and environmental investigations were developed. The dynamic development of this branch of science provides a number of

E. Jastrzębska (✉) · A. Rakowska · Z. Brzózka
Faculty of Chemistry, Warsaw University of Technology, Warsaw, Poland
e-mail: ejedrych@ch.pw.edu.pl

B. Buszewski et al. (eds.), *Electromigration Techniques*, Springer Series
in Chemical Physics 105, DOI: 10.1007/978-3-642-35043-6_14,
© Springer-Verlag Berlin Heidelberg 2013

publications, which describe new technologies or applications of microsystems. Since 2000 the number of publications focused on microsystems increased twice. Whereas in 1998 the number of patents related to microsystems was less than 24, six years later it reached about 300.

Miniature devices called *Lab-on-a-Chip* are an integrated micro-laboratory on the plate, in which it is possible to perform complex and multistep analysis. The origins of microchips are fields such as chemistry, biology, molecular biology, electronics [1]. Analytical chemistry and adaptation of analytical techniques in the microscale have the most important influence on the development of microfluidic systems. The next steps in microtechnology and microbioanalytics development were made after 1980 and were connected to molecular biology. In turn, the possibility of integration of microfluidic devices with electronic systems and utilization of electronic technologies became the consecutive steps leading to the construction of *Lab-on-a-Chip* system [1, 2]. The integration of these fields of science enabled to create a new interdisciplinary science called microbioanalytics. Microfluidic systems were employed in biotechnology, clinical diagnosis, pharmacy, nature science, tissue engineering, and nanomedicine [2, 3]. It is possible to construct a device, which is portable, with small dimensions and dedicated for analysis performed outside laboratory conditions. It is also possible to perform a flow analysis with extremely small reagent volumes and significant reduction of biological materials used. The application of microchips allows also for the reduction of analysis time and control in time and space [4].

14.2 Analytical Techniques in Miniaturization

Analytical chemistry is the main precursor of miniaturized systems. These applications of chips for multi-tests improve efficiently sample analysis. Small samples may be tested and simultaneously high sensitivity obtained. The best concept of microsystems used in analytical chemistry allows to perform a complex analysis which includes sampling, sample processing, component separation and detection, and sample removal. The automated microsystem allows to improve analytical procedures and to use advantages of miniaturization. A miniaturized gas chromatograph developed in silicon was the first microsystem dedicated for analytical chemistry. Microdevices are also applied for extraction, filtration, and separation of samples. Microsystems are miniature devices used in separation techniques such as capillary electrophoresis (CE), capillary iso-electric focusing (CIF), capillary gel electrophoresis (CGE), micellar electrokinetic capillary chromatography (Mecca), capillary electrochromatography (CEC), or high-performance liquid chromatography (HPLC). The usage of microsystems enable to perform series analytical processes on the single plate and elongation of separation way. Moreover, it is possible to design any gradient of mobile phase.

Different methods for analysis of samples are used depending on the research performed in the microchips. Electrochemical and optical detections are performed in microsystems: spectrophotometry, spectrofluorimetry, photothermal, and Raman spectroscopy are optical methods utilized in microscale. Likewise, amperometry, conductometry, and potentiometry belong to electrochemical methods which are used in microchips [5].

14.3 Cell Cultures

In the past decade, interest in researches related to cell culture and analysis of cell survival significantly increased. Cell cultures play a fundamental role in both biotechnology and pharmaceutical industries. They participate not only in the discovery and production of drugs, but also in the field of regenerative medicine [6]. The continuous increase in popularity of using culture to study the usefulness of new drugs is due to the fact that cells provide a fast and representative response to substances tested. It is also important that they limit the number of tests conducted on live animals, according to the 3Rs rule (Replacement, Reduction, Refinement), which assumes searching for alternative methods used to replace animal testing [7, 8].

Cell cultures are used both for studying biological processes and for application purposes such as tissue engineering. Cell cultures enable to culture selected type of cells and carry out the tests of cellular processes such as proliferation, cell adhesion to the substrate, protein secretion, and differentiation in tissues with specific phenotype [9]. A wide range of applications of cell cultures leads to the search for better methods of optimization which allow to improve the process of cell culturing and assimilation of its terms to the in vivo conditions. The cytotoxicity studies of anticancer drugs should be carried out also on normal cells. It is necessary to determine the extent to which drugs damage healthy cells in the human body [10]. There are many similarities and differences between cancer and normal cells. Understanding the mechanisms which regulate normal and tumor cells proliferation is essential to improve cancer research.

A noticeable difference between normal and cancer cell cultures is the tendency of healthy cells to achieve certain number of divisions and the possibility of programmable death after receiving a signal from the body. The distinguishing feature of the normal cells is also to serve their specific functions in the body. In vitro culture of these cells assume to maintain these functions, due to the fact that they are necessary to obtain material suitable for further research. Cancer cells are characterized by the imbalance between the numbers of dividing and dying cells, moreover, they exhibit a tendency to form multilayer aggregates which impede accurate analysis. Cancer cells can divide indefinitely to form three-dimensional structures. Furthermore, they have the ability to move and occupy the place of normal cells. Therefore, there are numerous publications that compare the cultures of normal and malignant cells [11].

14.4 Microsystems in Tissue Engineering

One of the latest techniques for biological research, which uses biological material, is miniaturized devices called *Lab-on-a-chip*. The utilization of such systems allows for elimination of research in highly specialized laboratories, which usually absorb huge costs. In addition, the volume used during the study is very small, so quantities of biological material used are also small. The usage of miniaturized analytical systems in such studies can also minimize the contribution of the human factor in the whole culturing process [12]. The *real-time* observation of the cultured cells allows to learn their behavior in different environmental conditions. It is especially significant for research carried out, *e.g.*, on cells. Another advantage is the ability to automate a microdevice, which allows for precise identification and elimination of errors in laboratory work. Moreover, these systems exhibit compatibility with most commonly used detection methods such as electrochemical methods, mass spectrometry, and optical methods, including absorption, chemiluminescence, and fluorescence [3]. In comparison with conventional cell culture methods, techniques based on miniature cell culturing systems have important advantages and enable to make an investigation in environment more similar to in vivo conditions [13]. In addition, during the flow of nutrients it is possible to control the shear stress as factor, which influences the cells.

Despite advanced technology, the designing of suitable microsystems dedicated to tissue engineering is still a complicated task. *Lab-on-a-chip* systems used in tissue engineering must fulfill several important functions. It is important that the manufactured chip allows cells to grow and control their behavior, *e.g.*, migration or adhesion to the substrate. It should also ensure a suitable oxygenation and flow of nutrients. Therefore, new miniaturized systems that allow for culturing different cell types and perform various types of analyses are still designed and developed.

14.5 Design and Development of Microsystems

14.5.1 Construction Materials

The structure of the microdevice used for cell analysis determines not only the properties of culture, but also ways of visualizing its results. This chapter presents the most often used materials to manufacture a microdevice dedicated to cellular engineering. The microsystem components should be biocompatible and non-toxic to cells. Furthermore, they should also allow to perform a desirable microstructures in them [14]. There are many currently used materials for production of microsystems, ranging from silicon, glass, and ceramics to various types of plastics. Polydimethylsiloxane (PDMS), polymethyl methacrylate (PMMA), poly [3-mercaptopropyl] methylsiloxane (PMMS), polycarbonate (PC), polycaprolactone

(PCL), polylactic acid (PLLA), poly [lactic-co-glycolic acid] (PLGA) [15–18] are polymers often used in microtechnology.

The most commonly used materials for construction of chip dedicated to tissue engineering are poly(dimethylsiloxane) and glass. The PDMS is nontoxic to cells, impermeable to water and permeable to gases. These are the main advantages of this material. Moreover, it is flexible, inexpensive, and optically transparent. In addition, PDMS is easily treated and allows to perform precise microstructures [19]. It also has the ability to create reversible and irreversible binding to other materials such as glass, silicon, other polymers, or to another layer of PDMS [20]. These bonds may be based on van der Waals forces (reversible) or covalent bonds (irreversible) between the surfaces of individual elements. The seal of microdevice is very important. During the flow of fluid through the microchip it allows suitable flow rates [15]. An important advantage of PDMS-based chips is their relatively simple structure, which allows to easily change the device design and optimization at minimal cost.

Glass is also often used material to create microchips. This material is nontoxic and chemically resistant. Microstructures in the glass are performed by wet etching with hydrofluoric acid [21]. In addition, the glass as a hydrophilic material ensures proper adhesion of cells to the substrate. On the other hand, a significant disadvantage is its limitations during fabrication of the microstructure. The geometry of the fabricated channels has influence on the culture conditions and other factors such as dispersion [22]. The fabrication of microchambers for cell culture in the glass enables to use various types of microscopy including fluorescence, confocal, and phase contrast to monitor cell culture processes.

14.5.2 Fabrication of Microsystem

The material used for construction of miniature analytical systems determines the method for producing a specific microstructures. Various types of materials are used to fabricate *Lab-on-a-chip* systems for cell culture and analysis. Glass and polymers are the most commonly used materials, so the methodology of fabrication of microstructures in these materials is discussed in this chapter. Microstructures in the glass are performed by using photolithography and wet etching techniques with hydrofluoric acid. Glass etching is isotropic, therefore the microstructure formed in the walls are curved. Microstructures in polymers are made by replication methods (hot embosing, injection molding, and casting), and by direct treatment (micromilling, laser micromachining) [2, 16]. The kind of material used to fabricate the microchip, microstructure parameters (geometry and size of the microchannels), and the availability of the components required are very important while choosing the fabrication technique [15]. The design of the geometry of microchannels and microchambers, which will be developed in the materials is the first step in microchip fabrication. Computer programs such as AutoCAD or Corel are used to design the model serving as the basis for manufacturing the microstructures in the material.

14.5.3 Casting

The first step in the casting process is to implement the so-called stamp (matrix). It can be made in different materials such as glass, quartz, nickel, or polymer. There are many techniques for manufacturing such stamps, *e.g.*, micromilling, wet etching, electroforming. The photolithography is the most often used method for stamps fabrication (particular in the case of microsystem made of PDMS) [2]. For this purpose a photoresist is deposited on a plate and exposed to UV light through a photomask [23]. Then, the exposed photoresist is developed using developer (for example water). In this step, it is obtained the stamp which contains the micro-structures and microchannel network [16]. There are two kind of photosensitive materials: the negative emulsion—a substance that becomes insoluble by ultraviolet radiation and positives emulsion—a substance soluble in the developer. The stamp is used for replication in PDMS. PDMS prepolimer is mixed with curing reagent and liquid PDMS is poured into the mold and allowed to cure for 1 h at 70 °C. After that, the structure is peeled of the master and the PDMS plate with precisely (with the accuracy of tens of micrometers) formed microchannel network is obtained. This method enables to reduce the time and cost of microchip fabrication process.

14.5.4 Hot Embossing

Hot embossing is a method in which polymer stamps are also used. To obtain microstructures using this technique, stamp and polymer plate (e.g. PMMA) should be placed in a vacuum chamber of hydraulic press and separately heated. The next step is to press the stamp in the polymer. After cooling a microstructure imprinted in the polymer is obtained [2, 16].

14.5.5 Injection Molding

In this method polymer is plasticized by increasing temperature. Then the polymer is injected into the special form containing microdesign by use of high pressure. This is a convenient way to perform the microstructure in the polymer plate allowing for accurate visualization of the design.

14.5.6 Laser Micromachining

Laser micromachining is a direct action of laser-generated UV or IR radiation. This method allows for fast obtaining of microstructures in the material fast. This type of various microstructures preparation is performed on the surface of the polymer or other materials. However, there are some limitations associated with high costs of equipment and a large surface roughness of the resulting shape [2, 16].

14.5.7 Micromilling

Micromilling is a mechanical processing of the material. Microstructures are performed in polymer by use of a microcutter. In this method we use micron-sized cutters covered with hard materials such as, e.g., titanium carbide. The technique enables to obtain a three-dimensional structures in a layer of a polymer. This method allows also to change the shape of the structure during project executing. Microstructures received in materials by micromilling are also used as stamps for structure molding, e.g., in PDMS [2, 16].

14.5.8 Microsystems Sealing

The combination of two plates with fabricated microstructures ensures their sealing. Different techniques can be used depending on the type of material. Polymers such as PMMA or PC can be connected by thermal bonding in thermal hydraulic press. In this method, the thermal properties of material are used. However, it is important not to exceed the temperature limit, since that may cause deformation in microstructures [16]. Another example of microstructures bonding is the use of oxygen plasma. This technique allows to connect materials to each other, e.g., PDMS with PDMS, PDMS with glass or PDMS with silicon.

14.6 Essential Parameters in Microsystem for Cell Culture

14.6.1 Geometry of Microsystem

The important thing which confirms advantages of microsystems is a possibility to manipulate their geometry. The microchannels and microchambers size respond to the cell dimension. The surface area to volume ratio (SAV) in the microsystem is higher than 1 being similar to in vivo conditions [24]. A large SAV ratio gives many advantages. First of all it allows for efficient mass transport of gases via diffusion [22, 25]. The experimental conditions mimicking in vivo environment are important for cell analysis because it increases the meaning of the obtained results. It is possible because in the microscale the cells are in biological inter-action between other cells. The microfluidic systems enable also to tests cells in carefully and precisely controlled conditions. Moreover, the microscale gives the possibility to design a suitable geometry and structure of cell culture systems.

14.6.2 Sterilization and Culture Condition

Effective sterilization is critical for maintaining the growth of long-term culture at both macro- and microscale. Autoclaving of microsystems is a fundamental method for sterilization, however, many materials cannot be exposed to high temperatures, because the microchannels developed in these materials can undergo deformation. If autoclaving is not possible the microfluidic systems can be exposed to UV light or oxygen plasma. Flushing the device with ethanol (70 %) is the next method to assure aseptic conditions inside microdevices. In the static cell culture, the microenvironment is uncontrolled. On the contrary, the changes of culture medium in the microsystems is controlled [26]. Suitable preparation of microsystems and in situ monitoring of culture conditions assure a proper cell growth and proliferation. The most important parameters, which must be controlled during system preparation and cell culture are flow rate value of all substances introduced into microdevices, procedure of cell seeding, density of cells introduced into microchamber, and cell-substrate adhesion. The aim of cell culture in the microfluidic systems in not only miniaturization of the conventional cell culture technique, but also usage of microscale benefits.

14.6.3 Gradient Generation

The control of flow rate of culture medium influences the cell culture in the microfluidic system. Unsuitable value of flow rate can indicate shear stress, which disadvantageously affects cell morphology and viability. Moreover, constant flow rate can flush paracrine (affecting other cells) and autocrine (affecting cells that produce it) signaling factors secreted by cells. Too high low rate can also remove cells from the microsystem. On the other hand, too small flow rate can reduce nutrients which leads to a decrease in cell viability [27]. Therefore, a proper flow rate of medium determines optimal cell growth with minimal damage to the cells. Liquid flow in the microscale have three physical phenomena: laminar flow, diffusion, and surface to volume ratio (SAV) [28, 29]. The Reynolds number is at a low level ($10 > Re > 0,001$), which is synonymous with the laminar flow. The ratio of width and height of microchannels is small, therefore the probability of any disturbances in the flow is also low [30]. Three kinds of flows are used in the microscale investigation: electroosmotic flow and flow propelled by centrifugal force or pressure [31]. During the laminar flow the mixing of substances proceeds through diffusion (Fig. 14.1). Scale reduction causes increase in the surface area to volume ratio (SAV). It contributes to increase in the rate of diffusion and heat conduction in microsystem. However, a high value of SAV results in faster adsorption of particles on the microchannel surface. Simultaneously, it reduces the efficiency of fluid flow [28].

Fig. 14.1 A mixing of stream X and Y. The diffusion rate increasing with time, according to [29]

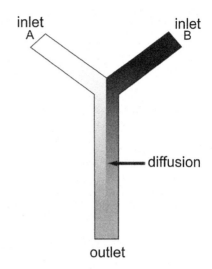

In miniature systems, it is possible to produce specific chemical concentration gradients similar to that in vivo, which are difficult to obtain at the macroscale. In the first type of gradient, laminar flow is utilized. Control of this kind of gradient can be done by a properly designed network of channels, whose inlets and outlets determine the concentrations obtained (Fig. 14.2) [28]. There is a second kind of gradient generator, in which chemical gradient is achieved by diffusion along the channel between the source and sink compartments (Fig. 14.3). Since mixing without laminar flow is utilized, it can be used for non-adherent cells. This type of gradient is more friendly to the cells and creates an environment with less shear stress on the cells [30].

Fig. 14.2 Concentration gradient generator where laminar flow is used. **a** Every microchannel has different concentration of reagent. **b** The outlet with different concentration of substance. Reprinted with permission from Breslauer et al. [28]. Copyright 2006 the Royal Society of Chemistry

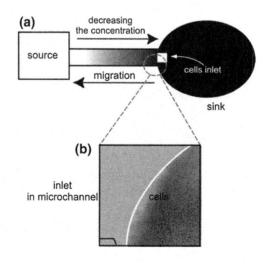

Fig. 14.3 Geometry of
chemical gradient generator
without flow. **a** The device
has source and compartment
sink covered with membrane.
The compounds diffused
through membrane create
chemical gradient. **b** Cells are
placed in the compartment
sink. Reprinted with
permission from Chung and
Choo [30]. Copyright 2010
the Royal Society of
Chemistry

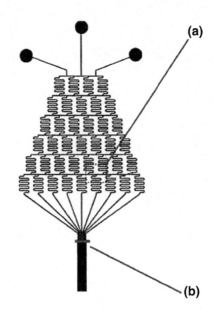

14.6.4 Other Parameters

Temperature, dampness, and concentration of carbon dioxide also influence on the
cells cultured in microsystems. These parameters are automatically regulated in
the incubator used for cell culture. However, microsystems integrated with heater
and devices for temperature and level of carbon dioxide control are also developed. The integrated microsystems assure optimal temperature and pH conditions
and allow to eliminate classic incubator during culture and analysis of cells in the
microchip.

14.7 Cell Culture in Microscale

The flow rate of cell suspension during introduction into the microsystem is an
important parameter. It determines the density of cells in separate microchambers.
A non-uniform placement of cells in the growth place is a meaningful problem
during cell seeding. Nutrients are consumed faster in microchambers with high
density of cells than in those with low cell density. Simultaneously, the secretion
of adventive products is increased. When adherent cells are used their density
should allow to play appropriate cell–cell and cell–matrix interactions [32]. After
introduction into the microsystem, adherent cells need a certain time to achieve an
appropriate level of adhesion to the substrate. In contrast, cells cultured in suspension require relatively less time to adapt to the new environment [25]. Another
important aspect of adherent cell culture is its response to the modification of

culture medium flow. Cell growth is directly dependent on the flow rate. At low flow rates cell growth is induced by exchange of mass, but there is a shortage of nutrients and excessive accumulation of metabolic products. However, providing large quantities of nutrients at high flow rates may result in a high hydrostatic pressure and a shear stress that inhibits cell growth [33]. Ensuring adequate adhesion of cells to substrate plays a very important role in cell culture. The behavior of the corresponding network structure of intercellular signals is basic for appropriate functions to maintain the biological test material. Adhesion factors are often used to ensure a proper attachment of cells to the substrate. These include proteins like fibronectin [34] or collagen [35]. Adhesion factors regulate several aspects of the cell behavior including proliferation, migration, and protein synthesis [36]. The adhesion of mouse embryonic fibroblasts to the above-mentioned proteins is described in the literature. The impact of factors facilitating the adhesion of cells and their migration iss also examined. The cells that adhered to the uncoated substrate surface have a greater tendency to migrate than in the surface with adhesion improving agent [34]. It should also be noted that the extent to which these factors will enhance bonding with the surface of cells depends on the type of biological material used. The investigation on influence of covering with polylysine on the surface of the PDMS showed that it increased the adhesion of fibroblasts by 32 %, while fibronectin did not exhibit such activity [37]. Park et al. showed that the adhesion affects both the size and the shape of cultured cells [38]. The shape and nature of substrate to which the cells stick are also meaningful. It was examined how the shape of adhesion surface influences growth of cells. For this purpose, adhesive islands of different sizes were coated with extracellular matrix. It was found that cells took the shape of the underlying substrate. Moreover, shape changes had an effect on the cytoskeleton function and cell apoptosis and proliferation [39]. Gaver and Kute [40] investigated the adhesion of the cells on different substrates. Parameters such as pressure, shear stress, flow rate, dimensions of microchannels, and fluid viscosity were examined. The shear stress increased linearly with increasing flow rate. It was also shown that when the ratio of cell size to channel dimensions is large, the shear stress experienced by cell may differ significantly from the stress acting on microchannel wall. This is the result of significant flow disturbances caused by the cells. These tests show that the presence of a cell changes the velocity distribution of substances. Figure 14.4 shows that the upper part of the cell is exposed to much higher shear stress than its lower part and that in the flow direction is also working on considerable force. The result of these forces is a visible change in cell shape.

Miniaturized analytical systems for cell culture have different geometry of microstructures. Microchips can be used for simultaneous observation and examination of proliferation and controlling the differentiation of one (Fig. 14.5) [19] or six different stem cell lines [41].

Cell adhesion is an important marker of the cell differentiation. Cells could be used as a functional tool for biological research and analysis of procedures used in different therapies, therefore, cell monitoring is very important. Cell growth and adhesion were observed in the microchambers obtained in PMMA (Fig. 14.6).

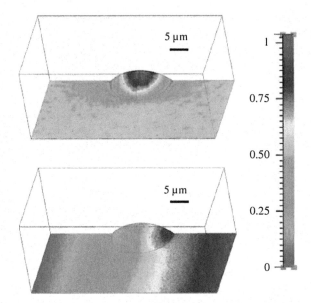

Fig. 14.4 3D model with distribution of forces, which have effect on the cell during the flushing of the microchip. Reprinted with permission from Lu et al. [34]. Copyright 2004 American Chemical Society

The microchambers were also connected to electrodes, used to determine cell-substrate adhesion [42].

14.8 Cytotoxicity Tests in Microsystems

A rapidly growing pharmaceutical industry requires faster and more efficient ways of finding, testing, and refinement of new drugs. Therefore, the main purpose of cytotoxicity testing became to examine the suitability of drugs in the early stages of their placing in the market. Conditions in the cell culture microsystem must be

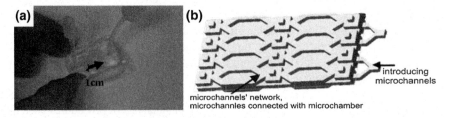

Fig. 14.5 Microsystem (a) and the geometry of the microchip (b). Reprinted with permission from Leclerc et al. [19]. Copyright 2006 Elsevier

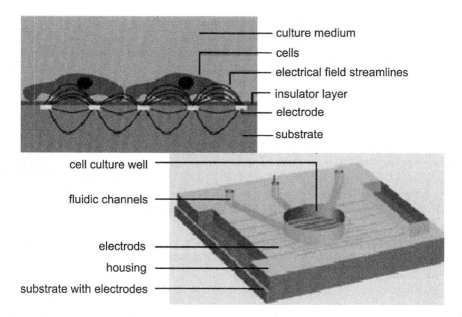

culture medium

cells

electrical field streamlines

insulator layer

electrode

substrate

cell culture well

fluidic channels

electrods

housing

substrate with electrodes

Fig. 14.6 The microsystem for monitoring of cells adhesion. Reprinted with permission from Maercker et al. [42]. Copyright 2008 Wiley–VCH

as close as possible to in vivo conditions for better prediction of the body's response to examined compound. Toxicity tests are carried out sequentially in vitro and in vivo to check whether they are sufficiently safe and effective for further clinical trials. The use of microchips for drug testing is the solution that gives the opportunity to limit not only costs and time-consuming researches, but also to reduce the amount of tests performed on animals.

Microsystems have a great advantage over static cultures, not only because they provide sterile and stable conditions [43], but also by creating a wide range of possibilities in controlling concentrations and dosage of tested substances. It is possible for example to develop a system that allows to simultaneously carry out the test for several concentrations of the tested agent [23]. Moreover, investigations in microscale offer the possibility of testing different types of cells and many substances. It gives the opportunity to perform a more detailed analysis of the effectiveness of tested drug and its safety profile. This is applicable in studying the synergistic effect of combinations of several drugs [44]. The volume of the material required for testing is low and it is especially important when tests are performed with cells derived from the patient. Not without significance are problems arising from two-dimensionality of cells cultured in microsystems. In the natural environment, mammalian cells are exposed to different stimuli including intercellular signals and the effect of extracellular matrix [45]. The ability to predict in vivo effects from in vitro responses of cells cultured in two-dimensional culture are limited. This is significant in the case of tumor cells tending to form heterogeneous three-dimensional structures [13]. Moreover, it is proved that cancer cells in the case of three-dimensional culture are more resistant to toxic compounds [46]. Despite the clear advantages of the conditions of 3-D cultures, two-dimensional cultures still constitute the majority of the toxicity studies. This is mainly due to relatively low costs of microsystems production and possibility of using simpler detection methods [13].

Studies using the microsystems may be useful in observing and analyzing the effects of drug on both cancer and normal cells [47]. Wang and Kim [48] have designed a microdevice that allowed to examine the response of cells to five different toxins at two different concentrations simultaneously. In these studies they used three types of cells: BALB/3T3, HeLa, and bovine endothelial cells. Another article describes tests carried out only on Balb/3T3 cells, taking into account the continuous monitoring of changes in metabolism and cell adhesion measured by sensors [47].

Microdevice to test the spatial structure cells (cells with this kind of structure better mimic the in vivo conditions) was performed. The microsystem was connected with pneumatic micropumps and microvalves. They enable the flow of nutrients and cells with agarose, which allows cells to achieve the correct spatial configuration and greatly simplifies the mechanism of entering cells to the chip. The microsystem geometry allowed to perform tests a few times in one step [48].

An example of a system used to study the cytotoxicity of chemical compounds can be a device equipped with a wireless magnetoelastic ribbon-like sensor. This solution gives the possibility to monitor the growth of tumor cells in situ and to assess the cytotoxicity of anticancer drugs [49]. The sensor is composed with thick, free-standing (long and narrow) magnetoelastic film, combined with chemical or biochemical layer of receptors. To transfer the energy to magnetoelastic sensor a magnetic pulse is applied. This pulse causes vibrations of the sensor with characteristic resonance frequency, which varies linearly in response to a small increase in weight or change in viscosity. Cell adhesion to the sensor surface causes changes in the resonance frequency and amplitude. Growth curves show the growth and death of cells, allowing for rapid determination of cytotoxicity of chemical compounds. 5+fluorouracil and cisplatin were chosen as model compounds. The addition of these

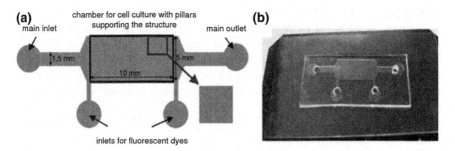

Fig. 14.7 The scheme (**a**) and the fabricated microchip (**b**). Reprinted with permission from Komen et al. [50]. Copyright 2008 Springer

cytostatic agents causes inhibition of cell proliferation. The study was conducted on breast cancer cell line MCF-7 [49]. To assess toxicity of chemical compounds also on the MCF-7 cells another microsystem made of PDMS was used. This device had a different geometry of channels (Fig. 14.7) than the previous one [50]. In this case, the cell viability was assessed after 24 h incubation with staurosporine.

Toxicity tests are also performed in hybrid microsystems built of glass and PDMS with different geometry, using many types of cell lines and various compounds tested [33, 51].

References

1. G.M. Whitesides, The origins and the future of microfluidics. Nature **442**, 368–373 (2006)
2. F. Gomez, *Biological Applications of Microfluidics* (Wiley, Hoboken, 2008)
3. P. Dittrich, A. Manz, Lab-on-a-chip: microfluidics in drug discovery. Nat. Rev. Drug Discov. **5**, 210–218 (2006)
4. L.Y. Yeo, H.C.H. Chang, P.P. Chan, J.R. Friend, Microfluidic devices for bioapplications. Small **7**, 12–48 (2011)
5. Z. Brzózka, Miniaturyzacja w analityce, Ofic. Wyd. Politechniki Warszawskiej, Warsaw (2005)
6. H. Andersson, A. van den Berg, Microfabrication and microfluidics for tissue engineering: state of art and future opportunities. Lab Chip **4**, 98–103 (2004)
7. J. Bożyk, Koniec testów na zwierzętach? Sprawy Nauki, 12 (2004)
8. S. Stokłosowa, Hodowla komórek i tkanek. PWN, Warsaw (2006)
9. D. Grabowski, Kultury komórkowe jako model eksperymentalny: Wybrane techniki frakcjonowania i hodowli komórek. Biologia Komórki, (2010)
10. D. Wlodkowic, J. Skommer, D. McGuinness, S. Faley, W. Kolch, Z. Darzynkiewicz, J.M. Cooperi, Chip-based dynamic real-time quantification of drug-induced cytotoxicity in human tumor cells. Anal. Chem. **81**, 6952–6959 (2009)
11. M. Shackleton, Normal stem cells and cancer stem cells: similar and different. Semin. Cancer Biol. **20**, 85–92 (2010)
12. B.S. Sekhon, S. Kamboj, Microfluidics technology for drug discovery and development. Int. J. Pharm. Tech. Res. **2**, 804–809 (2010)
13. M.H. Wu, S.B. Hung, G.B. Lee, Microfluidic cell culture systems for drug research. Lab Chip **10**, 939–956 (2010)
14. L. Kim, Y.C. Toh, J. Voldman, H. Yu, A practical guide to microfluidic perfusion culture of adherent mammalian cells. Lab Chip **7**, 681–694 (2007)
15. McJ Cooper, D. Duffy, J.R. Anderson, D.T. Chiu, H. Wu, O.J. Schueller, G.M. Whitesides, Fabrication of microfluidic systems in PDMS. Electrophor. **21**, 27–40 (2000)
16. Z. Brzózka, *Mikrobioanalityka Ofic. Wyd* (Politechniki Warszawskiej, Warsaw, 2009)
17. A. Prokop, Z. Prokop, D. Schaffer, E. Kozlov, J. Wikswo, D. Cliffel, F. Baudenbacher, NanoLiterBioReactor: long-term mammalian cell culture at nanofabricated scale. Biomed. Microdevices **6**, 325–339 (2004)
18. Y. Mi, Y. Chan, D. Trau, P. Huang, E. Chen, Micromolding of PDMS scaffolds and microwells for tissue culture and cell patterning: a new method of microfabrication by the self-assembled micropatterns of diblock copolymer micelles. Polymer **47**, 5124–5130 (2006)
19. E. Leclerc, B. David, L. Griscom, B. Lepioufle, T. Fujii, P. Layrolle, C. Legallaisa, Study of osteoblastic cells in a microfluidic environment. Biomaterials **27**, 586–595 (2006)
20. S.K. Sia, G.M. Whitesides, Microfluidic devices fabricated in poly(dimethylsiloxane) for biological studies. Electrophoresis **24**, 3563–3576 (2003)

21. V. Mokkapati, OM. Piciu, L. Zhang, J. Mollinger, J. Bastemeijer, A. Bossche, PDMS-glass bonded lab-on-a-chip device for single cell analysis, ASDAM 2008: Conference Proceedings of the 7th International Conference on Advanced Semiconductor Devices and Microsystems, pp. 211–214 (2008)
22. H.A. Stone, A.D. Stroock, A. Ajdari, Engineering flows in small devices. Microfluidics toward a lab-on-a-chip. Annu. Rev. Fluid Mech. **36**, 381–411 (2004)
23. H. Bang, W. Gu Lee, H. Yun, Ch. Chung, J.K. Chang, DCh. Han, A directly stackable microsystem onto the cultured cells for cytotoxicity tests. Microsyst. Technol. **14**, 719–724 (2008)
24. G.M. Walker, H.C. Zeringue, D. Beebe, Microenvironment design considerations for cellular scale studies. Lab Chip **4**, 91–97 (2004)
25. I. Barbulovic-Nad, A.R. Wheeler, Cell assays in microfluidics, in *Encyclopedia of Microfluidics and Nanofluidics*, ed. by D. Li (Springer, New York, 2007), pp. 209–216
26. L. Kim, M.D. Vahey, H.Y. Lee, J. Voldman, Microfluidic arrays for logarithmically perfused embryonic stem cell. Lab Chip **6**, 394–406 (2006)
27. K. Sato, K. Mawatari, K. Kitamori, Microchip-based cell analysis and clinical diagnosis system. Lab Chip **8**, 1992–1998 (2008)
28. D.N. Breslauer, J.P. Lee, L.P. Lee, Microfluidics-based systems biology. Mol. BioSyst. **2**, 97–112 (2006)
29. D. Beebe, G. Mensing, Physics and applications of microfluidics in biology. Annu. Rev. Biomed. Eng. **4**, 261–286 (2002)
30. B.G. Chung, J. Choo, Microfluidic gradient platforms for controlling cellular behavior. Lab Chip **31**, 3014–3027 (2010)
31. J. Clayton, Go with the microflow. Nat. Meth. **2**, 621–627 (2005)
32. H.A. Svahn, A. Berg, Single cells or large populations. Lab Chip **7**, 544–546 (2007)
33. P. Hung, P. Lee, P. Sabounchi, R. Lin, L.P. Lee, Continuous perfusion microfluidic cell culture array for high-throughput cell-based assays. Biotechnol. Bioeng. **89**, 1–8 (2005)
34. H. Lu, L.Y. Koo, W.M. Wang, D.A. Lauffenburger, L.G. Griffith, K.F. Jensen, Microfluidic shear devices for quantitative analysis of cell adhesion. Anal. Chem. **76**, 5257–5264 (2004)
35. G.M. Walker, N. Monteiro-Riviere, J. Rouse, A.T. O'Neill, A linear dilution microfluidic device for cytotoxicity assays. Lab Chip **7**, 226–232 (2007)
36. K. Lam, L. Zhang, K. Yamada, R. Lafrenie, Adhesion of epithelial cells to fibronectin or collagen I induces alterations in gene expression via a protein kinase C-dependent mechanism. J. Cell. Physiol. **189**, 79–90 (2001)
37. M.H. Wu, Simple poly(dimethylsiloxane) surface modification to control cell adhesion. Surf. Interf. Anal. **41**, 11–16 (2009)
38. T.H. Park, M.L. Shuler, Integration of cell culture and microfabrication technology. Biotechnol. Prog. **19**, 243–253 (2003)
39. ChS Chen, M. Mrksich, S. Huang, G.M. Whitesides, D.E. Ingber, Geometric control of cell life and death. Science **276**, 1425–1428 (1997)
40. D.P. Gaver III, S.M. Kute, A theoretical model study of the influence of fluid stresses on a cell adhering to a microchannel wall. Biophys. J. **75**, 721–733 (1998)
41. K. Kamei, S. Guo, Z. Tak, F. Yu, H. Takahashi, E. Gschweng, C. Suh, X. Wang, J. Tang, J. McLaughlin, O.N. Witte, K. Lee, H. Tseng, An integrated microfluidic culture device for quantitative analysis of human embryonic stem cells. Lab Chip **9**, 555–563 (2009)
42. C. Maercker, T. Rogge, H. Mathis, H. Ridinger, K. Bieback, Development of live cell chips to monitor cell differentiation processes. Eng. Life Sci. **8**, 33–39 (2008)
43. M.H. Wu, J.P. Urban, Z. Cui, Z. Cui, Development of PDMS microbioreactor with well-defined and homogenous culture environment for chondrocyte 3-D culture. Biomed. Microdevices **8**, 331–340 (2006)
44. A. Dodge, E. Brunet, S. Chen, J. Goulpeau, V. Labas, J. Vinh, P. Tabeling, PDMS-based microfluidics for proteomic analysis. Analyst **131**, 1122–1128 (2006)
45. P. Lee, P. Hung, V. Rao, L.P. Lee, Nanoliter scale microbioreactor array for quantitative cell biology. Biotechnol. Bioeng. **5**, 5–14 (2006)

46. Y. Torisawa, H. Shiku, T. Yasukawa, M. Nishizawa, T. Matsue, Multi-channel 3-D cell culture device integrated on a silicon chip for anticancer drug sensitivity test. Biomaterials **26**, 2165–2172 (2005)

47. L. Ceriotti, A.I. Kob, S. Drechsler, J. Ponti, E. Thedinga, P. Colpo, R. Ehret, F. Rossi, Online monitoring of BALB/3T3 metabolism and adhesion with multiparametric chip-based system. Anal. Biochem. **371**, 92–104 (2007)

48. Z. Wang, H. Kim, M. Marquez, T. Thorsen, High-density microfluidicarrays for cell cytotoxicity analysis. Lab Chip **276**, 1425–1428 (2007)

49. X. Xiaoa, M. Guoa, Q. Li, Q. Caia, S. Yaoa, C.A. Grimes, In-situ monitoring of breast cancer cell (MCF-7) growth and quantification of the cytotoxicity of anticancer drugs fluorouracil and cisplatin. Biosens. Bioelectron. **24**, 247–252 (2008)

50. J. Komen, F. Wolbers, H.R. Franke, H. Andersson, I. Vermes, A. Berg, Viability analysis and apoptosis induction of breast cancer cells in a microfluidic device: effect of cytostatic drugs. Biomed. Microdevices **10**, 727–737 (2008)

51. K. Ziółkowska, E. Jedrych, R. Kwapiszewski, J.M. Lopacinska, M. Skolimowski, M. Chudy, PDMS/glass microfluidic cell culture system for cytotoxicity tests and cells passage. Sens. Actuator B: Chem. **145**, 533–542 (2010)

Chapter 15
Application of Electromigration Techniques: Metabolomics– Determination of Potential Biomarkers Using Electromigration Techniques

Michał J. Markuszewski, Małgorzata Waszczuk-Jankowska,
Wiktoria Struck and Piotr Kośliński

Abstract Over the last decade the systems biology has developed into a new research platform that is based on a multidisciplinary approach to the analysis of complex biological systems (e.g., molecules, cells, organisms, or specific species). In addition, technological advances combined with the rapid development of the advanced analytical equipment allowed the study of genes, transcriptomes, proteins, and metabolites at the global level. The complement of knowledge about the observed relationship between genome, transcriptome, proteome, and both patho- and physiological processes is the analysis of metabolites—intermediate products of the genetic code. While the human body is composed of 30,000–50,000 genes, 150,000–300,000 transcriptomes, and about 1 million proteins, it has only 3.5,000–10,000 metabolites. Compared to the transcriptome and proteome, the number of metabolites in the body is relatively small, but the number of dependencies that affect the final metabolic profile is multi-dimensional. This makes metabolomics research a field of interest. In this chapter we describe metabolomics in the context of its role in cancer diagnosis. As an example we compiled various approach for determination of urinary nucleosides and pterins using electromigration techniques as well as advanced bioinformatics methods of data processing and analysis.

M. J. Markuszewski (✉) · M. Waszczuk-Jankowska · W. Struck
Faculty of Pharmacy, Medical University of Gdańsk, Gdańsk, Poland
e-mail: markusz@gumed.edu.pl

M. Waszczuk-Jankowska
e-mail: waska@gumed.edu.pl

W. Struck
e-mail: wiktoria.struck@gumed.edu.pl

P. Kośliński
Faculty of Pharmacy Ludwig Rydgier Collegium Medicum in Bydgoszcz, Nicolaus
Copernicus University in Toruń, Toruń, Poland
e-mail: piotr.koslinski@cm.umk.pl

B. Buszewski et al. (eds.), *Electromigration Techniques*, Springer Series
in Chemical Physics 105, DOI: 10.1007/978-3-642-35043-6_15,
© Springer-Verlag Berlin Heidelberg 2013

One of the characteristic properties of living beings is their ability to carry out metabolism, which amounted to simultaneously perform a wide variety of chemical reactions. In some reactions of low molecular weight organic compounds (amino acids, sugars, nucleotides, and lipids) are decomposed or modified to provide other compounds of low molecular weight needed for the cell. In subsequent reactions of the same, low molecular weight compounds are used for the synthesis of proteins, nucleic acids, and other macromolecules, which are suitable for living systems. In each cell, the reaction is controlled by a specialized protein—enzymes, which catalyze one of many possible types of reactions. Enzymatically, catalyzed reactions are usually combined in sequences so that the product of one reaction becomes the starting material (substrate) for the next. Sequences form successive reaction pathways for cell survival, growth, and reproduction. In every living cell, the information concerning the above dependencies is stored in the form of genes. One can suggest that genes are the instructions, by which organisms develop and operate. Human genome contains nearly a thousand times more DNA particles than the genome of Escherichia coli. Reading genomic sequences of individual organisms (including human) led to a flowering of research on genes and proteins such as genomics, transcriptomics, proteomics, and metabolomics. Metabolomics analyze metabolites which results from metabolic processes occurring in living organisms. The exact number of metabolites present in the human body is not known and probably varies from several thousand to about 20,000. Comparatively, the estimated number of genes are 30,000 and proteins—even one to two million. Metabolomics investigate the metabolic profiles in biological samples such as urine, blood, and saliva. The term metabolome is understood as the set of all metabolites present in a cell which reflects its physiological state at the time of the study. In the human body there are many different types of cells, each with potentially different levels of metabolomes. A number of genes and proteins determine the sequence of events that happens in the cell, as well as catalyze nearly all the important chemical reactions that determine the course of chemical transformations in biological systems. The level of metabolites is usually dependent on the functional state of the cell. In fact, signaling pathways in the cell or energy metabolism and communication between cells, all these processes are regulated by metabolites. Genomics and proteomics may inform what happens in the body, while metabolomics primarily provides information about what is actually happening in the body. By rapid technological development in bioanalitics and in consequence of metabolomic advancement of research, it is possible to obtain answers to key questions that could not be fully reflected in the other "-omics". Obtaining a picture of all processes in the cell requires a holistic approach to the analysis of phenomena observed in living organisms. In the longer term, researchers are looking at metabolomics that complement the existing gaps in biological systems, research paradigm that focuses on all the interconnected molecular pathways in the cell, and thus throughout the body. The objectives of clinical research include a search for

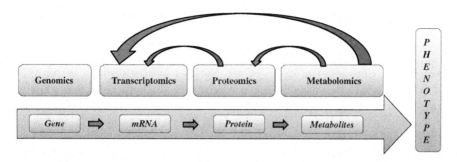

Fig. 15.1 Schematic representation of the relationship between the main directions of research in systems biology

molecular biomarkers or indicators of disease states. The concept of using the various metabolites as biomarkers of diseases is known and used for many years. Elevated blood glucose levels is an example of the indicator of a metabolic disease that is diabetes. Similarly, the measurement of the concentration of individual fractions of cholesterol and triglycerides in the blood can provide information about the risk of cardiovascular disease as well as measurement of serum creatinine level used as an indicator of kidney functional status.

Metabolomics, in contrast to the traditional approach based on individual markers, offers search and identification of many biomarker compounds simultaneously from whole profiles of metabolites. The term metabolomics complements genomics and proteomics (Fig. 15.1). Metabolomics can be regarded as an integral part of metabonomics. Metabonomics deals with the quantitative determination of the analyzed concentrations of metabolites and tracking their changes in a dynamic state with the passage of time. It is not inconceivable that metabolomics (metabonomics) will find clinical application earlier than genomics or proteomics. This may be due to economic factors: generation of metabolic profiles is cheap (assuming that the cost of purchased equipment, in particular NMR, are subtracted), and fast. In DNA microarray analysis, the technology used in genomics, the cost of the study is a much more expensive which automatically limits its wide availability for clinical use. Analysis of proteins is time-consuming and often difficult because of the much greater molecular weight and complexity of the chemical structure (functional elements). In addition, most of the functions of genes and proteins remain unknown, while the metabolites can often be attributed to specific tissues and metabolic pathways in a particular disease, which makes it relatively easy to predict their functions. The decisive advantage of metabolomics is that the most often used biological material is urine which puts it in a number of non-invasive methods, allowing for the repeated, multiple sampling over time, without significant complications for the patient or the person examined.

15.1 Determination of Nucleosides in Biological Samples as Markers of Carcinogenesis

According to the estimate made by the National Cancer Institute in the U.S., mortality from cancer in 2007 in the world was 7.6 million people, which is equivalent to about 20,000 deaths a day. Still the main factor responsible for increased mortality is late detection of cancer and, consequently, ineffective treatment. This is due to, among other things, the lack of an effective marker compounds that could be used to diagnose the disease early, often clinically difficult to diagnose, stage of disease. In addition to prostate specific antigen (PSA), which is characterized by a sensitivity of 70 % and a specificity of 59–97 %, and breast cancer antigen (CA 15-3), sensitivity 63 %, specificity 80–88 %, still lacks the specific compounds diagnosing disease at an early stage of its development, i.e., before it will be visible by macroscopic changes at the cellular level [1–3]. In addition, these markers are used only to predict specific types of cancer, and screening tests are of little use. Based on literature data it is known that a group of potential biomarkers for predicting cancer are nucleosides. This group includes both the basic nucleosides such as adenosine, guanosine, cytidine, uridine, and their modified derivatives: methylated (e.g., 5-methyluridine), acetylated (e.g., N4-acetylcitidine), reduced (dihydrouridine) and others such as: 8-hydroxydeoxyguanozine. A special hope for the diagnostic value is associated with modified nucleoside derivatives. Elevated levels of these compounds in urine are associated with the process of increased turnover of ribonucleic acid molecules which occurs in pathological states of the body, such as inflammation, cancer, and the acquired immune disorders (AIDS). Usually, in the physiologically normal conditions nucleosides are metabolized into β-alanine and uric acid or are re-enabled for the biosynthesis of nucleotides and RNA triphosphates. However, in pathological conditions modified nucleoside derivatives are not included in the cycle of re-synthesis of RNA, and are excreted unchanged in urine. There is a correlation between the concentration levels of modified nucleosides and the pathological condition of the body [4, 5].

The interest in nucleosides as potential markers of carcinogenesis is due to a fact that in cancer, besides the increased turnover of RNA molecules there is also hyperactivity of tRNA methyltransferases found, which may also affect the levels of nucleosides in urine [6]. At the same time the concentration of modified nucleosides in urine is influenced by factors that cause damage to cellular RNA, or cause cell death. In the metabolomic studies it is therefore necessary to take into account pathological states of the organism, the severity of the disease, and the fact of exposing the patient to side effects of chemo- or radiotherapy. Also important factors are the impact of diet, exposure to stress, alcoholism, and/or smoking. Thus, the various processes that determine the increased excretion of modified nucleosides in the urine causes difficulties in the interpretation of designated profiles. By using advanced bioinformatic data processing methods it is possible to extract relevant information from the collected raw data and, subsequently, identify the potential biomarkers of disease entities.

On the basis of changes in concentrations of nucleosides in the urine of healthy persons and patients with cancer, increased urinary excretion of nucleosides to urine was observed in diseases such as leukemia [7], lymphoma [8], breast cancer [9–11], colon cancer [12, 13], lung cancer [14], kidney and bladder cancers [15–18], and liver cancer [19].

Significant differences in concentration of the modified nucleosides in cancer diseases, confirmed by many publications, together with the physicochemical properties of these metabolites, encourages to mark them in biological fluids with the help of both High Performance Liquid Chromatography (HPLC) [7, 10, 12, 13] and electromigration techniques (i.e., Capillary Electrophoresis, CE) [6, 11, 15–18, 20–29] coupled with various types of detection methods.

Capillary electrophoresis technique has an evident advantage over high performance liquid chromatography as to the volume of the dosed sample (nL) and small amount of the produced organic waste. Moreover, it is characterized by better resolution and a generally shorter time of analysis. Bare fused silica capillary, used in this technique as a separation column, is cheaper and less problematic in storing. Besides, if properly conditioned, it retains the repeatability of analyses. This technique is also suitable for vast series of analyses of urine samples and could be effectively used in the clinical laboratory. Low cost of analyses together with a short time of analysis as well as a high resolution make capillary electrophoresis technique significant in the analysis of the series of both exogenous and endogenous analytes in biological fluids.

So far as electromigration techniques are concerned, the following methods are used in determination of nucleosides: micellar electrokinetic chromatography (MECK), capillary zone electrophoresis (CZE), and capillary electrochromatography (CEC).

15.2 Micellar Electrokinetic Chromatography

Micellar electrokinetic chromatography is widely applied in determination of nucleosides in biological samples. In this method a background electrolyte (BGE) consists of a mixture of borate and phosphate buffers with the addition of sodium dodecyl sulfate (SDS) in the pH range between 6.7 and 6.9. Sodium tetraborate is a complexing factor in a background electrolyte, which creates complexes with hydroxyl groups of nucleosides. Thanks to this, it changes their charge and influences their electrophoretic properties. In turn, SDS, by creating micelles, enables nucleosides' separation which reveal affinity with them. Table 15.1 summarizes the electrophoretic parameters currently available and applied in the literature concerning determination of nucleosides in biological samples. Modifications are generally related to differences in the proportions of specific substrates of BGE. For example, Liebich et al. [23, 30] applied 25 mmol/L sodium

Table 15.1 Examples of modifications of electrophoretic methods applied for determination of non-modified and modified nucleosides in urine

Number of analytes	Type of extraction	Background electrolyte (BGE)	Capillary	Detection method	Total analysis time	Ref.
13	SPE	25 mmol/L borate/50 mmol/L phosphate/300 mmol/L SDS; pH 6.7	Fused silica capillary 50 cm × 50 μm	UV–VIS $\lambda = 254$ nm	45 min	[5]
16	SPE	25 mmol/L borate/42.5 mmol/L phosphate/300 mmol/L SDS; pH 6.7	Fused silica capillary 50 cm × 50 μm	UV–VIS $\lambda = 210$, 260 nm	30 min	[20]
11	SPE	1 M acetic acid/methanol/ethanol (1:1:1) (v/v/v)	Fused silica capillary 60 cm × 50 μm	ESI/MS	18 min	[16]
12	SPE	100 mmol/L borate/72.5 mmol/L phosphate/160 mmol/L SDS; pH 6.7	Fused silica capillary 70 cm × 50 μm	UV–VIS $\lambda = 254$ nm	25 min	[17, 18]
13	SPE	25 mmol/L borate/50 mmol/L phosphate/300 mmol/L SDS; pH 6.9	Fused silica capillary 50 cm × 50 μm	UV–VIS $\lambda = 254$ nm	45 min	[11]
15	SPE	25 mmol/L borate/50 mmol/L phosphate/200 mmol/L SDS; pH 6.7	Fused silica capillary 50 cm × 50 μm and 56.5 cm × 50 μm	UV–VIS $\lambda = 210$, 260 nm	45 min	[30]
15	SPE	25 mmol/L borate/42.5 mmol/L phosphate/200 mmol/L SDS; pH 6.7	Fused silica capillary 57 cm × 50 μm	UV–VIS $\lambda = 190$–390 nm	30 min	[21]
14	SPE	25 mmol/L borate/50 mmol/L phosphate/300 mmol/L SDS; pH 6.7	Fused silica capillary 50 cm × 50 μm	UV–VIS $\lambda = 260$ nm and MALDI-TOF–MS	40 min	[23]

tetraborate, 50 mmol/L phosphate, and 300 mmol/L SDS at pH 6.7 as BGE. In turn, in Zheng et al. [5, 11] modification of BGE consisted of changing pH from 6.7 to 6.9, which did not have any significant impact on the results of analyses. Kim et al. [20] and La et al. [21] applied buffer that consisted of 25 mmol/L sodium tetraborate, 42.5 mmol/L phosphate and 200 mmol/L sodium dodecyl sulfate at pH 6.7. Also, Szymańska et al. [17, 18, 26] used 100 mmol/L sodium tetraborate, 72.5 mmol/L phosphate, and 160 mmol/L SDS at pH 6.7 to separate nucleosides in urine. An entirely different background electrolyte was presented in the work of Wang et al. [16]. The authors applied a mixture of 150 mmol/L acetic acid with 15 % methanol and 15 % ethanol. Regardless of the applied background electrolyte, the proposed determination methods enabled a separation from 13 to 16 nucleosides in the urine samples of both healthy volunteers and/or cancer patients. Thus, the applicability of the electromigration techniques in metabolomic research (metabolomic profiling) was confirmed.

15.2.1 Selected Methods of Detection

In spite of the wide applicability and many advantages, electromigration techniques are characterized by lower sensitivity in comparison with HPLC. This refers mainly to the results of the analyses obtained using spectrophotometric detection. In most reported cases, the analyses were conducted with UV detection, where the wavelength was set at 254 nm [6, 11, 17, 18, 20–22, 26, 27, 30]. Also, when applying spectrophotometric detection, the reference substances must be used in order to confirm the identity of the nucleosides determined in biological samples. Determinations with the help of mass spectrometry are conducted on the 10^{-9}–10^{-12} mol/L level of concentrations. Furthermore, thanks to fragmentation of ions it is possible to identify the analytes based on the relation of their mass to their charge. In this part of the analysis proper databases are of utmost help. Thanks to intensive development of the analytical tools the nucleosides are gradually more often determined through the use of the combined techniques of CE-MS or CE-MS/MS.

Wang et al. [16] proposed a CE-MS technique to determine nucleosides in urine, wherein the total time of analysis was 18 min. The accuracy of the method calculated as a recovery was equal to 93.5–118.2 %. The limit of detection (LOD) of the applied method was in the range 0.0086–3.82 μmol/L for individual nucleosides. In turn, when applying spectrophotometric detection, LOD was 3.1–74.2 μmol/L [5].

Liebich et al. [23] performed analysis of nucleosides with the use of capillary electrophoresis hyphenated with mass spectrometry. They applied matrix-assisted laser desorption ionization (MALDI) together with the time-of-flight analyzer (TOF) [23]. The authors tested two matrices: DHB (2,5-dihydroxybenzoic acid) and CHCA (α-cyano-4-hydroxycinnamic acid), which absorbed laser energy ($\lambda = 337$ nm). In case of MALDI method, a way of preparing a sample before measurement has a great impact on the final results. The authors prepared the solutions of nucleosides in a volatile solvent, and then mixed it with a matrix solution that was added in excess. The newly prepared solution was subsequently put on a plate to make the solvent evaporate and to crystallize the mixture of analytes and matrix. Because the quality of the spectrum depends on the regular dispersion of the analytes in the matrix material, the procedure of mixing two solutions (analytes and matrix) is extremely important. Moreover, the measurements were performed in the post source decay mode (PSD), which resulted in obtaining a spectrum of nucleosides wherein the sugar moiety and purine/pirimidine fragments were separated from each other. The determinations were characterized by the limit of detection between 100 femtomoles and 10 pmol with the application of CHCA matrix. A very high sensitivity is an unquestionable advantage of this technique, in spite of a basic disadvantage—namely, a time-consuming sample preparation procedure.

15.3 Capillary Electrochromatography

In a techinque of capillary electrochromatography, that is, a hybrid of chromatographic and electrophoretic approach, the analyses are performed with the use of a column of length from 25 to 50 cm and diameter 50–150 µm under high voltage. Thanks to this technique Helboe et al. developed a method for determination of 6 nucleosides [27]. A background electrolyte consisted of a mixture of 5 mmol/L acetic acid with 3 mmol/L triethylamine at pH 5.0 and acetonitrile 92:8 (v/v). Because of a small number of the determined nucleosides (n=6), the method does not meet the expectations of complex metabolomic analyses focused on determination of the largest possible amounts of metabolites.

However, the method proposed by Xie et al. [15] is worth noticing. The authors applied a pressurized capillary electrochromatography (pCEC) for determination of nucleosides. The method, based on applying pressure together with a proper voltage, is characterized by high resolution, sensitivity, and efficacy. The authors performed determination with the use of C18 column and mobile phase that consisted of acetonitrile and water with the addition of trifluoroacetic acid. The urine samples were collected from rats submitted to a highly caloric diet as well as from the control group. As a result, a significant differences in the metabolic profiles were observed between these two groups, which confirms the possibility of applying this method in metabolomics.

15.4 Bioinformatic Method of Analysis of the Electrophoretic Data

Electrophoretic analytical data comprise only a part information about processes in living organism achieved in metabolomics. This is because besides metabolites (e.g., nucleosides) concentration levels, additional factors, such as clinical data (medical history, organ dysfunction), physiological state (age, sex, BMI), and environment (work, place of residence) have also significant impact on data interpretation. In order to obtain valuable data on metabolic profiles in metabolomics it is crucial to apply bioinformatic tools which take into account all available variables.

Zheng et al. [5] observed elevated levels of 11 from 13 nucleosides in patients with thyroid cancer in comparison with control group ($p > 0.001$) (Table 15.2). Further studies with advanced bioinformatics tools showed that pseudouridine, 1-methylinosine, N4-acetylcytydine, 1-methylguanosine, 2-methylguanosine are the most frequently elevated nucleoside levels. To confirm the differences between the studied groups Principal Component Analysis (PCA) method was applied. The authors used 13 nucleosides concentrations as a vector data. As a result of the calculations, a model allowing the distinction between thyroid cancer patients and control group with 72 % efficiency was obtained.

Obtained electrophoretic data (especially from biological sample) include a lot of "noise" resulting from analytic interferences. It is necessary to apply bioinformatic methods for data preparation (preprocessing) in order to remove random analytical variation without significant information lost. Szymańska et al. [17, 18, 26] applied respectively baseline correction, denoising and alignment of migration time obtained by established CE method. To compensate analyte migration times three methods were tested: COW (Correlation Optimized Warping), DTW (Dynamic Time Warping), PTW (Parametric Time Warping). COW provided best results in the shortest time. Analytical data were compared with raw data using PCA method. Classification and distinction between healthy and patients was better in the case of analytical results obtained after bioinformatics data preprocessing in comparison to raw analytical data.

In order to compare the obtained electrophoretic data from biological samples, classification methods like k-NN method (k-Nearest Neighbor) and p-PLS-DA method (Probabilistic Partial Least Squares Discriminant Analysis) were applied [31]. Classification accuracy was approximately 80 % which allows to establish the potential diagnostic value of nucleosides in cancer disease. A comprehensive approach to metabolomics data including analytical, clinical, and bioinformatic information offers new prospects for diagnostic and cancer prevention (Fig. 15.2).

Table 15.2 Mean urinary nucleoside concentrations (µmol nucleoside/mmol/L creatinine) in urine of healthy volunteers and cancer patients. n denotes a number of analyzed samples and in squared brackets are given reference numbers

Analyte	Non-cancer controls						Cancer patients				
	Zheng et al. [5]; n=70	Zheng et al. [11]; n=41	Liebich et al. [30]; n=24	Kim et al. [20]; n=10	La et al. [21]; n=12	Szymańska et al. [31]; n=96	Zheng et al. [5]; n=70	Zheng et al. [11]; n=41	Kim et al. [20]; n=10	La et al. [21]; n=12	Szymańska et al. [31]; n=160
Pseudouridine	17.73 (±4.42)	19.25 (±5.82)	25.32 (±10.32)	14.80 (±3.36)	17.15 (±4.91)	37.18 (±12.8)	44.4 (±4.42)	55.12 (±34.63)	30.47 (±27.24)	25 (±2.83)	60.12 (±47.44)
Uridine	0.34 (±0.16)	0.35 (±0.14)	4.25 (±1.105)	0.23 (±0.11)	0.19 (±0.11)	0.65 (±0.22)	0.77 (±0.58)	0.94 (±0.64)	0.38 (±0.21)	0.79 (±0.62)	1.17 (±0.98)
5-Methyluridine	–	–	–	0.14 (±0.04)	0.17 (±0.11)	0.74 (±0.26)	–	–	0.39 (±0.46)	0.33 (±0.10)	1.23 (±1.08)
Cytidine	0.31 (±0.19)	0.33 (±0.21)	0.07 (±0.09)	0.1 (±0.05)	0.099 (±0.073)	0.23 (±0.16)	0.79 (±0.89)	0.98 (±0.85)	0.12 (±0.09)	0.48 (±0.27)	0.37 (±0.81)
N4-Acetylcytidine	0.49 (±0.14)	0.53 (±0.18)	0.6 (±0.384)	0.23 (±0.16)	0.51 (±0.33)	0.84 (±0.33)	1.31 (±0.93)	1.55 (±0.89)	0.7 (±0.85)	0.57 (±0.24)	1.27 (±0.98)
Inosine	0.37 (±0.19)	0.35 (±0.19)	0.14 (±0.1)	0.09 (±0.07)	0.086 (±0.12)	0.31 (±0.16)	0.70 (±0.67)	0.82 (±0.67)	0.51 (±0.38)	1.15 (±1.12)	0.59 (±0.51)
Guanosine	0.12 (±0.07)	0.12 (±0.08)	0.01 (±0.021)	–	–	1.06 (±0.46)	0.3 (±0.36)	0.5 (±0.48)	–	–	1.49 (±1.21)
N2-N2-Dimethylguanosine	–	–	–	1.74 (±0.6)	0.38 (±0.26)	2.54 (±0.98)	–	–	3.19 (±2.88)	1.11 (±0.33)	4.10 (±3.28)
Xanthosine	0.82 (±0.39)	0.86 (±0.38)	0.45 (±0.26)	0.62 (±0.25)	0.83 (±0.57)	1.35 (±0.8)	1.41 (±1.12)	1.68 (±1.13)	0.83 (±0.63)	1.63 (±0.89)	2.13 (±1.56)
Adenosine	0.41 (±0.12)	0.44 (±0.17)	0.18 (±0.17)	0.21 (±0.08)	0.22 (±0.14)	0.46 (±0.21)	1.02 (±0.77)	1.19 (±0.83)	0.36 (±0.29)	0.52 (±0.18)	0.57 (±0.44)

(continued)

Table 15.2 (continued)

Analyte	Non-cancer controls					Cancer patients				
1-Methyladenosine	–	–	–	–	2.14 (±0.78)	–	–	–	–	3.37 (±4.23)
6-Methyladenosine	0.09 (±0.11)	0.01 (±0.023)	0.05 (±0.05)	0.093 (±0.1)	0.41 (±0.24)	0.19 (±0.40)	0.22 (±0.43)	0.21 (±0.28)	0.57 (±1.21)	0.68 (±0.84)
1-Methylinosine	1.02 (±0.32)	1.27 (±0.457)	–	–	–	2.45 (±1.80)	2.95 (±1.79)	–	–	–
1-Methylguanosine	0.86 (±0.22)	0.82 (±0.298)	0.23 (±0.16)	0.48 (±0.31)	–	1.97 (±1.54)	2.32 (±1.41)	0.78 (±0.65)	1.33 (±0.33)	–
2-Methylguanosine	0.49 (±0.19)	0.39 (±0.197)	–	–	–	1.34 (±0.97)	1.47 (±0.83)	–	–	–

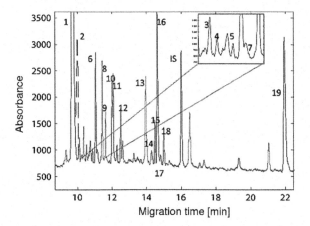

Fig. 15.2 Electropherogram of urinary nucleoside profile from urine extract. Capillary electrophoretic conditions are: 100 mmol/L borate, 72.5 mmol/L phosphate, 160 mmol/L SDS, pH 6.7; 25 kV voltage, 30 °C temperature during analysis; injection 5 s × 0.5 psi; capillary: untreated fused-silica, 70 cm length to detector, 50 μm I.D. Peaks: *1* pseudouridine (pU), *2* dihydrouridine, *3* uridine (U), *4* cytidine (C), *5* 5-methyluridine (5 mU), *6* unidentified metabolite, *7* inosine (I), *8* unidentified metabolite, *9* N4-acetylcytidine (acC), *10* guanosine (G), *11* unidentified metabolite, *12* adenosine (A), *13* unidentified metabolite, *14* unidentified metabolite, *15* unidentified metabolite, *16* N2,N2-dimethylguanosine (dmG), *17* 6-methylade-nosine (6 mA), *18* xanthosine (X), *19* 1-methyladenosine (1 mA). Reprinted with permission from Szymańska et al. (2010) [31] Copyright 2010 Elsevier

15.5 Determination of Pteridines Compounds in Cancer Diagnosis

Pteridines are important cofactors in the process of cell metabolism and their levels have significant importance in clinical diagnosis. They are excreted by humans in urine. Levels of pteridines have been elevated significantly when the cellular immune system is activated by certain diseases such as cancer [32]. In vivo studies have shown that neopterin (NP), one of the pteridines, is a useful marker for monitoring the activation of cellular immunity in patients [33]. What is crucial for the diagnostic function is the fact that normal somatic cells metabolize pterins differently from cancer cells. Each type of tumor shows its own pattern in changes of pteridine concentrations, since different pteridine derivatives may play various roles in different tumor-related disease [32]. Xanthopterin, isoxanthop-terin, biopterin, and neopterin urinary concentrations, seem to vary independently from each other [34].

Neopterin was first isolated from human urine in the 1960s. In 1979, Wachter identified NP as one of the molecules responsible for the fluorescence of urine in cancer patients [35]. Neopterin is excreted from the organism in an unchanged form mainly through the kidneys. Increased concentration of NP is found in various types of malignancies, correlating positively with tumor stage and poor prognosis. For example in the studies performed by Weiss et al. [36] one could observe that NP is strongly associated with the prognosis of adenocarcinoma of the colon. The problem of breast cancer and the possible use of neopterin in the diagnosis was studied by Yuksel et al. [37]. They observed the relationship between the levels of neopterin among patients with benign and malignant breast cancer disease and the relation with the stage of the malignant process. They noticed that there is no significant difference between the control group and between patients with benign disorders. However, NP levels in patients with malignant breast cancer were significantly higher than patients with benign disorders.

Cha et al. [38] used CE coupled with UV spectrophotometric detection in the determination of pteridines standards. This application due to the insufficient sensitivity of the UV detector and complexity of the urine matrix, had no diagnostic value. Han et al. [39] also employed CE but with laser-induced fluorescence (LIF) detection. LIF detection is recognized to be an extremely sensitive detection method sufficient for capillary electrophoresis because it can detect 10^{-14} mol/L of samples, which is about 1,000 times more sensitive than the typical spectrophotometric UV detector. They observed the lowest detection limit for pteridine compounds when phosphate buffer was used. Authors compared this value with 0.1 mol/L Tris/0.1 mol/L borate/2 mmol/L EDTA.

Although this work has shown particular promise in its feasibility for clinical laboratories using CE, only a limited number of parameters were investigated. In addition, because of the complexity of urine samples, the method encountered significant challenges in avoiding interferences from other compounds during the separation and quantification of pteridines in real urine samples. Gibson et al. [40] attempted to optimize CE method for determination of pterins in urine sample. The CE instrument setup, method, and sample preparation were investigated. The optimized electrophoretic conditions were: running buffer 0.1 mol/L Tris/0.1 mol/L borate/2 mmol/L EDTA, pH 9.63; capillary: 50 μm × 70 cm, running voltage: 371 V/cm; LIF detection at 325/445 nm (ex/em).

The CE-LIF method was proven to be fast, simple, and a more reliable method as compared to other electromigration-based methods. In fact this is also a sensitive analytical detection suitable for metabolomic analyses.

References

1. U. Manne, R.G. Srivastara, S. Srivastara, Recent advances in biomarkers for cancer diagnosis and treatment. Drug Discov. Today **14**, 965–976 (2005)
2. O. Stoss, T. Henkel, Biomedical marker molecules for cancer—current status and perspectives. Drug Discov. Today **3**, 228–237 (2004)
3. J.A.M. Gimenez, G.T. Saez, R.T. Seisdedos, On the function of modified nucleosides in the RNA world. J. Theo. Biol. **194**, 485–490 (1998)
4. H.M. Liebich, S. Müller-Hagedorn, M. Bacher, H.G. Scheel-Walter, X. Lu, A. Frickenschmidt, B. Kammerer, K.R. Kim, H. Gérard, Age-dependence of urinary normal

and modified nucleosides in childhood as determined by reversed-phase high-performance liquid chromatography. J. Chromatogr. B **814**, 275–283 (2005)

5. Y.F. Zheng, G.W. Xu, D.Y. Liu, J.H. Xiong, P.D. Zhang, C. Zhang, Q. Yang, S. Lv, Study of urinary nucleosides as biological marker in cancer patients analyzed by micellar electrokinetic capillary chromatography. Electrophoresis **23**, 4104–4109 (2002)

6. K.H. Schram, Urinary nucleosides. Mass Spectrom. Rev. **17**, 131–251 (1998)

7. C.G. Zambonin, A. Aresta, F. Palmisano, G. Specchia, V. Liso, Liquid chromatographic determination of urinary 5-methyl-2′-deoxycytidine and pseudouridine as potential biological markers for leukaemia. J. Pharm. Biomed. Anal. **21**, 1045–1051 (1999)

8. T. Rasmuson, G.R. Björk, Urinary excretion of pseudouridine and prognosis of patients with malignant lymphoma. Acta Oncol. **34**, 61–67 (1995)

9. A.J. Sasco, F. Rey, C. Reynaud, J.Y. Bobin, M. Clavel, A. Niveleau, Breast cancer prognostic significance of some modified urinary nucleosides. Cancer Lett. **108**, 157–162 (1996)

10. Y. Zheng, G. Xu, J. Yang, X. Zhao, T. Pang, H. Kong, Determination of urinary nucleosides by direct injection and coupled-column high-performance liquid chromatography. J. Chromatogr. B **819**, 85–90 (2005)

11. Y.F. Zheng, H.W. Kong, J.H. Xiong, S. Lv, G.W. Xu, Clinical significance and prognostic value of urinary nucleosides in breast cancer patients. Clin. Biochem. **38**, 24–30 (2005)

12. B. Feng, M.H. Zheng, Y.F. Zheng, A.G. Lu, J.W. Li, M.L. Wang, J.J. Ma, G.W. Xu, B.Y. Liu, Z.H. Zhu, Normal and modified urinary nucleosides represent novel biomarkers for colorectal cancer diagnosis and surgery monitoring. J. Gastroenterol. Hepatol. **20**, 1913–1919 (2005)

13. Y.F. Zheng, J. Yang, X.J. Zhao, B. Feng, H.W. Kong, Y.J. Chen, S. Lv, M.H. Zheng, G.W. Xu, Urinary nucleosides as biological markers for patients with colorectal cancer. World J. Gastroenterol. **11**, 3871–3876 (2005)

14. T.P. Waalkes, M.D. Abeloff, D.S. Ettinger, Modified ribonucleosides as biological markers for patients with small cell carcinoma of the lung. Eur. J. Cancer Clin. Oncol. **18**, 1267–1274 (1982)

15. G. Xie, M. Su, P. Li, X. Gu, C. Yan, Y. Qiu, H. Li, W. Jia, Analysis of urinary metabolites for metabolomic study by pressurized CEC. Electrophoresis **28**, 4459–4468 (2007)

16. S. Wang, X. Zhao, Y. Mao, Y. Cheng, Novel approach for developing urinary nucleosides profile by capillary electrophoresis-mass spectrometry. J. Chromatogr. A **1147**, 254–260 (2007)

17. E. Szymańska, M.J. Markuszewski, X. Capron, A.M. van Nederkassel, Y. Vander Heyden, M. Markuszewski, K. Krajka, R. Kaliszan, Evaluation of different warping methods for the analysis of CE profiles of urinary nucleosides. J. Pharm. Biomed. Anal. **43**, 413–420 (2007)

18. E. Szymańska, M.J. Markuszewski, K. Bodzioch, R. Kaliszan, Development and validation of urinary nucleosides and creatinine assay by capillary electrophoresis with solid phase extraction. J. Pharm. Biomed. Anal. **44**, 1118–1126 (2007)

19. J. Yang, G. Xu, Y. Zheng, H. Kong, T. Pang, S. Lv, Q. Yang, Diagnosis of liver cancer using HPLC-based metabonomics avoiding false-positive result from hepatitis and hepatocirrhosis diseases. J. Chromatogr. B **813**, 59–65 (2004)

20. K.R. Kim, S. La, A. Kim, J.H. Kim, H.M. Liebich, Capillary electrophoretic profiling and pattern recognition analysis of urinary nucleosides from uterine myoma and cervical cancer patients. J. Chromatogr. B **754**, 97–106 (2001)

21. S. La, J.H. Cho, J.H. Kim, K.R. Kim, Capillary electrophoretic profiling and pattern recognition analysis of urinary nucleosides from thyroid cancer patients. Anal. Chim. Acta **486**, 171–182 (2003)

22. Y. Ma, G. Liu, M. Du, I. Stayton, Recent developments in the determination of urinary cancer biomarkers by capillary electrophoresis. Electrophoresis **25**, 1473–1484 (2004)

23. H.M. Liebich, S. Müller-Hagedorn, F. Klaus, K. Meziane, K.R. Kim, A. Frickenschmidt, B. Kammerer, Chromatographic, capillary electrophoretic and matrix-assisted laser desorption ionization time-of-flight mass spectrometry analysis of urinary modified nucleosides as tumor markers. J. Chromatogr. A **1071**, 271–275 (2005)

24. P. Iadarola, G. Cette, M. Luisetti, L. Annovazzi, B. Casado, J. Baraniuk, C. Zanone, S. Viglio, Micellar electrokinetic chromatographic and capillary zone electrophoretic methods for screening urinary biomarkers of human disorders: a critical review of the state-of-the-art. Electrophoresis **26**, 752–766 (2005)

25. Y.X. Zhang, Artificial neural networks based on principal component analysis input selection for clinical pattern recognition analysis. Talanta **73**, 68–75 (2007)

26. E. Szymańska, M.J. Markuszewski, X. Capron, A.M. van Nederkassel, Y. Vander Heyden, M. Markuszewski, K. Krajka, R. Kaliszan, Evaluation of different warping methods for the analysis of CE profiles of urinary nucleosides. Electrophoresis **28**, 2861–2873 (2007)

27. T. Helboe, S.H. Hansen, Separation of nucleosides using capillary electrochromatography. J. Chromatogr. A **836**, 315–324 (1999)

28. Y. Mao, X. Zhao, S. Wang, Y. Cheng, Urinary nucleosides based potential biomarker selection by support vector machine for bladder cancer recognition. Anal. Chim. Acta **598**, 34–40 (2007)

29. A.S. Cohen, S. Terabe, J.A. Smith, B.L. Karger, High-performance capillary electrophoretic separation of bases, nucleosides, and oligonucleotides: retention manipulation via micellar solutions and metal additives. Anal. Chem. **59**, 1021–1027 (1987)

30. H.M. Liebich, R. Lehmann, G. Xu, H.G. Wahl, H.U. Häring, Application of capillary electrophoresis in clinical chemistry: the clinical value of urinary modified nucleosides. J. Chromatogr. B **745**, 189–196 (2000)

31. E. Szymańska, M.J. Markuszewski, M. Markuszewski, R. Kaliszan, Altered levels of nucleoside metabolite profiles in urogenital tract cancer measured by capillary electrophoresis. J. Pharm. Biomed. Anal. **53**, 1305–1312 (2010)

32. Y. Ma, G. Liu, M. Du, I. Stayton, Recent developments in the determination of urinary cancer biomarkers by capillary electrophoresis. Electrophoresis **10**(11), 1473–1484 (2004)

33. A. Berdowska, K. Zwirska-Korczala, Neopterin measurement in clinical diagnosis. J. Clin. Pharm. Ther. **5**, 319–329 (2001)

34. J.L. Lord, ADe Peyster, P.J.E. Quintana, R.P. Metzger, Cytotoxicity of xanthopterin and isoxanthopterin in MCF-7 cells. Cancer Lett. **1**, 119–124 (2005)

35. B. Melichar, D. Solichová, R.S. Freedman, Neopterin as an indicator of immune activation and prognosis in patients with gynecological malignancies. Int J Gynecol Cancer **1**, 240–252 (2006)

36. G. Weiss, P. Kronberger, F. Conrad, E. Bodner, H. Wachter, G. Reibnegger, Neopterin and prognosis in patients with adenocarcinoma of the colon. Cancer Res. **2**, 260–265 (1993)

37. O. Yuksel, T.T. Sahin, G. Girgin, H. Sipahi, K. Dikmen, O. Samur, A. Barak, E. Tekin, T. Baydar, Neopterin, catalase and superoxide dismutase in females with benign and malignant breast tumors. Pteridines **4**, 132–138 (2007)

38. K.W. Cha, S.I. Park, Y.K. Lee, J.J. Yim, Capillary electrophoretic separation of pteridine compounds. Pteridines **4**, 210–213 (1993)

39. F. Han, B.H. Huynh, H. Shi, B. Lin, Y. Ma, Pteridine analysis in urine by capillary electrophoresis using laser-induced fluorescence detection. Anal. Chem. **7**, 1265–1269 (1999)

40. S.E. Gibbons, I. Stayton, Y. Ma, Optimization of urinary pteridine analysis conditions by CE-LIF for clinical use in early cancer detection. Electrophoresis **20**, 3591–3597 (2009)

Chapter 16
Applications of Electromigration Techniques: Electromigration Techniques in Detection of Microorganisms

Ewelina Dziubakiewicz and Bogusław Buszewski

Abstract The detection and identification of microbes is a challenge and an important aspect in many fields of our lives from medicine to bioterrorism defense. However, the analysis of such complex molecules brings a lot of questions mainly about their behavior. Bacteria are biocolloid, whose surface charge originates from the ionization of carboxyl, phosphate, or amino groups and the adsorption of ions from solution. Consequently, the charged cell wall groups determine the spontaneous formation of the electrical double layer. In this chapter application of electromigration techniques for microorganism's identification and separation are described. This approach represents the possibility to apply electromigration techniques in medical diagnosis, detection of food contamination, and sterility testing.

One of the examples of applying of electromigration techniques in the area of '-omics' is a possibility to apply CE in the determination of microorganisms for the purposes of medical analysis, environmental analysis, or food analysis.

Both structurally and chemically, a bacterial cell is dynamic and non-homogeneous. Differences in the construction of a cell wall allow to single out two different groups of bacteria—Gram-negative and Gram-positive (Fig. 16.1).

The cell wall of Gram-positive bacteria consists of a multi-layer murein into which are combined polysaccharides and teichoic acids. Additionally, there are also found molecules of lipoteichoic acids packed with their lipid part in a cytoplasmatic membrane. In Gram-negative bacteria, teichoic acids have not been

E. Dziubakiewicz · B. Buszewski (✉)
Faculty of Chemistry, Department of Environmental Chemistry,
Nicolaus Copernicus University, Toruń, Poland
e-mail: bbusz@chem.umk.pl

E. Dziubakiewicz
e-mail: ewelina.dziubakiewicz@gmail.com

B. Buszewski et al. (eds.), *Electromigration Techniques*, Springer Series
in Chemical Physics 105, DOI: 10.1007/978-3-642-35043-6_16,
© Springer-Verlag Berlin Heidelberg 2013

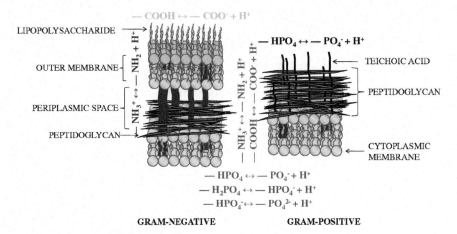

Fig. 16.1 Scheme of gram-positive and gram-negative bacteria cell wall structure, according to [3]

found, and their cell wall is surrounded by an external membrane whose integral part is lipopolysaccharide (LPS) [1, 2].

The surface of a bacterial cell has a charge coming from ion adsorption from a solution and proteolysis of ionic phosphate groups, carboxyl, amine groups located on its surface (Table 16.1). As a result, this charge determines a spontaneous formation of an electrical double layer (Fig. 16.2).

A charge on the surface, typical properties of membranes and cell walls, and self-mobility in an electrical field under special conditions, allow to observe the behavior of live bacterial cells using electromigration techniques [3].

A pioneer of electrophoretic separation of microorganisms was Hjertén [4]. In 1987 he proved that in the conditions of capillary electrophoresis under the influence of applied electrical field, both *Nicotiana virus* and *Lactobacillus casei* bacteria migrated together with the electroosmotic flow along silica capillary covered with methylcellulose whose task was to prevent their adhesion to the capillary walls.

In 1993, Ebersole and McCormick [5] partially separated five strains of bacteria using capillary zone electrophoresis (CZE): *Entercoccus faecalis, Streptococcus pyogenes, Streptococcus agalactiae, Streptococcus pneumoniae,* and *Staphylococcus*

Table 16.1 Groups present on the surface of a bacterial cell influencing a charge, according to [19]

Reaction	Molecule–groups	pK$_a$
—COOH $\leftrightarrow$ —COO$^-$ + H$^+$	Polysaccharide	2.8
	Protein peptidoglycan	4.0–5.0
—NH$_3^+$ $\leftrightarrow$ —NH$_2$ + H$^+$	Protein peptidoglycan	9.0–9.8
—HPO$_4$ $\leftrightarrow$ —PO$_4^-$ + H$^+$	Teichoic acids	2.1
—H$_2$PO$_4$ $\leftrightarrow$ —HPO$_4^-$ + H$^+$	Phospholipids	2.1
—HPO$_4^-$ $\leftrightarrow$ —PO$_4^{2-}$ + H$^+$	Phospholipids	7.2

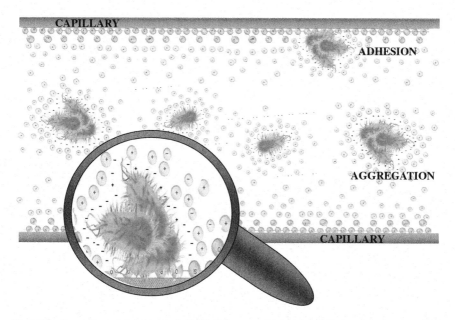

Fig. 16.2 Formation of the electrical double layer on a bacterial cell during the electrophoretic analysis

aureus. However, bacteria created agglomerates, bands obtained were wide, and the time of analysis was long, which reduced the efficiency of separation.

The available literature sources confirm a wide interest in electromigration techniques in microorganisms identification and separation (Table 16.2). The conducted research is aimed at shortening the analysis time and improvement of selectiveness of electrophoretic separation of bacteria by using a number of substances modifying an internal wall of capillary (Fig. 16.3) [6, 7], as well as additions to the buffer solution, e.g., poly(ethylene oxide)—PEO.

A breakthrough in the application of capillary electrophoresis is the identification and separation of pathogenic bacteria contained in urine (Fig. 16.4) [8, 9] or faeces [10], which in future may constitute a basis for CZE application as a quick, economical, and precise identification of the health condition of a patient.

Due to its high selectiveness and sensitiveness, CZE provides an opportunity to separate various bacteria strains in the scope of the same species (Fig. 16.5) [11, 12]. In addition, it allows to make a lot of determinations with a small consumption of a sample, in a short time and with low financial expenses. This gives an opportunity to introduce this technique in hospitals for routine screening examinations.

CZE was applied in the identification and quantitative determination of probiotics [13–16]. Probiotics are live microorganisms which, as active components of food, are necessary for a human organism. This made CE a tool for automatic analyses in the food and pharmaceutical industry.

Table 16.2 Examples of determination of microorganisms using electromigration techniques

Microorganisms	CE mode	Background electrolyte (BGE)	CE conditions	Detection method	Ref.
Pseudomonas species, Pseudomonas putida, Methanobrevibacter smithii, Rhodococcus erythropolis, Alcaligenes eutrophus	CZE	TBE (4.5 mmol/L Tris/ 4.5 mmol/L borate/ 0.1 mmol/L EDTA); pH 8.4	Fused-silica capillary 250 × 250, 144.5 cm × 250 μm, U = 25–30 kV	UV	[20]
Escherichia coli, Pseudomonas putida, Pseudomonas fluorescens Serratia rubidae, Enterobacter aerogenes, Micrococcus luteus, Saccharomyces cerevisiae	CZE CIEF	TBE (4.5 mmol/L Tris/ 4.5 mmol/L borate/ 0.1 mmol/L EDTA); pH 8.4; 0.00125 % PEO 20 mmol/L NaOH; 20 mmol/L H_3PO_4	Fused-silica capillary 27 cm × 100 μm, U = 10 kV methylcellulose-modified capillary U = 20 kV	UV UV	[21]
Escherichia coil, Paracoccus denitrificans, Pseudomonas fluorescens, Bifidobacterium longum, Acetobacter pasteurianus, Acetobacter pasteurianus, Serratia marcescens, Saccharomyces cerevisiae	CZE	10 mmol/L phosphate (pH 7.0; 7.8) with NaCl (I = 0.019–0.227 mol/L)	Fused-silica capillary 80 cm × 50 μm, U = 10 kV	UV	[22]
Escherichia coli, Staphylococcus saprophyticus	CZE	TBE (4.5 mmol/L Tris/ 4.5 mmol/L borate/ 0.1 mmol/L EDTA); pH 8.4; 0.00125 % PEO	Fused-silica capillary 27 cm × 100 μm, U = 10 kV	UV	[8]
Lactobacillus acidophilus, Bifidobacterium infantis, Saccharomyces cerevisiae	CZE	TBE (4.5 mmol/L Tris/ 4.5 mmol/L borate/ 0.1 mmol/L EDTA); pH 8.4; 0.00125 % PEO	Fused-silica capillary 30 cm × 100 μm, U = 15 kV,	LIF	[23]

(continued)

Table 16.2 (continued)

Microorganisms	CE mode	Background electrolyte (BGE)	CE conditions	Detection method	Ref.
Cellulomonas cartae KYM-7, *Agrobacterium tumefaciens* KYM-8	CZE CGE	20 mmol/L borate (pH 9.3) 10 mmol/L phosphate (pH 7.0); dextran	Fused-silica capillary 80.5 cm × 100 µm, U = 10 kV	UV	[24]
Bifidobacteria infantis, Lactobacillus acidophilus, Saccharomyces cerevisiae	CZE	TBE (0.5–9.0 mmol/L Tris/0.5–9.0 mmol/L borate/0.011 mmol/L L–0.2 mmol/L EDTA); pH 8.4; PAA, PAcA, PEO, PVP	Fused-silica capillary 58 cm × 100 µm, U = 15 kV	LIF	[13]
Salmonella enteritidis, Salmonella typhimurium	CZE	TBE (0.1 mol/L Tris/ 0.1 mol/L borate/ 2 mol/L EDTA); pH 8.4; 0.1 % sodium alginate and/or 10 % NaCl	Fused-silica capillary 31 cm × 75 µm, U = 10 kV	LIF	[25]
Escherichia coli, Proteus vulgaris, Bacillus cereus, Pseudomonas fluorescens	CZE	5 mmol/L phosphate Tris (pH 8.5)	Acrylamide-modified capillary 33.5 cm × 75 µm, U = 20 kV	UV	[6]
Aerobacter aerogenes, Pseudomonas fluorescens, Micrococcus lysodeikticus	CZE	TBE (4.5 mmol/L Tris/ 4.5 mmol/L borate/ 0.1 mmol/L EDTA); pH 8.4; 0.00125 % PEO	Fused-silica capillary 27 cm × 75 µm, U = 10 kV	UV	[26]

(continued)

Table 16.2 (continued)

Microorganisms	CE mode	Background electrolyte (BGE)	CE conditions	Detection method	Ref.
Bifidobacterium infantis	CZE	TBE (394 mmol/L Tris/56 mmol/L borate/1.3 mmol/L EDTA); pH 9.1; 0.05 % PEO 0.1–10 mmol/L phosphate (pH 7.0)	Fused-silica capillary 30 cm × 75 µm,	LIF	[14]
Escherichia coli, Salmonella enteriditis, Leuconostoc mesenteroides, Listeria monocitogenes, Yersinia enterocolitica, Enterococcus faecium, Lactobacillus plantarum, Staphylococcus aureus	CZE	25 mmol/L phosphate (pH 7.0), 25 µmol/L CaCl$_2$, 35 µmol/L myo-inositol hexaphosphatez	Fused-silica capillary 47 cm × 75 µm, U = 15 kV	UV	[27]
Streptococcus thermophilus, Lactobacillus delbrueckii subsp. bulgaricus, Saccharomyces cerevisiae	µ-czip CE	TBE (0.1 mol/L Tris/0.1 mol/L borate/2 mol/L EDTA); pH 8.4; 0.1 % sodium alginate and/or 10 % NaCl	µ-czip Type D (110 × 50 µm) Type T (110 × 50 µm)	LIF	[17]
Escherichia coil, Proteus vulgaris, Bacillus megaterium, Micrococcus species, Arthorobacter globiformis	CZE	TBE (4.5 mmol/L Tris/4.5 mmol/L borate/0.1 mmol/L EDTA); pH 8.53; 0.00125 % PEO	Trimethylchlorosilane-modified and divinylbenzen e-modified capillary 33.5 cm × 75 µm, U = −15 kV	UV	[7]
Escherichia coli, Staphylococcus aureus, Pseudomonas aeruginosa, Klebsiella pneumoniae	CZE	TBE (4.5 mmol/L Tris/4.5 mmol/L borate/0.1 mmol/L EDTA); pH 8.5; 0.00125 % PEO	Acrylamide-modified capillary 27 cm × 100 µm, U = −15 kV	UV, LIF	[10]

(continued)

Table 16.2 (continued)

Microorganisms	CE mode	Background electrolyte (BGE)	CE conditions	Detection method	Ref.
Escherichia coli, Candida albicans, Candida parapsilosis, Candida krusei, Candida glabrata, Candida tropicalis, Proteus vulgaris, Saccharomyces cerevisiae, Enterococcus faecalis, Staphylococcus epidermidis, Stenotrophomonas maltophilia	CIEF	20 mmol/L NaOH; 100 mmol/L H_3PO_4; ethanol; PB-PEG	Fused-silica capillary 32 cm × 100 μm, U = −20 kV	LIF	[28]
Saccharomyces cerevisiae, Escherichia coli, Candida albicans, Candida parapsilosis, Candida krusei, Staphylococcus aureus, Streptococcus agalactiae, Enterococcus faecalis, Staphylococcus epidermidis, Stenotrophomonas maltophilia	CIEF	20 mmol/L NaOH; 100 mmol/L H_3PO_4; PEG	Fused-silica capillary 27 cm × 100 μm, U = −20 kV	UV	[29]
Escherichia coli, Helicobacter pylori, Serratia marcescens	CZE	TBE (4.5 mmol/L Tris/ 4.5 mmol/L borate/ 0.1 mmol/L EDTA); pH 8.53; 5 mmol/L MES (pH 6.1); 0.00125 % PEO	Fused-silica capillary 33.5 cm × 50 μm, U = 20 kV	UV	[9]
Escherichia coli, Salmonella subterreanea, Listeria innocua, Brevibacterium tapei, Corynebacterium acetoacidophilum, Aerococcus viridans, Pseudomonas fluorescens, Escherichia blattae, Staphylococcus aureus	CZE	1 mmol/L T Tris/ 0.33 mmol/L citric acid (pH 7.0); CTAB	Fused-silica capillary 30 cm × 100 μm, U = −2 kV	UV, LIF	[30]
Escherichia coli, Staphylococcus epidermidis, Candida albicans, bakteriofag ΦX174	CIEF	60 mmol/L NaOH; 100 mmol/L H_3PO_4; ethanol; PEG	Fused-silica capillary 35 cm × 100 μm, U = −20 kV	UV, LIF	[31]

(continued)

Table 16.2 (continued)

Microorganisms	CE mode	Background electrolyte (BGE)	CE conditions	Detection method	Ref.
Brevibacterium taipei, Corynebacterium acetoacidophilum, Escherichia blattae, Bacillus cereus, Bacillus subtilis, Candida albicans, Rhodotorula, Bacillus megaterium	CZE	1 mmol/L Tris/ 0.33 mmol/L citric acid (pH 7.0); CTAB; sulphobetaine	Fused-silica capillary 30 cm × 100 μm, $U = -2$ kV	LIF	[32]
Trzy szczepy Staphylococcus aureus	CZE	5 mmol/L MES (pH 6.5)	Fused-silica capillary 33.5 cm × 75 μm, $U = 20$ kV	UV	[11]
Lactobacillus delbrueckii subsp. bulgaricus, Streptococcus thermophilus	CZE	TBE (4.5 mmol/L Tris/ 4.5 mmol/L borate/ 0.1 mmol/L EDTA); pH 8.5; 0–0.05 % PEO	Fused-silica capillary 56 cm × 100 μm, $U = 20$ kV	UV	[15]
Pseudomonas corrugate, siedem szczepów Pseudomonas syringae	CIEF CZE	40 mmol/L NaOH; 100 mmol/L H_3PO_4; 1.5 mmol/L taurine-Tris (pH 8.4); ethanol, PEG	Fused-silica capillary 35 cm × 50 μm, $U = -20$ kV	UV	[33]
Brevibacterium taipei, Bacillus cereus, Bacillus subtilis, Candida albicans, Bacillus megaterium	CIEF	1 mmol/L T Tris/ 0.33 mmol/L citric acid (pH 7.0); CTAB; sulphobetaine	Fused-silica capillary 30 cm × 100 μm, $U = -3$ kV	LIF	[34]

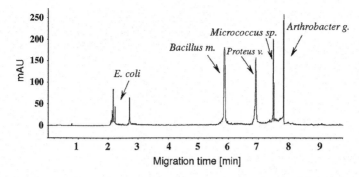

Fig. 16.3 Electropherogram obtained for a sample containing five species of bacteria. Conditions of the analysis: capillary modified with trimethylchlorosilane, buffer TBE (pH 8.53), L_{eff} = 8.5 cm, λ = 210 nm, U = −15 kV, dosing 100 mbar·s. Reprinted with permission from Szumski et al. [7]. Copyright 2005 Elsevier

From the biological, medical, and pharmaceutical points of view, the miniaturization is more and more important due to its unique benefits, i.e., short time and low costs. Seemingly identical cells are often non-homogeneous in respect of their chemical composition, biological activity, and synchronization and response to external impulses. The electrophoretic separation of microorganisms conducted on a microchip gives a new look on the identification of pathogenic cells in physiologic fluids for an early detection of a disease [17, 18] (Table 16.2).

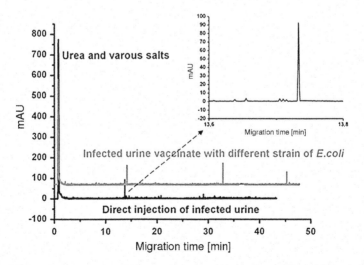

Fig. 16.4 Electropherogram obtained for a sample of infected urine. Conditions of the analysis: buffer TBE (pH 8.53), c = 5 mmol/l with the addition of PEO, L_{eff} = 25 cm, λ = 210 nm, U = −20 kV. Reprinted with permission from Kłodzińska et al. [9]. Copyright 2006 Wiley-VCH

Fig. 16.5 Electropherogram obtained for a sample containing three strains of the same species of *S. aureus* and a picture from a stereoscopic fluorescent microscope (visible spots represent migrating bacteria aggregates). Conditions of the analysis: buffer MES (pH 6.5), $c = 5$ mmol/l, $L_{eff} = 25$ cm, $\lambda = 214$ nm, $U = 20$ kV. Reprinted with permission from Kłodzińska et al. [11]. Copyright 2009 Wiley-VCH

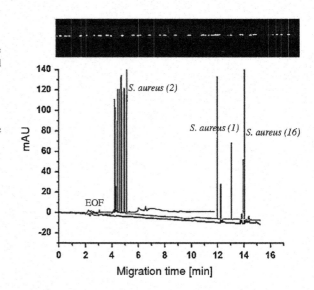

References

1. W. Kilarski, *Strukturalne podstawy biologii komórki* (PWN, Warsaw, 2005)
2. G.M. Fuller, D. Shields, Podstawy molekularne biologii komórki: aspekty medyczne. PZWL (2005)
3. E. Dziubakiewicz, E. Kłodzińska, B. Buszewski, Elektryczne mikroby. Analityka **2**, 11–14 (2009)
4. S. Hjertén, K. Elenbring, F. Kilar, J.L. Liao, A.J. Chen, C.J. Siebert, M.D. Zhu, Carrier-free zone electrophoresis, displacement electrophoresis and isoelectric focusing in a high-performance electrophoresis apparatus. J. Chromatogr. **403**, 47–61 (1987)
5. R.C. Ebersole, R.M. McCormick, Separation and isolation of viable bacteria by capillary zone electrophoresis. Biotechnology **11**, 1278–1282 (1993)
6. B. Buszewski, M. Szumski, E. Klodzinska, H. Dahm, Separation of bacteria by capillary electrophoresis. J. Sep. Sci. **26**, 1045–1049 (2003)
7. M. Szumski, E. Kłodzińska, B. Buszewski, Separation of microorganisms using electromigration techniques. J. Chromatogr. A **1084**, 186–193 (2005)
8. D.W. Armstrong, J.M. Schneiderheinze, Rapid identification of the bacterial pathogens responsible for urinary tract infections using direct injection CE. Anal. Chem. **72**, 4474–4476 (2000)
9. E. Kłodzińska, H. Dahm, H. Rożycki, J. Szeliga, M. Jackowski, B. Buszewski, Rapid identification of Escherichia coli and Helicobacter pylori in biological samples by capillary zone electrophoresis. J. Sep. Sci. **29**, 1180–1187 (2006)
10. P. Gao, G. Xu, X. Shi, K. Yuan, J. Tian, Rapid detection of staphylococcus aureus by a combination of monoclonal antibody-coated latex and capillary electrophoresis. Electrophoresis **27**, 1784–1789 (2006)
11. E. Kłodzińska, M. Szumski, K. Hrynkiewicz, E. Dziubakiewicz, M. Jackowski, B. Buszewski, Differentiation of *Staphylococcus aureus* strains by CE, zeta potential and coagulase gene polymorphism. Electrophoresis **30**, 3086–3091 (2009)

12. B. Buszewski, E. Kłodzińska, H. Dahm, H. Różycki, J. Szeliga, M. Jackowski, Rapid identification of *Helicobacter pylori* by capillary electrophoresis: an overview. Biomed. Chromatogr. **21**, 116–122 (2007)

13. M. Girod, D.W. Armstrong, Monitoring the migration behavior of living microorganisms in capillary electrophoresis using laser-induced fluorescence detection with a charge-coupled device imaging system. Electrophoresis **23**, 2048–2056 (2002)

14. J. Zheng, E.S. Yeung, Mechanism of microbial aggregation during capillary electrophoresis. Anal. Chem. **75**, 818–824 (2003)

15. O. Lim, W. Suntornsuk, L. Suntornsuk, Capillary zone electrophoresis for enumeration of Lactobacillus delbrueckii subsp bulgaricus and Streptococcus thermophilus in yogurt. J. Chromatogr. B **877**, 710–718 (2009)

16. D.W. Armstrong, J.M. Schneiderheinze, J.P. Kullman, L. He, Rapid CE microbial assays for consumer products that contain active bacteria. FEMS Microbiol. Lett. **194**, 33–37 (2001)

17. T. Shintani, M. Torimura, H. Sato, H. Tao, T. Manabe, Rapid separation of microorganisms by quartz microchip capillary electrophoresis. Anal. Sci. **21**, 57–60 (2005)

18. P.C.H. Li, D.J. Harrison, Transport, manipulation, and reaction of biological cells on-chip using electrokinetic effects. Anal. Chem. **69**, 1564–1568 (1997)

19. A.T. Poortinga, R. Bosa, W. Nordea, H.J. Busscher, Electric double layer interactions in bacterial adhesion to surfaces. Surf. Sci. Rep. **47**, 1–32 (2002)

20. A. Pfetsch, T. Welsch, Determination of the electrophoretic mobility of bacteria and their separation by capillary zone electrophoresis Fresenius. J. Anal. Chem. **359**, 198–201 (1997)

21. D.W. Armstrong, G. Schulte, J.M. Schneiderheinze, D.J. Westenberg, Separating microbes in the manner of molecules 1. Capillary electrokinetic approaches. Anal. Chem. **71**, 5465–5469 (1999)

22. M. Torimura, S. Ito, K. Kano, T. Ikeda, Y. Esaka, T. Ueda, Surface characterization and on-line activity measurements of microorganisms by capillary zone electrophoresis. J. Chromatogr. B **721**, 31–37 (1999)

23. D.W. Armstrong, L. He, Determination of cell viability in single or mixed samples using capillary electrophoresis laser-induced fluorescence microfluidic systems. Anal. Chem. **73**, 4551–4557 (2001)

24. K. Yamada, M. Torimura, S. Kurata, Y. Kamagata, T. Kanagawa, K. Kano, T. Ikeda, T. Yokomaku, R. Kurane, Application of capillary electrophoresis to monitor populations of Cellulomonas cartae KYM-7 and Agrobacterium tumefaciens KYM-8 in mixed culture. Electrophoresis **22**, 3413–3417 (2001)

25. T. Shintani, K. Yamada, M. Torimura, Optimization of a rapid and sensitive identification system for salmonella enteritidis by capillary electrophoresis with laser-induced fluorescence. FEMS Microbiol. Lett. **210**, 245–249 (2002)

26. B.G. Moon, Y.-I. Lee, S.H. Kang, Y. Kim, Capillary electrophoresis of microbes. Bull. Korean Chem. Soc. **24**, 81–85 (2003)

27. B. Palenzuela, B.M. Simonet, R.M. Garcia, A. Ríos, M. Valcarcel, Monitoring of bacterial contamination in food samples using capillary zone electrophoresis. Anal. Chem. **76**, 3012–3017 (2004)

28. M. Horka, F. Ruzicka, J. Horky, V. Hola, K. Slais, Capillary isoelectric focusing and fluorometric detection of proteins and microorganisms dynamically modified by poly(ethylene glycol) pyrenebutanoate. Anal. Chem. **78**, 8438–8444 (2006)

29. M. Horka, F. Ruzicka, J. Horky, V. Hola, K. Slais, Capillary isoelectric focusing of proteins and microorganisms in dynamically modified fused silica with UV detection. J. Chromatogr. B **841**, 152–159 (2006)

30. M.A. Rodriguez, A.W. Lantz, D.W. Armstrong, Capillary electrophoretic method for the detection of bacterial contamination. Anal. Chem. **78**, 4759–4767 (2006)

31. M. Horka, O. Kubicek, F. Ruzicka, V. Hola, I. Malinovska, K. Slais, Capillary isoelectric focusing of native and inactivated microorganisms. J. Chromatogr. A **1155**, 164–171 (2007)

32. W. Lantz, Y. Bao, D.W. Armstrong, Single-cell detection: test of microbial contamination using capillary electrophoresis. Anal. Chem. **79**, 1720–1724 (2007)

33. M. Horká, J. Horký, A. Kubesova, K. Mazanec, H. Matouskova, K. Lais, Electromigration techniques: a fast and economical tool for differentiation of similar strains of microorganisms. Analyst **135**, 1636–1644 (2010)
34. M.Y. Tong, C. Jiang, D.W. Armstrong, Fast detection of Candida albicans and/or bacteria in blood plasma by sample-self-focusin using capillary electrophoresis-laser-induced fluorescence. J. Pharm. Biomed. Anal. **53**, 75–80 (2010)

Chapter 17
Applications of Electromigration Techniques: Applications of Electromigration Techniques in Food Analysis

Piotr Wieczorek, Magdalena Ligor and Bogusław Buszewski

Abstract Electromigration techniques, including capillary electrophoresis (CE), are widely used for separation and identification of compounds present in food products. These techniques may also be considered as alternate and complementary with respect to commonly used analytical techniques, such as high-performance liquid chromatography (HPLC), or gas chromatography (GC). Applications of CE concern the determination of high-molecular compounds, like polyphenols, including flavonoids, pigments, vitamins, food additives (preservatives, antioxidants, sweeteners, artificial pigments) are presented. Also, the method developed for the determination of proteins and peptides composed of amino acids, which are basic components of food products, are studied. Other substances such as carbohydrates, nucleic acids, biogenic amines, natural toxins, and other contaminations including pesticides and antibiotics are discussed. The possibility of CE application in food control laboratories, where analysis of the composition of food and food products are conducted, is of great importance. CE technique may be used during the control of technological processes in the food industry and for the identification of numerous compounds present in food. Due to the numerous advantages of the CE technique it is successfully used in routine food analysis.

Determination of qualitative and quantitative composition of ready food products, as well as raw plant materials used for their production, allows to obtain the quality food accepted by consumers, dieticians, and also medical doctors and scientists examining the influence of food components on human health. Also, the problem of bioavailability of biologically active compounds present in raw materials and food

P. Wieczorek (✉)
Faculty of Chemistry, University of Opole, Opole, Poland
e-mail: piotr.wieczorek@uni.opole.pl

M. Ligor · B. Buszewski
Faculty of Chemistry, Nicolaus Copernicus University, Toruń, Poland

B. Buszewski et al. (eds.), *Electromigration Techniques*, Springer Series
in Chemical Physics 105, DOI: 10.1007/978-3-642-35043-6_17,
© Springer-Verlag Berlin Heidelberg 2013

products is an important issue. The food proposed by producers is often characterized by an unsuitable quality. Only high quality food is valuable for consumers, mainly due to the issue of hunger satisfying, and thus supplying an organism with products conditioning energy formation in metabolic processes, and also exhibiting health promoting properties and suitable taste and aroma features.

Electromigration techniques, including capillary electrophoresis (CE), are used for mixtures separation and for identification of components present in food products. A variety of CE applications allows their utilization in determination of high-molecular compounds, especially peptides, proteins, nucleic acids, and also compounds of lower molecular mass such as: amino acids, carbohydrates, vitamins, polyphenols, including flavonoids and pigments, pesticides, inorganic ions, organic acids and others [1–4]. The possibility of CE application in food control laboratories, where analysis of the composition of food and food products, are conducted is of great importance.

17.1 Polyphenols

It is commonly known that polyphenols have a variety of valuable biological properties. Among them the ability of free radicals scavenging, thus their properties labelled as antioxidative, anti-inflammatory, and anticarcinogenic [5–9]. These compounds may also play a significant role in prevention of heart and circulatory system diseases formation.

Determination of polyphenols using electromigration techniques concerns not only ready food products, but also raw materials, mainly plant ones, since plants are the most abundant source of these compounds. The available literature sources confirm large possibilities of CE application in polyphenols determination [10]. The studies conducted concerned mainly polyphenols determining in grapes and wine [11–16], plant material [17–23], various kinds of tea [13, 24, 25], and olive oil [26–28]. As was demonstrated, the valuable sources of polyphenols are also berries [29], fruits [30, 31], soya [32], algae and microalgae [33, 34], and cocoa and chocolate [35].

Solid phase extraction (SPE) has been used for the separation of numerous polyphenolic compounds from red wine samples. They included apigenin, luteolin, naringenin, hesperidin, kaempferol, quercetin, as well as myricetin and their potential precursor—caffeic acid. The components of the mixture containing polyphenolic compounds may be separated using quartz capillaries, basic electrolyte 35 mM borax of pH 8.9 and UV–Vis detector (wave length $\lambda = 250$ nm) [11]. Whereas when applying CE with electrochemical detector (ED), it was possible to determine *trans*-resveratrol, (−)-epicatechin and (+)-epicatechin in the samples of red wine [12]. An application of borate buffer of a concentration of 100 mmol/L and pH 9.2 was required to conduct the analysis; with capillary diameter of 300 μm, which allowed the concurrent determination of all mentioned compounds. The detection limit for *trans*-resveratrol, (−)-epicatechin, and (+)-epicatechin was within the range from 0.2 to 0.5 μg/mL.

An interesting proposition is an application of microemulsion electrokinetic capillary chromatography (MEEKC) and micellar electrokinetic chromatography (MEKC) for determination of phenolic acids in grapes and tea: syringic, p-coumaric, vanillic, caffeic, gallic, 3,4-dihydroxybenzoic, 4-hydroxybenzoic and (+)-catechin, (−) epigallocatechin, (−)-epicatechin gallate, (−)-epigallocatechin gallate, (−)-epicatechin and (−)-gallocatechin, and also methylxanthines: caffeine and theophylline. An addition of surface-active substance in a form of sodium dodecyl sulfate (SDS) did not affect in a significant manner separation in MECK technique, as opposed to MEEKC where increased SDS concentration clearly improved separation results. An addition of organic modifier (methanol or acetonitrile) demonstrates a substantial influence on separation and selectivity both in MEEKC and MEKC. In turn, an increase in voltage and temperature in the capillary affects the effectiveness of separation of compounds present in analyzed mixture, without deterioration of separation in MEEKC technique. However, in the same conditions the separation in the case of MEKC decreases distinctly. In both techniques used, the selectivity according to analyzed polyphenolic compounds is different [13].

The electromigration techniques allowed also to detect the presence of procyanidins in grapes. Adding 40 mmol/L of sodium cholate and 10 mmol/L of SDS to buffer solution containing 50 mmol/L of phosphate (pH 7.0), the compounds from that group may be determined in a time shorter than 15 min [17]. The method proposed may be useful in determination of flavonoids and anthocyanins in postproduction wastes, peels, and seed of grapes (wine industry cakes), which are potential sources of these compounds.

The range of anthocyanins in a form of glucosides and galactosides (malvidins, pelargonidins, and cyanidins) was identified in grape peel extracts whose purpose was wine production. The highest selectivity of separation was obtained when the buffer of a following composition was used: 30 mmol/L of potassium phosphate, 400 mmol/L of sodium borate and Tris, with an addition of 50 mmol/L of SDS, the mixture of pH 7.0 was obtained. High concentration of borate profitably influenced separation of malvidin diastereomers. That method is characterized by linearity in the range of concentrations of 10–100 μg/ml for diglycoside derivatives, and 25–100 μg/mL for monoglycoside derivatives. Among anthocyanins separated from grape peels, the most important is malvidin 3-glucoside [15].

Connection of capillary zone electrophoresis (CZE) with isotacho- phoresis (ITP) allowed to obtain higher sensitivity and separation ability of the system with respect to 14 selected flavonoids and polyphenolic acids separated from red wine [16]. Preliminary analysis using ITP and suitable composition of electrolytes for ITP and CZE allowed to determine flavonoids and polyphenols; additionally, the analysis time did not exceed 45 min. The detection limit for phenolic acids, phenolic acids, and rutin was 30 ng/mL, for kaempferol and epicatechin it was 100 ng/mL, and for catechin −250 ng/mL [16].

Extracts obtained from plants contain numerous biologically active compounds. The seeds of *Garcinia kola* plant, called *bitter kola* in English, are used in herbal medicine by Northern American Indians. Due to its bitter taste, they may also be used in beer and wine production. The seeds of *bitter kola* are the source of biflavones (GB1, GB2, glucoside GB1, and colaflavone). Due to the application of the CE-UV method, the quantitative determination of these compounds was possible within only 12 min. The electrolyte contained 100 mmol/L of borate, pH 9.5. The detection limit for biflavones was from 3 to 6 µg/L [17]. The range of active flavonoids: epicatechins, catechins, rutin, kaempferol, quercetin were separated from common sea-buckthorn (*Hippophae rhamnoides*) fruits. The analysis of extracts using CE with amperometric detection (AD) allowed successful identification of these compounds [19].

Furthermore, MEKC has found an application in determination of iridoid glucosides and flavonoids in extracts of verbena leaves (*Verbena officinalis*) [20]. The solution of sodium borate of a concentration of 50 mmol/L (pH 9.3) with an addition of 50 mmol/L of SDS was used during analyses. Due to different maxima of analytes absorption, the detection was conducted with two wavelengths: $\lambda = 205$ and $\lambda = 235$ nm.

The leaves of rosemary (*Rosmarinus officinalis*) are the source of numerous antioxidants. Using extraction with overheated water steam and capillary electrophoresis with UV detection or in connection with mass spectrometry (MS), rosmarinic acid, carnosic acid, and carnosol in extracts from rosemary leaves were identified and determined quantitatively [21–23].

An interesting issue is variable content of (−)-epigallocatechin gallate and (−)-epicatechin gallate in tea leaves depending on the degree of plant development. In order to investigate the content of particular components electromigration techniques were used. Concentrations of compounds mentioned above are higher in young leaves. Leaves of fully developed plant contain higher amounts of free catechins [24]. Other sources confirm healthy properties of green tea since when compared to black tea it has 137-fold higher concentration of (−)-epigallocatechin gallate [25].

Olive oil is a rich source of polyphenolic compounds, polyphenolic acids, tyrosol, and hydroksytyrosol. Using CE, where 45 mM sodium borate (pH 9.6) was used as a buffer, the determination of 21 polyphenolic compounds was conducted in a time of 10 min [26]. In order to separate polyphenolic acids from olive oil, liquid–liquid extraction (LLE) was used. Concentrations of polyphenols in olive oil are usually on ppb level. The extracts obtained were subject to CE analysis with UV detection. The time of analysis of 16 min was sufficient to confirm the presence of fourteen polyphenolic acids in olive oil samples [27, 28]. Examples electropherograms are presented in Fig. 17.1. In that kind of analysis, an addition of polycationic surfactant (hexadimethrine bromide) to electrolyte may be used, which evenly covers the walls of capillary and may positively influence anodic EOF acceleration [27].

MEKC technique has found application in determination of biologically active compounds in organisms of *S. Platensis* microalgae [33]. In vitro studies confirmed, that compounds separated from these organisms exhibit antioxidative properties. Another algae species, brown algae *Fucus vesiculosus*, also contain the range of bioactive compounds [34]. When analyzing extracts from these organisms using CE with UV detection ($\lambda = 210$ nm) with borate and acetate buffer with an

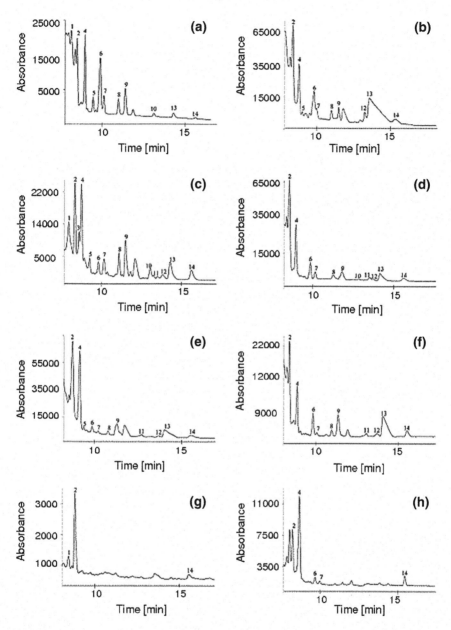

Fig. 17.1 Example CZE electropherograms of phenolic fraction obtained after LLE of real olive oil sample: **a** Arbequina. **b** Lechin de Sevilla. **c** Picual. **d** Hojiblanca. **e** Lechin de Granada. **f** Cornicabra. **g** Refined olive oil. **h** Mixture of refined and virgin olive oils. Separation conditions: capillary, 57 cm × 675 mm; applied voltage, 25 kV; applied temperature, 25 °C; buffer, 25 mM sodium borate pH 9.60; hydrodynamic injection, 0.5 psi for 8 s. Detection was performed at 210 nm. Peaks: *1 trans*-cinnamic acid, *2* 4-hydroxyphenylacetic acid, *3* sinapinic acid, *4* gentisic acid, *5* (+)-taxifolin, *6* ferulic acid, *7 o*-coumaric acid, *8 p*-coumaric acid, *9* vanillic acid, *10* caffeic acid, *11* 4-hydroxybenzoic acid, *12* 3,4 dihydroxyphenylacetic acid, *13* gallic acid, *14* protocatechuic acid. Reprinted with permission from Carrasco-Pancorbo [28]. Copyright 2004 American Chemical Society

addition of methanol and acetonitrile, the presence of polyphenols in samples examined was confirmed.

The methods developed for the determination of selected polyphenolic compounds based on data reported in Ref. [2] are presented in Table 17.1.

Electromigration techniques are a perfect tool in isoflavones determination [51]. Especially interesting is an application of electrochromatography (CEC) with various kinds of packings, including monolithic ones. Unlike in high-performance liquid chromatography (HPLC), in case of CEC, the separation in capillary column takes place after application of electric field. Using monolithic bed based on lauryl acrylate as a stationary phase for CEC, the determination of daidzein, genistein, glycitein, and others in products for infants feeding based on soya was conducted [52]. The compounds of isoflavones group may be identified using various kinds of detection systems, mainly UV and ED. Also, an application of devices like mass spectrometer (MS) or fluorescence detector is suitable. These devices allow the selective determination of selected compounds from that group [53]. Additionally, MS allows to investigate the structure of the compound analyzed. In case of isoflavones the electromigration techniques allow not only their determination in plant material, by also examination of chemism, e.g., determination of ionisation constants [54], study of stability effected by UV-B radiation activity [55], and others [56, 57].

An important analytical problem is enantiomeric purity of flavonoids biosynthesized by transgenic plants [58, 59]. Using an addition of two cyclodextrins (HP-β-CD and HP-γ-CD) to electrolyte and methanol as organic modifier the components of the mixture containing R and S enantiomers of westitone and their metabolites (+)-medikarpin and (−)-medikarpin were successfully separated [59]. Catechin and epicatechin enantiomers in tea and tea drinks were separated using 6-O-α-d-glucosyl-β-cyclodextrin (6G-β-CD) and HP-γ-CD. However, only the application of 6G-β-CD allowed selective determination of catechin and epicatechin against other catechins [60]. The separation of *cis*- and *trans*-resveratrol, while only *trans*- isomer has the practical meaning, requires an application of MEKC [61, 62].

17.2 Pigments

The range of pigments having different functions exists in the world of plants. These compounds play mainly protective function contributing in the process of plant photo-screening against destructive activity of UV radiation, and also take part in photosynthesis process. The plant pigments include carotenoids, anthraquinones, curcumins, and the previously described anthocyans and flavonoids. They may be obtained from the plants in the form of extracts or in a synthetic manner. In that form they are a perfect addition to food products. The aim of such activities is an improvement in the aesthetic features of ready products. The special position in pigments determination is occupied by electromigration techniques, both CZE and MEKC [63, 64].

Table 17.1 Example methodologies applied for the determination of polyphenols in food products by CE

Application	CE mode	Background electrolyte (BGE)	CE conditions	Detection method	LOD^a (µg/mL)	References
Resveratrol in wine, herbs, and health food	CZE	100 mmol/L borate (pH 9.24)	65 cm × 25 µm ID fused-silica capillary, 30 kV	Electrochemical	0.06	[36]
Resveratrol in red wine	MEKC	20 mmol/L sodium tetraborate, 25 mmol/L polyethylene glycol 400, 25 mmol/L SDS, 10 % methanol	57 cm × 75 µm ID fused-silica capillary, 28 kV, temp. 25 °C	DAD^b	0.05	[37]
Anthocyanins in wine	CZE	50 mmol/L borate (pH 8.4), 15 % methanol	46 cm × 75 µm ID fused-silica capillary, 25 kV, temp. 10 °C	VIS absorption, λ = 599 nm	1	[38]
Resveratrol, catechin, rutin, quercetin, myricetin, caffeic acid, chlorogenic acid, gallic acid in plants	CZE	25 mmol/L borate (pH 9.4)	75 cm × 50 µm ID fused-silica capillary, 18 kV	UV absorption, λ = 240 nm	–	[39]
Rutin and quercetin in plants	MEKC	20 mmol/L borate (pH 8.8), 40 mmol/L SDS, 10 % acetonitrile	50 cm × 50 µm ID fused-silica capillary, 12 kV	Electrochemical	From 0.05 to 400	[40]
Quercetin, rutin, kaempferol, catechin, galic acid plants	CZE	100 mmol/L borate (pH 10.0)	51 cm × 50 µm ID fused-silica capillary, 15 kV, temp. 32 °C	UV absorption, λ = 270 nm	0.86	[41]
Procyanidins after thiolysis	MEKC	50 mmol/L phosphate (pH 7.0), 40 mmol/L sodium cholate, 10 mmol/L SDS	47 cm × 50 µm ID fused-silica capillary, 15 kV, temp. 25 °C	UV absorption, λ = 214 nm	From 0.62 to 11.2	[42]

(continued)

Table 17.1 (continued)

Application	CE mode	Background electrolyte (BGE)	CE conditions	Detection method	LOD[a] (μg/mL)	References
Catechins in green tea	MEEKC	Microemulsion of 50 mmol/L phosphate (pH 2.5), SDS, n-heptane, 2-hexanol or cyklohexanol	24 cm × 50 μm ID fused-silica capillary, – 10 kV, temp. 40 °C	UV absorption, λ = 230 nm	From 0.30 to 6.61	[43, 44]
Catechins in green tea	MEKC	3 parts 20 mmol/L phosphate, 1 part 50 mmol/L borate, 2 parts 200 mmol/L SDS (pH 7.0)	47 cm × 50 μm ID fused-silica capillary, 30 kV, temp. 29 °C	UV absorption, λ = 200 nm	From 0.0011 to 0.0051	[45, 46]
Theaflavin composition of black tea	NACE[c]	71 % v/v acetonitrile, 25 % v/v methanol, 0.1 mol/L potassium hydroxide, 4 % glacial acetic acdid, 90 mmol/L ammonium acetate	40 cm × 50 μm ID fused-silica capillary, 22.5 kV, temp. 18.5 °C	UV absorption, λ = 380 nm	From 21 to 25	[47, 48]
Chlorogenic acid, ferulic acid, vanilic acid, caffeic acid, catechol in coffee extract	CZE	50 mmol/L borate (pH 9.5)	48.5 cm × 50 μm ID fused-silica capillary, 20 kV, temp. 25 °C	DAD	–	[49, 50]

[a] Limit of detection (LOD)
[b] UV diode array detection (DAD)
[c] Non-aqueous capillary electrophoresis (NACE)

Table 17.2 Results of the determination of natural pigments in food products

Food product	Source of pigment	Compound	Concentration (μg/mL)	RSD (%)	References
Jelly	Cochineal	Carmic acid	639	0.37	[65]
	Elderberry	Cyanidin-3-sambubioside	905	3.24	
		Cyanidin-3-glucoside	439	4.62	
Juice	Cochineal	Carmic acid	850	0.47	[66]
	Safflower	Safflomin A	427	2.36	
Candies	Safflower	Safflomin A	1,253	0.94	[67]
	Gardenia	Crocin	590	2.54	
	Monascus purpureus	Xanthomonasin A	105	2.38	[68]
	Elderberry	Xanthomonasin B	76	2.45	[69]
		Cyanidin-3-sambubioside	863	2.36	
		Cyanidin-3-glucoside	421	3.54	
Noodle	Gardenia	Crocetin	67	2.38	[67]

Selected results of quantitative analysis concerning pigments determination in food products are presented in Table 17.2.

17.3 Vitamins

Electromigration techniques have found an application, e.g., in investigation of riboflavin and its derivatives, flavin nucleotides, and flavin adenine dinucleotides [70–74]. Flavines are soluble in water, demonstrate fluorescence and thus it is profitable to use LIF detector, which is characterized by suitable sensitivity towards the analytes and concurrently allows the selective identification. Trace amounts of flavines were detected when CZE with LIF detector were used with phosphate buffer of a concentration of 30 mmol/L (pH 9.8), quartz capillary of a length of 84 cm, and internal diameter of 75 μm, with voltage of 30 kV, temperature of 15 °C [70]. The possibilities of trace amounts of riboflavins determination presented in the literature concern mainly wine [71] and vegetables, wheat flour, and tomatoes [72].

17.4 Food Additives

The range of food additives like preservatives, antioxidants, sweeteners, pigments, and others are currently used. Their introduction to food products serves first of all durability elongation, improvement of taste, and aroma features while addition of pigments increase product attractiveness for consumers. Law regulations concerning the kind of additives used and their amounts are established individually by state offices. That situation creates the possibility of numerous divergent conclusions non-necessarily in consumers' interest. Thus, only those food additives that have been well recognized in terms of human health influence should be used.

Analytical methods currently available are entirely adequate for qualitative and quantitative assessment of food products containing additives [75]. Examples of application of electromigration techniques in preservatives, antioxidants, sweeteners, and pigments determination are presented in Tables 17.3, 17.4, 17.5, and 17.6.

17.5 Amino Acids, Peptides, Proteins

Proteins and peptides composed of amino acids are basic components of food products determining their nutritional properties and food quality. Taking into consideration the fact of the presence of many various proteins, peptides, and amino acids of differentiated and complicated structure and different range of particular components concentrations, determination of these compounds in food samples is one of the most significant problems in food analysis . Due to the possibility of determination of numerous compounds of similar structures present in different concentrations, the wide application of capillary electrophoresis is observed in analysis of amino acids, peptides, and proteins in food samples [104–106].

An important role in food analysis is played by an analysis of free proteinogenic amino acids included in the composition of all proteins and peptides. It provides suitable information concerning quality, composition, and food production process, and also possible food products falsifications and adulteration. For that reason, the methods of precise determination of amino acids in food are still being developed, including those using capillary electrophoresis. Due to the lack of chromophore groups in most amino acids molecules, in order to determine their quantity it is necessary, like in the case of determination using HPLC method, to transform them in adequate derivatives. These are most often the derivatives absorbing in the range of UV or allowing the detection using fluorescence, which enables the determination with much better sensitivity. The possibility of estimation of seven various neutral amino acids (Ala, Asn, Leu, Lys, Pro, Tyr and Val) using CE-DAD was demonstrated in samples of wine, beer, and fruit juices without the need for previous preparation and condensation of samples, except filtration and possible dilution. The separation was conducted in borate buffer of pH 10 with an addition of 10 % ACN. As derivatization agent, 1,2-naphthoquinone-4-sulfonate was used [107]. For determination of as many as 18 amino acids from marine algae, transformed in derivatives using phenylisothiocyanate, the same method was used (CE-DAD) [108]. In that case 0.4 M borate buffer of pH 9.5 with an addition of 1 mmol/L ethylenediamine was used as a basic electrolyte. That allowed to obtain the wide range of linearity of determination,

Table 17.3 Example methodologies applied for the determination of preservatives in food products

Food sample	Compounds	CZE conditions	Detection method	LOD	References
Wine, soft drinks, fruit and juice concentrates, margarine, marmalade	Sorbic acid	100 mmol/L MES, 10 mmol/L Bis–TRIS, 0.2 % PEG (pH 5.2)	UV $\lambda = 254$ nm	0.06 µg/mL	[76]
Fruit juices, cordials, soft drinks, wine, jam, dips, cheese slices	Sorbic acid, benzoic acid	50 mmol/L SDS, 20 mmol/L disodium hydrogen phosphate (pH 9.2)	UV $\lambda = 230$ nm	0.1 µg/mL	[77]
Soft drinks, soy sauce	Sorbic acid, benzoic acid	25 mmol/L sodium phosphate (pH 10.0)	UV $\lambda = 248$ nm	0.27 µg/mL (benzoic acid) 0.47 µg/mL (sorbic acid)	[78]
Plum preserves, bean curd, soy sauce	Benzoic acid, sorbic acid, p-hydroxybenzoates, p-hydroxybenzoic acid	35 mmol/L borax-NaOH (pH 10.0), with 2 mmol/L α-cyceodextrin	UV DAD	0.4–2.2 µg/ml	[79]
Cordials, vegetables (cabbage, onions), seafoods, processed foods (dried apricot, onion slice)	Sulfate	5 mmol/L sodium chromate, 0.5 mmol/L OFM Anion-BT reagent (pH 8.0)	UV $\lambda = 254$ nm	5 µg/g	[80]
Wine, beer	Sulfite	6 mmol/L chromate, 0.3 mmol/L boric acid, 23 µmol/L CTAB (pH 8.75)	–	–	[81]

Table 17.4 Example methodologies applied for the determination of antioxidants in food products

Food sample	Compounds	CZE conditions	Detection method	LOD (μg/mL)	References
Sesame oil, wine	*Tert*-butylhydroquinone (BHQ), butylated hydroxyanisole (BHA), butylated hydroxytoluene (BHT)	10 mmol/L borate, 50 mmol/L SDS (pH 9.5)	UV $\lambda = 214$ nm	–	[82]
	BHQ, BHA, BHT, propyl gallate, dodecyl gallate, octyl gallate, ascorbic acid, erythrobic acid	40.0 mmol/L sodium cholate, 15 mmol/L SDS, 10 % methanol, 10 mmol/L borate (pH 9.3)	UV $\lambda = 214$ nm	1–10	[83]
	BHA, BHT, gallic acid, derivatives of gallic acid	30 mmol/L SDS, 30 mmol/L phosphate (pH 7.0)	UV DAD	–	[84]
	BHQ, BHT, BHA, propyl gallate, dodecyl gallate, octyl gallate	20 mmol/L boric acid, 20 % acetonitrile, 20 mmol/L bis (2-ethylohexyl) sodium sulfosuccinate	UV $\lambda = 280$ nm	–	[85]
Fruit juices, wine	Ascorbic acid	100 mmol/L N-[tris(hydroxymethyl) methyl] glycin (*Tricine*) (pH 8.8)	UV $\lambda = 254$ nm	1.6	[86]
Beer, wine, fruit, drinks	Ascorbic acid, erythrobic acid	0.05 mol/L sodium deoxycholate, 0.02 mol/L potasium dihydrogen *ortho*-phosphate, 0.02 mol/L sodium tetraborate (pH 8.6)	UV $\lambda = 254$ nm	–	[87]

Table 17.5 Example methodologies applied for the determination of sweeteners in food products

Food sample	Compound	CZE conditions	Detection method (nm)	LOD	References
Soft drinks, solid sweeteners, tomato sauce, jam	Aspartame, saccarin, acesulfame-K, dulcin (with benzoic acid, sorbic acid, caffeine)	10 mmol/L borate, 10 mmol/L potassium dihydrogenorthophosphate, 50 mmol/L sodium deoxycholate	UV λ = 220	–	[88]
Cordials, soft drinks, jam	Cyclamate	10 mmol/L sodium benzoate, 1 mmol/L hexadecyltrimethylammonium hydroxide (pH 6.6)	Indirect UV λ = 254	–	[89]
Soft drinks, iced tea, solid sweetener, diet dessert	Aspartame	30 mmol/L phosphate, 19 mmol/L Tris (pH 2.14)	UV λ = 211	–	[90]
Soft drinks, coffee containing added solid sweetener	Aspartame and its degradation products	25 mmol/L phosphate, 25 mmol/L borate (pH 9.0)	UV λ = 214	–	[91]
Stevia extract (*Stevia rebaudiana*)	Stevia glycosides	50 mmol/L sodium tetraborate, 55 % acetonitrile (pH 9.3)	UV λ = 210	–	[92]
Candies, chewing gum	Alditols (mannitol, sorbitol, xylitol, lactitol, maltitol, glycerol)	50 mmol/L or 65 mmol/L (for chewing gum analysis) borate (pH 9.4)	UV λ = 200	1–3 mmol	[93]

Table 17.6 Example methodologies applied for the determination of artificial pigments in food products

Food sample	Compound	CZE conditions	Detection method	LOD	References
Fruit drinks, ice cream	Erythrosine, ploxine, rose bengal, acid red, amaranth, new coccine, allura red	(i) 25 mmol/L sodium phosphate, 25 mmol/L borate (pH 8.0), 10 mmol/L SDS (ii) 25 mmol/L sodium phosphate, 25 mmol/L borate (pH 8.0), 10 mmol/L β-cyclodextrin	DAD	–	[94]
	Erythrosine, fast green, light green SF yellow, sunset yellow, amaranth, new coccine, tartrazine	10 mmol/L sodium tetraborate (pH 7.5), 15 mmol/L β-cyclodextrin	UV $\lambda = 254$ nm	–	[95]
	Tartrazine, new coccine, amaranth, brilliant black, sunset yellow, indigo carmine, chromotrope FB, patent blue V, brilliant blue, erythrosine, quinoline yellow	30 mmol/L N-tris(hydroxymethyl) methyl-2-aminoethane sulfonic acid, 8 mmol/L imidazole 0.2 % (w/v) PEG, 3–9 mmol/L β-cyclodextrin	UV $\lambda = 254$ nm	11–300 ppb	[96]
	New coccine, erythrosine, allura red, tartrazine sunset yellow, brilliant blue, indigo carmine, fast green	25 mmol/L borax (pH 9.5), 10 mmol/L β-cyclodextrin	DAD	–	[97]
	Brilliant blue, tartrazine, indigo carmine, ponceau 4R, sunset yellow, amaranth	20 mmol/L borate (pH 9)	DAD	>2 ppm	[98]
Ice lolly, fruit syrups	Erythrosine, sunset yellow, carmic acid, amaranth, ponceau, carmoisine, red 2G	15 mmol/L borate (pH 10.5)	UV $\lambda = 216$ nm	0.4–2.2 μg/mL	[99]

(continued)

Table 17.6 (continued)

Food sample	Compound	CZE conditions	Detection method	LOD	References
Confectionery, cordials	Green S, brilliant blue, erythrosine, allura red, indigo carmine, sunset yellow, azorubine, amaranth, ponceau, tartrazine	15 % acetonitrile, 85 % 50 mmol/L sodium deoxycholate/5 mmol/L phosphate/5 mmol/L borate (pH 8.6)	UV λ = 214 nm		[100]
Candies, juice, jelly	Anthocyanins	30 mmol/L phosphate, 60 mmol/L borate, 30 mmol/L SDS (pH 7.0)	UV λ = 560 nm	0.5 µg/mL	[69]
Soft drinks	Anthocyanins	150 mmol/L sodium borate (pH 8.0)	UV λ = 580 nm	–	[101]
	Caramel (class III)	30 mmol/L phosphate (pH 1.9)	UV λ = 275 nm	–	[102]
	Caramels (class IV)	50 mmol/L carbonate (pH 9.5)	UV λ = 200, 280 nm	–	[103]

from 50 to 1,000 μmol/L. Derivatives obtained as a result of reaction with fluorescein isothiocyanate [109] and *o*-phthalic aldehyde [110] was used in determination of amino acids using capillary electrophoresis with fluorescence detection. That allowed to obtain high sensitivity, and in the latter case, with in-capillary derivatization, the detection limit obtained was 0.02 μmol/L for alanine and 0.004 μmol/L for *γ*-aminobutyric acid.

Except 20 amino acids included in proteins of living organisms, and thus in food, hundreds of other compounds of a similar structure may be found (amino acids and their derivatives), which are commonly called non-protein amino acids. Non-protein amino acids are present in food usually as the products formed as a result of food products processing, as products of metabolic processes in animal tissues, or as additives introduced to food in order to improve its quality and nutritional properties [111–116]. Therefore, determination of their presence in food may provide interesting information concerning quality and ways of food quality improvement, and also possible adulteration of some products. These compounds are usually classified, taking into consideration their structure, to: selenium amino acids, sulfone amino acids, aliphatic monoamino acids containing nitrogen atom in side chain, derivatives of aromatic amino acids, cycloalkane, hydroxyl, isoxaline, and heterocyclic amino acids. For these compounds determination, except HPLC, the capillary electrophoresis is used more and more often, which is characterized by better separation and enables quick determination of numerous substances in complicated matrices, which are food samples. For non-protein amino acids, CZE is used the most often with an application of buffer of given pH, with or without additives facilitating dissolution, or MEKC with an addition of SDS of a concentration higher than CMC and/or CD as micellar pseudophase enabling separation of these compounds in complicated matrices. More often the detection is conducted using UV absorption, more rarely LIF, ESI–MS or ICP–MS. Examples of an application of electromigration techniques in non-protein amino acids determination are presented in Table 17.7.

Electrophoretic methods have also found wide application in analysis of proteins and peptides in food products. This involves mainly determination of composition and content of particular proteins in protein extracts that allows estimation of components origin and authenticity of given food, the way of food processing, monitoring and optimization of technological process, or finally the comparison of high quality products with products of poorer quality or adulterated products.

Capillary electrophoresis appeared to be a useful technique for analysis and comparison of protein composition of cow, sheep, and goat milk [124]. Determination of protein composition and differences in its composition resulting from protein composition polymorphism and genetic differences, in particular, animals species is of high significance in the dairy industry, since casein content and coagulation time determine cheese production yield. For example, capillary electrophoresis with UV detection (CE-UV) was used for a comparison of protein composition of milk from different sheep varieties [125] and for characterization

Table 17.7 Methods of determination of non-protein amino acids in selected victuals

Kind of samples	Compound	CE conditions	Detection method	LOD (M)	References
Human milk	Selenomethionine (Se-Met), Selenocystine (Se-Cys)	CZE: (i) 100 mM phosphoric buffer (pH 2.5), (ii) 100 mM acetic buffer (pH 1.9)	$\lambda = 200$ nm	1.0×10^{-5}	[114]
	Selenocystine (Se-Cys)	CZE: 10 mM carbonate buffer (pH 11.5), CIEF: Leading buffer: NaOH (pH 2) Closing buffer: phosphoric acid (pH 10)	ICP-MS	$4–6 \times 10^{-4}$ $1–4 \times 10^{-4}$	[115]
Powder milk	Taurine	50 mM boric buffer (pH 9.2)	Electrochemical	1.0×10^{-7}	[116]
Orange juice	γ-aminobutyric acid (GABA)	EKC: 100 mM ammonium acetate (pH 6.0), 5 mM β-CD	ESI–MS	2.0×10^{-6}	[117]
Dietetic supplements	Carnitine	CZE: 5 mM Tris, 7 mM phosphoric acid, 0.5 mM quinine	$\lambda = 254$ nm	2.0×10^{-5}	[118]
Tea	Teanine	EKC: 25 mM phosphoric buffer (pH 7.0), 100 mM SDS, 5–6 % methanol	$\lambda = 200$ nm	2.0×10^{-5}	[119]
Duck eggs	Lysilalanine (LysAla)	EKC: 250 mM boric buffer (pH 9.5), 200 mM SDS, 75 mM β-CD (methanol)	$\lambda = 254$ nm	4×10^{-4}	[120]

(continued)

Table 17.7 (continued)

Kind of samples	Compound	CE conditions	Detection method	LOD (M)	References
Beans	3,4-Dihydroxyphenylalanine (DOPA)	CZE: 35 mM phosphoric buffer (pH 4.55),	$\lambda = 210$ nm	3.0×10^{-6}	[121]
Lentil	2-Carboxymethyl-3-izoksazoline-5-on (CMI)	CZE: 25 mM phosphoric buffer (pH 7.5), 8 % 1-propanol	$\lambda = 254$ nm	3.0×10^{-5}	[122]
	β-(izoksazoline-5-on-2-yl)alanine (BIA)	CZE: 25 mM phosphoric buffer (pH 7.5), 8 % 1-propanol	$\lambda = 254$ nm	2.0×10^{-5}	[122]
	γ-Glutamyl- β-(izoksazolin-5-on-2-yl)alanine (γ-Glu-BIA)	CZE: 25 mM phosphoric buffer (pH 7.5), 8 % 1-propanol	$\lambda = 254$ nm	1.0×10^{-5}	[123]

of a few varieties of fresh cheese, e.g., quark and mozarella cheese depending on the kind of milk and its processing (heating, ultrafiltration) [126]. Capillary gel electrophoresis (CE-SDS) appeared to be a useful method for determination of protein profiles of fractions soluble in water, obtained for examined beef samples and ostrich meat, which enabled identification of particular proteins [127].

17.6 Carbohydrates

Carbohydrates (sugars, saccharides) are differentiated group of organic compounds, which includes monosaccharides, oligosaccharides, and polysaccharides. Sugars may also be included in other bimolecules, such as glycoproteins. Since they are present in all living organisms, including certainly food, therefore their determination in food samples is also an essential issue. Gas (GC) or liquid (LC) chromatography is used the most often for sugars determination. Due to numerous advantages, such as simplicity of performance, not too much complicated procedures of samples preparation and purification, and the possibility of many compounds determination without columns change necessary in chromatographic methods, the capillary electrophoresis has been used more and more often in sugars analysis recently. Due to the lack of chromophore group in sugars' molecules, the main problem in these compounds analysis is detection manner. Therefore, the indirect detection with an addition of a substance absorbing UV radiation in carrying buffer is often used. In this manner the concentration of various sugars, including, e.g., glucose, fructose, lactose, maltose, and galactose in cereal and dairy products was determined [128]. It is also possible to determine sugars using direct UV detection with wavelength of 196 nm. Using that method of detection glucose was determined, and LOQ obtained was not too low and was on a level of 0.66 mg/ml [129]. Much better detection level is obtained analyzing sugars derivatives, which absorb in UV range. Several mono and disaccharides were transformed in unsaturated diol derivatives in alkaline environment in a solution inside the capillary. As a result of mutarotation, enolization and isomerisation reaction proceeding at those conditions, it was possible to obtain LOD of 0.02 mmol/L (detection UV $\lambda = 270$ nm) [130].

17.7 Nucleic Acids

Capillary electrophoresis is also used for determination of deoxyribonucleic acids (DNA) in food, which allows its application for investigation of food authenticity, pathogens determination, and most of all is useful in detection of genetically modified organisms (GMO) present in food. The analysis of specific markers, DNA fragments, using capillary electrophoresis allows to determine food authenticity. Using capillary gel electrophoresis with fluorescence detection (CGE-LIF) the presence of different varieties of grapes was noted in the composition of the most examined [131]. Rice varieties in a mixture were distinguished by multiplication of markers using polymerase chain reaction (PCR) and determining with CGE-LIF method [132], and also varieties of apples contained in a juice [133] and olives in olive oil [134]. The same method was also applied for an identification of pathogens present in food, such as *Escherichia*, *Clostridium*, *Campylobacter*, *Salmonella*, or *Bacillus* [135], and also for examination of the gene encoding *Clostridium botulinum* and neurotoxin C in DNA extract from that bacteria [136]. Starting from 2002, when the first papers concerning PCR-CGE-LIF application in GMO determining in food samples were published [136, 137], the method gained more significance and is used mostly for determination of different genetically modified maize varieties [138–140].

17.8 Biogenic Amines, Natural Toxins, and Other Contaminations

Both the production and processing, and also conditions of food storage, may influence the presence and content of numerous organic compounds contaminating the food, which often demonstrates a harmful activity. Such compounds include both biogenic amines formed in food as a result of an activity of microorganisms causing amino acids decarboxylation, and also natural toxins produced by fungi, algae, bacteria, and plants, and also these used in the industry for production of machinery, devices, containers, and packages. That group may also include synthetic organic compounds like phthalates and acrylates, and finally compounds intentionally added to food where the most spectacular example is melamine.

Melamine is used in the industry for production of adhesives and plastics, and may penetrate into food from these products. Due to its toxic properties its application is prohibited and should not be present in food. In China the substance was added to many dairy products for children which were sold to many countries worldwide, in order to increase the hypothetical protein content measures using standard Kieldahl and Dumas method involving determination of total nitrogen content. As a result, a few hundred thousands of children were poisoned, including 50,000 who were hospitalized and at least 6 died. For that reason, quick and simple analytical methods allowing to determine melamine in food are essential. Besides chromatographic methods, electromigration methods (CZE-DAD, CZE-MS, and MEKC-DAD) are also used for that purpose. Using CE with MS detection, after suitable procedure of preparation of milk, dairy products, and fish samples, and concentration using SPE, the limit of quantitation (LOQ) in a range of 0.05–0.25 µg/g for liquid samples, and 0.10–0.50 µg/g for solid samples, was obtained [141]. Slightly worse LOD (0.25–0.50 µg/g) for melamine determination in dairy products, fish, eggs, and meat was obtained using UV detection after concentration using SPE [142, 143]. Capillary electrophoresis with DAD detection was also used for determination of acryloamide commonly used in a production of polyacrylamide polyelectrolytes classified by the International Agency for Research on Cancer as substances of possible carcinogenic activity. The limit of detection obtained using that method was 0.041 mg/L [144].

Biogenic amines in turn (spermine, spermidine, putrescine, cadaverine, histamine, tyramine, and many others) are a group of low-molecular compounds present commonly in many food products, mainly in meat, fish, cheese, and fermented food. However, their high concentration in an organism negatively influences physiological processes and may form carcinogenic nitrosamines in reactions with nitro-compounds. Due to the lack of ionophoric groups in molecules of biogenic amines, in order for their determination, it is necessary to transform them into derivatives or an application of mass detector. Capillary electrophoresis with UV detection (CZE-UV) was used for determination of a few biogenic amines after derivatization conducted using 1,2-naphthoquinone sulfonate in samples of various kinds of red wine obtaining LOD in the range of 0.02–0.91 mg/L [145]. Similar LOD was obtained when estimating amines in samples of fish, meat, and meat products [146]. Much lower limits of detection were obtained using fluorescence detection (CZE-LIF). When analyzing amines in beer, the LOD obtained was in the range of 5–198.3 µg/L [147]. Recently, in detection of biogenic amines using CE, mass detectors are used more often. In order to determine amines in wine, CE-ESI–MS and CE-ESI-TOF systems were used without previous derivatization step, obtaining significantly better results in the second case and LOD in the range of 0.01–0.071 µg/mL [148].

Natural toxins present in food are mainly mycotoxins, phytotoxins, plant toxins, and those produced by bacteria. Due to their often considerable toxicity, the allowable level of these toxins in food is limited in most countries, and the allowable content both in EU and USA is within the range of 15–20 ng/g. The consequence of that situation is the necessity of elaboration of cheap and simple methods of finding of these compounds in food, and also electromigration methods are used for that purpose. The samples of an application of electromigration techniques in determination of selected natural toxins in food are presented in Table 17.8.

17.9 Pesticides

Activities aimed at food production increase, and production and storage losses decrease involve first of all the common application of pesticides whose use increased rapidly after the Second World War. From many groups of pesticides, herbicides (weeds killing agents), fungicides (anti-fungal agents), and insecticides (insect exterminants) are used in the highest amounts. They are mostly low-molecular organic compounds of very differentiated structure, polarity, and durability. Due to their common application of usually many agents on the same cultivations, they are present in environmental samples (water, soil) and their

Table 17.8 Methods of determination of natural toxins in selected victuals

Kind of samples	Component	Preconcentration method	Detection method	LOD	References
Coffee, wine	Ochratoxine A	Extraction on magnetic nanoparticles coated with albumin	CE-MS	4 µg/L	[149]
Apple juice	Patuline	Liquid–liquid extraction	MEKS-UV	0.7–3 µg/L	[150]
Maze, corn	Zearalenon	SFE	CZE-AD	<35 µg/L	[151]
	Aflatoxins (AFB1, AFB2, AFG1, AFG2)	Affinity SPE	MEKC-LIF	1 ng/g	[152]
	Fuminisins (FB1, FB2)	C18 SPE	CE-LIF	50 ng/g	[153]
	Mykotoxin ZEN	Liquid–liquid extraction, Immunoaffinity SPE	CE-LIF	5 ng/g	[154]
Wine	Ochratoxin A	SLM-C18SPE	CE-LIF	30 µg/L	[155]
Crustaceans	Yesotoxins (YTX)	SPE	CE-MS	0.02–0.08 µg/L	[156]
Drinking water	Mikrocystins (MC-YR, MC-LR, MC-RR)	SPE	CZE-UV	0.076–0.110 µg/L	[157]

residues and metabolites in cultivated plans and food. Taking into consideration the fact that they are usually compounds of toxic activity the suitable agencies of most countries over the world, including European Union and USA, determined their allowable concentration on a level of 0.1 μg/L for single pesticide and 0.5 μg/L for total content of pesticides in water samples, and in a food in the range of 100–200 μg/L (kg). Therefore, quick and cheap analytical methods allowing determination of low concentrations both of single pesticides and mixtures of these compounds in various food samples (plants, fruits, juices etc.) are necessary. Capillary electrophoresis, due to the possibility of determination of a wide range of compounds, simplicity of performance, and small cost of analysis is more and more often used in determination of pesticides in food samples. However, due to limited sensitivity of that method, the elaboration of new procedures of pesticides determination is focused mainly on the strategy of analytes purification and concentration. In order to achieve required detection limits, after pesticides separation from the matrix using solid–liquid or liquid–liquid extraction, the liquid–liquid extraction and solid phase extraction (SPE) are used the most often for purification and concentration using different sorbents, and more rarely membrane methods or extraction using liquid membranes (Table 17.9).

17.10 Antibiotics

In veterinary medicine, and especially in the process of animal production and industrial fish breeding, huge amounts of antibiotics are used, both in treatment of diseases of microbiological background and for preventive purposes. Among many groups of antibiotics, the most often and in the highest amounts used are the following ones: β-lactames, sulphonamides, quinolones, and also aminoglycosides and nitrofurane antibiotics. The result of this situation is presence of residues of different antibiotics and their metabolites in animal meat and also in milk, and its products as well. The separate problem is an occurrence of microorganisms resistant to activity of particular antibiotics, which may result in serious health problems. Therefore, the European Union and many other countries determined, similar like for pesticides, maximum allowable limits of antibiotics content in food. Chromatographic methods and also electromigration ones are commonly used for these compounds residues determination in food. In order to improve selectivity and sensibility of determination methods, the new procedures have been elaborated with an application of capillary electrophoresis.

CZE with UV detection was used to determine four different penicillins (ampicillin, penicillin G, amoxicillin, and cloxacillin), β-lactam antibiotics in milk obtaining LOD in the range of 0.5–1.1 mg/L after previous purification and concentration using C18-SPE columns [173]. Much lower LOD was obtained using CZE with DAD detection for determination of eight various antibiotics from that group, including penicillins, in milk. In that case the LOD were two orders lower and were from 2 to 10 μg/L, after previous concentration using SPE [174].

Table 17.9 Methods of determination of selected pesticides in victuals

Pesticide		Type of samples	Preconcentration method	Detection method	LOD	References
Herbicides	Chlorosulfuron, diuron, monuron and others	Oranges, tomatoes	SPE	MEKC-UV	50 µg/L	[158]
	Linuron, diuron, metobromuron	Soy-beans, potatoes	SPE	RP-CEC-UV	50 µg/L	[159]
	Atrazine	Fruit juices	SLM-SPE	MEKC-UV	50 µg/L	[160]
	Cyromazine	Milk, eggs	Liquid–liquid extraction	CZE-DAD	0.42 mg/L	[161]
	Triazines	Potatoes	SPE (carbon nanotubes)	NACE-UV	10–15 µg/kg	[162]
	Glyphosate and metabolites	Wheat	Liquid–liquid extraction	CE-MS	<5 mg/L	[163]
	Glyphosate and metabolites	Different plants	Liquid–liquid extraction	RP-CEC-UV	0.1 µg/mL	[164]
Fungicides	Carbendazim, o-phenylphenol, tiabendazol	Grapes, oranges, tomatoes	SPE	MEKC-UV	0.1–1 mg/kg	[165]
	Tiabendazol, carbendazim	Different fruits	SPE	LVSS-CZE-MS	5–50 µg/kg	[166]
	Picloram, o-phenylphenol	Oranges, tomatoes	SPME	CE-MS	0.05–5 mg/kg	[167]
	Pyrifenox, carbendazim	Mineral water	SPE (carbon nanotubes)	NACE-UV	0.058 µg/L	[168]
Insecticides	Carbamates	Vegetables	SPE	CEC-UV	0.05–1.6 mg/kg	[169]
	Phosphoroorganic	Potatoes	SPE	MEKC-UV	0.03 mg/kg	[170]
	Phosphoroorganic	Vegetables	Liquid–liquid extraction	CE-QD-LIF	0.05–0.18 mg/kg	[171]
	Pyretroides	Chinese cabbage	Liquid–liquid extraction, SPE	CEC-UV	50–80 µg/kg	[172]

The same CZE-DAD technique was used for determination of nine various sulphonamides in meat, after concentration using SPE, obtaining LOD in the range of 5–10 μg/kg [175]. Slightly lower LOD (0.1–0.5 μg/L) was obtained for 15 various sulphonamides determined in beef and poultry meat and in samples of honey using CZE-UV method. In that case, food samples were prepared and purified using only the series of liquid–liquid extraction [176]. Also, capillary electrophoresis with mass detector (CE-MS) is used for these antibiotics determination, and LOD are close to previous ones and are 0.6–1.1 μg/L [177]. Whereas quinolines are determined using CZE with an application of various detection methods, UV, LIF, MA, and potential gradient detection (PGD). The respective LOD after concentration using SPE of liquid–liquid extraction are as follows: 10–25 μg/kg in chicken meat using UV detection [178], 0.2–1.0 mg/L in fish samples (CEC-UV) [179], 23–65 mg/L in milk using LLE-CE-PGD [180] method, and finally 6 μg/L in samples of cow's milk determined using CZE-MS/MS [181]. Also, aminoglycoside antibiotics are determined in cow's milk obtaining LOD in the range of 0.5–1.5 μg/L using MEKC-LIF method [182], and tetracyclines residues using CE-MS method with LOD on a level of 7.14–14.9 μg/L [183]. While nitrofuran antibiotics and their metabolites were determined in shrimps obtaining LOD in the range of 0.19–2 mg/L using MEKC-UV method and concentration using SPE [184]. Very low LOD for 12 different antibiotics (4 sulphonamides, 4 β-lactams and 4 quinolones) ranging from 18 to 50 ng/kg were obtained when determining them in fish and meat using CE-MS after extraction and concentration using C18-SPE [185].

17.11 Summary

Capillary electrophoresis as an analytical technique is widely used in food analysis It may be also considered as alternate and complementary with respect to commonly used analytical techniques, such as high-performance liquid chromatography (HPLC) or gas chromatography (GC). An application of CE allows conducting determination of compounds of similar structures present on similar concentrations level. The CE technique may be used during the control of technological processes in the food industry and for identification of numerous compounds present in food. Due to numerous advantages of CE technique their common application in routine food analysis seems to be a real issue.

References

1. A. Cifuentes, Recent advances in the application of capillary electromigration methods for food analysis. Electrophoresis **27**, 283–303 (2006)
2. R.A. Frazier, A. Papadopoulou, Recent advances in the application of capillary electrophoresis for food analysis. Electrophoresis **24**, 4095–4105 (2003)

3. R.A. Frazier, Recent advances in capillary electrophoresis methods for food analysis. Electrophoresis **22**, 4197–4206 (2001)

4. R.A. Frazier, J.M. Ames, H.E. Nursten, The development and application of capillary electrophoresis methods for food analysis. Electrophoresis **20**, 3156–3180 (1999)

5. C.T. da Costa, D. Horton, S.A. Margolis, Analysis of anthocyanins in foods by liquid chromatography, liquid chromatography-mass spectrometry and capillary electrophoresis. J. Chromatogr. A **881**, 403–410 (2000)

6. H. Wang, G. Cao, R.L. Prior, Oxygen radical absorbing capacity of anthocyanins. J. Agric. Food Chem. **45**, 304–309 (1997)

7. M.T. Satue-Garcia, M. Heinonen, E.N. Frankel, Anthocyanins as antioxidants on human low-density lipoprotein and lecithin-liposome systems. J. Agric. Food Chem. **45**, 3362–33676 (1997)

8. P.M. Abuja, M. Murkovic, W. Pfannhauser, Antioxidant and pro-oxidant activities of elderberry (Sambucus nigra) extract in low-density lipoprotein oxidation. J. Agric. Food Chem. **46**, 4091–4096 (1998)

9. A.S. Meyer, O.S. Yi, D.A. Pearson, A.L. Waterhouse, E.N. Frankel, Inhibition of human low-density lipoprotein oxidation in relation to composition of phenolic antioxidants in grapes (Vitis vinifera). J. Agric. Food Chem. **45**, 1638–1643 (1997)

10. M. Herrero, E. Ibáñez, A. Cifuentes, A analysis of natural antioxidants by capillary electromigration methods. J. Sep. Sci. **28**, 883–897 (2005)

11. S.P. Wang, K.J. Huang, Determination of flavonoids by high-performance liquid chromatography and capillary electrophoresis. J. Chromatogr. A **1032**, 273–279 (2004)

12. Y. Peng, Q. Chu, F. Liu, J. Ye, Determination of phenolic constituents of biological interest in red wine by capillary electrophoresis with electrochemical detection. J. Agric. Food Chem. **52**, 153–156 (2004)

13. H.Y. Huang, W.C. Lien, C.W. Chiu, Comparison of microemulsion electrokinetic chromatography and micellar electrokinetic chromatography methods for the analysis of phenolic compounds. J. Sep. Sci. **28**, 973–981 (2005)

14. J.M. Herrero-Martínez, C. Ràfols, M. Rosés, J.L. Torres, E. Bosch, You have full text access to this content Mixed micellar electrokinetic capillary chromatography separation of depolymerized grape procyanidins. Electrophoresis **24**, 707–713 (2003)

15. P. Bednar, A.V. Tomassi, C. Presutti, M. Pavlikova, K. Lemr, S. Fanali, Separation of structurally related anthocyanins by MEKC. Chromatographia **58**, 283–287 (2003)

16. R. Hamoudova, M. Urbanek, M. Pospisilova, M. Polasek, Assay of phenolic compounds in red wine by on-line combination of capillary isotachophoresis with capillary zone electrophoresis. J. Chromatogr. A **1032**, 281–287 (2004)

17. C.O. Okunji, T.A. Ware, R.P. Hicks, M.M. Iwu, D.J. Skanchy, Capillary electrophoresis determination of biflavanones from Garcinia kola in three traditional African medicinal formulations. Planta Med. **68**, 440–444 (2002)

18. X.H. Sun, C.L. Gao, W.D. Cao, X.R. Yang, Capillary electrophoresis with amperometric detection of curcumin in Chinese herbal medicine pretreated by solid-phase extraction. J. Chromatogr. A **62**, 117–125 (2002)

19. Q.C. Chu, W.Q. Qu, Y.Y. Peng, Q.H. Cao, J.N. Ye, Determination of flavonoids in Hippophae rhamnoides L and its phytopharmaceuticals by capillary electrophoresis with electrochemical detection. Chromatographia **58**, 67–71 (2003)

20. A. Müeller, M. Ganzera, H. Stuppner, Analysis of the aerial parts of Verbena officinalis L by micellar electrokinetic capillary chromatography. Chromatographia **60**, 193–197 (2004)

21. A.L. Crego, E. Ibáñez, E. García, R. de Pablos Rodríguez, F.J. Señoráns, G. Reglero, A. Cifuentes, Capillary electrophoresis separation of rosemary antioxidants from subcritical water extracts. Eur. Food Res. Technol. **219**, 549–555 (2004)

22. M. Herrero, D. Arráez-Román, A. Segura, E. Kenndler, B. Gius, M.A. Raggi, E. Ineza, A. Cifuentes, Pressurized liquid extraction–capillary electrophoresis–mass spectrometry for the analysis of polar antioxidants in rosemary extracts. J. Chromatogr. A **1084**, 54–62 (2005)

23. M. Bonoli, M. Pelillo, G. Lercker, Fast separation and determination of carnosic acid and rosmarinic acid in different rosemary (Rosmarinus officinalis) extracts by capillary zone electrophoresis with ultra violet-diode array detection. Chromatographia **57**, 505–512 (2003)

24. C.H.N. Chen, C. Liang, J. Lai, Y. Tai, J.S. Tsay, J.K. Lin, Capillary electrophoretic determination of theanine, caffeine, and catechins in fresh tea leaves and oolong tea and their effects on rat neurosphere adhesion and migration. J. Agric. Food Chem. **51**, 7495–7503 (2003)

25. D.J. Weiss, C.R. Anderton, Determination of catechins in match a green tea by micellar electrokinetic chromatography. J. Chromatogr. A **1011**, 173–180 (2003)

26. M. Bonoli, M. Montanucci, T.G. Toschi, G. Lercker, Fast separation and determination of tyrosol, hydroxytyrosol and other phenolic compounds in extra-virgin olive oil by capillary zone electrophoresis with ultraviolet-diode array detection. J. Chromatogr. A **1011**, 163–172 (2003)

27. A. Carrasco-Pancorbo, A. Segura-Carretero, A. Fernandez-Gutierrez, Co-electroosmotic capillary electrophoresis determination of phenolic acids in commercial olive oil. J. Sep. Sci. **28**, 925–934 (2005)

28. A. Carrasco-Pancorbo, C. Cruces-Blanco, A. Segura-Carretero, A. Fernandez-Gutierrez, Sensitive determination of phenolic acids in extra-virgin olive oil by capillary zone electrophoresis. J. Agric. Food Chem. **52**, 6687–6693 (2004)

29. D.J. Watson, A.A. Bushway, R.J. Bushway, Separation of peonidin and cyanidin, two anthocyanidins, in cranberries by capillary electrophoresis. J. Liq. Chromatogr. Rel. Technol. **27**, 113–121 (2004)

30. L.J. Juang, S.J. Sheu, T.C. Lin, Determination of hydrolyzable tannins in the fruit of *Terminalia chebula* Retz by high-performance liquid chromatography and capillary electrophoresis. J. Sep. Sci. **27**, 718–724 (2004)

31. Y. Peng, F. Liu, Y. Peng, J. Ye, Determination of polyphenols in apple juice and cider by capillary electrophoresis with electrochemical detection. Food Chem. **92**, 169–175 (2005)

32. Y. Peng, Q. Chu, F. Liu, J. Ye, Determination of isoflavones in soy products by capillary electrophoresis with electrochemical detection. Food Chem. **87**, 135–139 (2004)

33. M. Herrero, E. Ibáñez, J. Señoráns, A. Cifuentes, Pressurized liquid extracts from Spirulina platensis microalga: Determination of their antioxidant activity and preliminary analysis by micellar electrokinetic chromatography. J. Chromatogr. A **1047**, 195–203 (2004)

34. K. Truus, M. Vaher, M. Koel, A. Mähar, I. Taure, Analysis of bioactive ingredients in the brown alga Fucus vesiculosus by capillary electrophoresis and neutron activation analysis. Anal. Bioanal. Chem. **379**, 849–852 (2004)

35. R. Gotti, J. Fiori, F. Mancini, V. Cavrini, Modified micellar electrokinetic chromatography in the analysis of catechins and xanthines in chocolate. Electrophoresis **25**, 3282–3291 (2004)

36. L. Gao, Q. Chu, J. Ye, Determination of trans-Resveratrol in wines, herbs and health food by capillary electrophoresis with electrochemical detection. Food Chem. **78**, 255–260 (2002)

37. V. Brandolini, A. Maietti, P. Tedeschi, E. Durini, S. Vertuani, S. Manfredini, Capillary electrophoresis determination, synthesis, and stability of Resveratrol and related 3–O–β–d–Glucopyranosides. J. Agric. Food Chem. **50**, 7407–7411 (2002)

38. R. Saenz-Lopez, P. Fernandez-Zurbano, M.T. Tena, Development and validation of a capillary zone electrophoresis method for the quantitative determination of anthocyanins in wine. J. Chromatogr. A **990**, 247–258 (2003)

39. M. Vaher, M. Koel, Separation of polyphenolic compounds extracted from plant matrices using capillary electrophoresis. J. Chromatogr. A **990**, 225–230 (2003)

40. X. Li, Y. Zhang, Z. Yuan, Separation and determination of rutin and quercetin in the flowers of Sophora japonica L by capillary electrophoresis with electrochemical detection. Chromatographia **55**, 243–246 (2002)

41. L. Suntornsuk, S. Kasemsook, S. Wongyai, Quantitative analysis of aglycone quercetin in mulberry leaves (*Morus alba* L) by capillary zone electrophoresis. Electrophoresis **24**, 1236–1241 (2003)

42. J.M. Herrero-Martinez, C. Rafols, M. Roses, E. Bosch, C. Lozano, J.L. Torres, Micellar electrokinetic chromatography estimation of size and composition of procyanidins after thiolysis with cysteine. Electrophoresis **24**, 1404–1410 (2003)

43. R. Pompanio, R. Gotti, N.A. Santagati, V. Cavrini, Analysis of catechins in extracts of *Cistus* species by microemulsion electrokinetic chromatography. J. Chromatogr. A **990**, 215–223 (2003)

44. R. Pompanio, R. Gotti, B. Luppi, V. Cavrini, Microemulsion electrokinetic chromatography for the analysis of green tea catechins: Effect of the cosurfactant on the separation selectivity. Electrophoresis **24**, 1658–1667 (2003)

45. M. Bonoli, M. Pelillo, T.G. Toschi, G. Lercker, Analysis of green tea catechins: comparative study between HPLC and HPCE. Food Chem. **81**, 631–638 (2003)

46. M. Bonoli, P. Colabufalo, M. Pelillo, T.G. Toschi, G. Lercker, Fast determination of catechins and xanthines in tea beverages by micellar electrokinetic chromatography. J. Agric. Food Chem. **51**, 1141–1147 (2003)

47. L.P. Wright, J.P. Aucamp, Z. Apostolides, Analysis of black tea theaflavins by non-aqueous capillary electrophoresis. J. Chromatogr. A **919**, 205–213 (2001)

48. L.P. Wright, N.I.K. Mphangwe, H.E. Nyirenda, Z. Apostolides, Analysis of the theaflavin composition in black tea for predicting the quality of tea produced in Central and Southern Africa. J. Sci. Food Agric. **82**, 517–525 (2002)

49. M.D. del Castillo, J.M. Ames, M.H. Gordon, Effect of roasting on the antioxidant activity of coffee brews. J. Agric. Food Chem. **50**, 3698–3703 (2002)

50. P. Charurin, J.M. Ames, M.D. del Castillo, Antioxidant activity of coffee model systems. J. Agric. Food Chem. **50**, 3751–3756 (2002)

51. J. Vacek, B. Klejdus, L. Lojková, V. Kubán, Current trends in isolation, separation, determination and identification of isoflavones: a review. J. Sep. Sci. **31**, 2054–2067 (2008)

52. J.A. Starkey, Y. Mechref, C.K. Byun, R. Steinmetz, J.S. Fuqua, O.H. Pescovitz, M.V. Novotny, Determination of trace isoflavone phytoestrogens in biological materials by capillary electrochromatography. Anal. Chem. **74**, 5998–6005 (2002)

53. M.A. Aramendia, I. Garcia, F. Lafont, J.M. Marinas, Determination of isoflavones using capillary electrophoresis in combination with electrospray mass spectrometry. J. Chromatogr. A **707**, 327–333 (1995)

54. G.S. McLeod, M.J. Shepherd, Determination of the ionisation constants of isoflavones by capillary electrophoresis. Phytochem. Anal. **11**, 322–326 (2000)

55. G. Dinelli, I. Aloisio, A. Bonetti, I. Marotti, A. Cifuentes, Compositional changes induced by UV-B radiation treatment of common bean and soybean seedlings monitored by capillary electrophoresis with diode array detection. J. Sep. Sci. **30**, 604–611 (2007)

56. Y. Cao, Ch. Lou, X. Zhang, Q. Chu, Y. Fang, J. Ye, Determination of puerarin and daidzein in *Puerariae radix* and its medicinal preparations by micellar electrokinetic capillary chromatography with electrochemical detection. Anal. Chim. Acta **452**, 123–128 (2002)

57. P. Li, S.P. Li, Y.T. Wang, Optimization of CZE for analysis of phytochemical bioactive compounds. Electrophoresis **27**, 4808–4819 (2006)

58. P. Jáč, M. Polášek, M. Pospišilová, Recent trends in the determination of polyphenols by electromigration methods. J. Pharma Biomed. Anal. **40**, 805–814 (2006)

59. D.J. Allen, J.C. Gray, N.L. Paiva, J.T. Smith, An enantiomeric assay for the flavonoids medicarpin and vestitone using capillary electrophoresis. Electrophoresis **21**, 2051–2057 (2000)

60. S. Kodama, A. Yamamoto, A. Matsunaga, H. Yanai, Direct enantioseparation of catechin and epicatechin in tea drinks by 6-O-α-D-glucosyl-β-cyclodextrin-modified micellar electrokinetic chromatography. Electrophoresis **25**, 2892–2898 (2004)

61. J.J.B. Nevado, A.M.C. Salcedo, G.C. Penalvo, Simultaneous determination of cis- and trans-resveratrol in wines by capillary zone electrophoresis. Analyst **124**, 61–66 (1999)

62. X. Gu, L. Creasy, A. Kester, M. Zeece, Capillary electrophoretic determination of resveratrol in wines. J. Agric. Food Chem. **47**, 3223–3227 (1999)
63. T. Watanabe, S. Terabe, Analysis of natural food pigments by capillary electrophoresis. J. Chromatogr. A **880**, 311–322 (2000)
64. X. Sun, X. Yang, E. Wang, Chromatographic and electrophoretic procedures for analyzing plant pigments of pharmacologically interests. Anal. Chim. Acta **547**, 153–157 (2005)
65. T. Watanabe, N. Hasegawa, A. Yamamoto, S. Nagai, S. Terabe, Capillary electrophoresis of anthraquinone pigments for food. Bunseki Kagaku **45**, 765–770 (1996)
66. T. Watanabe, N. Hasegawa, A. Yamamoto, S. Nagai, S. Terabe, Separation and determination of yellow and red safflower pigments in food by capillary electrophoresis. Biosci. Biotech. Biochem. **61**, 1179–1183 (1997)
67. T. Watanabe, A. Yamamoto, S. Nagai, S. Terabe, Separation and determination of yellow gardenia pigments for food and iridoid constituents in gardenia fruits by micellar electrokinetic chromatography. Food Sci. Technol. Int. Tokyo **4**, 54–58 (1998)
68. T. Watanabe, A. Yamamoto, S. Nagai, S. Terabe, Separation and determination of monascus yellow pigments for food by micellar electrokinetic chromatography. Anal. Sci. **3**, 571–575 (1997)
69. T. Watanabe, A. Yamamoto, S. Nagai, S. Terabe, Analysis of elderberry pigments in commercial food samples by Micellar electrokinetic chromatography. Anal. Sci. **14**, 839–840 (1998)
70. T.R.I. Cataldi, D. Nardiello, G.E. De Benedetto, S.A. Bufo, Optimizing separation conditions for riboflavin, flavin mononucleotide and flavin adenine dinucleotide in capillary zone electrophoresis with laser-induced fluorescence detection. J. Chromatogr. A **968**, 229–239 (2002)
71. T.R.I. Cataldi, D. Nardiello, L. Scrano, A. Scopa, Assay of riboflavin in sample wines by capillary zone electrophoresis and laser-induced fluorescence detection. J. Agric. Food Chem. **50**, 6643–6647 (2002)
72. T.R.I. Cataldi, D. Nardiello, V. Carrara, R. Ciriello, G.E. De Benedetto, Assessment of riboflavin and flavin content in common food samples by capillary electrophoresis with laser-induced fluorescence detection. Food Chem. **82**, 309–314 (2003)
73. P. Britz-McKibbin, K. Otsuka, S. Terabe, On-line focusing of flavin derivatives using dynamic pH junction-sweeping capillary electrophoresis with laser-induced fluorescence detection. Anal. Chem. **74**, 3736–3743 (2002)
74. P. Britz-McKibbin, M.J. Markuszewski, T. Iyanagi, K. Matsuda, T. Nishioka, S. Terabe, Picomolar analysis of flavins in biological samples by dynamic pH junction-sweeping capillary electrophoresis with laser-induced fluorescence detection. Anal. Biochem. **313**, 89–96 (2003)
75. M.C. Boyce, Determination of additives in food by capillary electrophoresis. Electrophoresis **22**, 1447–1459 (2001)
76. D. Kaniansky, M. Masar, V. Madajova, J. Marak, Determination of sorbic acid in food products by capillary zone electrophoresis in a hydrodynamically closed separation compartment. J. Chromatogr. A **677**, 179–185 (1994)
77. I. Pant, V. Trenerry, The determination of sorbic acid and benzoic acid in a variety of beverages and foods by micellar electrokinetic capillary chromatography. Food Chem. **53**, 219–226 (1995)
78. K. Waldron, J. Li, Investigation of a pulsed-laser thermo-optical absorbance detector for the determination of food preservatives separated by capillary electrophoresis. J. Chromatogr. B **683**, 47–54 (1996)
79. K. Kuo, Y. Hsieh, Determination of preservatives in food products by cyclodextrin-modified capillary electrophoresis with multiwavelength detection. J. Chromatogr. A **768**, 334–341 (1997)
80. V. Trenerry, The determination of the sulphite content of some foods and beverages by capillary electrophoresis. Food Chem. **55**, 299–303 (1996)

81. P. Kuban, B. Karlberg, On-line coupling of gas diffusion to capillary electrophoresis. Talanta **45**, 477–484 (1998)

82. C. Hall III, A. Zhu, G. Zeece, Comparison between capillary electrophoresis and high-performance liquid chromatography separation of food grade antioxidants. J. Agric. Food Chem. **42**, 919–921 (1994)

83. M.C. Boyce, E.E. Spickett, Separation of food grade antioxidants (synthetic and natural) using mixed micellar electrokinetic capillary chromatography. J. Agric. Food Chem. **47**, 1970–1975 (1999)

84. J. Summanen, H. Vuorela, R. Hiltunen, H. Siren, M. Riekkola, Determination of phenolic antioxidants by capillary electrophoresis with ultraviolet detection. J. Chromatogr. Sci. **33**, 704–711 (1995)

85. M. Delgado-Zamarreno, A. Sanchez-Perez, I. Maza, J. Hemandez-Mendez, Micellar electrokinetic chromatography with bis(2-ethylhexyl)sodium sulfosuccinate vesicles: Determination of synthetic food antioxidants. J. Chromatogr. A **871**, 403–414 (2000)

86. E. Koh, M. Bissell, R. Ito, Measurement of vitamin C by capillary electrophoresis in biological fluids and fruit beverages using a stereoisomer as an internal standard. J. Chromatogr. A **633**, 245–250 (1993)

87. P.A. Marshall, V.C. Trenerry, C.O. Thompson, The determination of total ascorbic acid in beers, wines, and fruit drinks by micellar electrokinetic capillary chromatography. J. Chromatogr. Sci. **33**, 426–432 (1995)

88. C.O. Thompson, V.C. Trenerry, B. Kemmery, Micellar electrokinetic capillary chromatographic determination of artificial sweeteners in low-Joule soft drinks and other foods. J. Chromatogr. A **694**, 507–514 (1995)

89. C. Thompson, V.C. Trenerry, B. Kemmery, Determination of cyclamate in low joule foods by capillary zone electrophoresis with indirect ultraviolet detection. J. Chromatogr. A **704**, 203–210 (1995)

90. J.J. Pesek, M.T. Matyska, Determination of aspartame by high-performance capillary electrophoresis. J. Chromatogr. A **781**, 423–428 (1997)

91. H.Y. Aboul-Enein, S.A. Bakr, Comparative study of the separation and determination of aspartame and its decomposition products in bulk material and diet soft drinks by HPLC and CE. J. Liq. Chromatogr. Rel. Technol. **20**, 1437–1444 (1997)

92. J. Liu, S.F.Y. Li, Separation and determination of stevia sweeteners by capillary electrophoresis and high performance liquid chromatography. J. Liq. Chromatogr. Rel. Technol. **18**, 1703–1719 (1995)

93. C. Corradini, A. Cavazza, Application of capillary zone electrophoresis (CZE) and micellar electrokinetic chromatography (MEKC) in food analysis. Ital. J. Food Sci. **10**, 299–316 (1998)

94. S. Suzuki, M. Shirao, M. Aizawa, H. Nakazawa, K. Sasa, H. Sasagawa, Determination of synthetic food dyes by capillary electrophoresis. J. Chromatogr. A **680**, 541–547 (1994)

95. S. Razee, A. Tamura, T. Masujima, Improvement in the determination of food additive dyestuffs by capillary electrophoresis using β-cyclodextrin. J. Chromatogr. A **715**, 179–1881 (1995)

96. M. Masar, D. Kaniansky, V. Madajova, Separation of synthetic food colourants by capillary zone electrophoresis in a hydrodynamically closed separation compartment. J. Chromatogr. A **724**, 327–336 (1996)

97. L.K. Kuo, H.Y. Huang, Y.Z. Hsieh, High-performance capillary electrophoretic analysis of synthetic food colorants. Chromatographia **47**, 249–256 (1998)

98. H. Lui, T. Zhu, Y. Zhang, S. Qi, A. Huang, Y. Sun, Determination of synthetic colourant food additives by capillary zone electrophoresis. J. Chromatogr. A **718**, 448–453 (1995)

99. J.J. Berzas Nevado, C. Guiberteau Cabanillas, A.M. Contento Salcedo, Method development and validation for the simultaneous determination of dyes in foodstuffs by capillary zone electrophoresis. Anal. Chim. Acta. **378**, 63–71 (1999)

100. C. Thompson, V. Trenerry, Determination of synthetic colours in confectionery and cordials by micellar electrokinetic capillary chromatography. J. Chromatogr. A **704**, 195–201 (1995)

101. P. Bridle, C. Garcia-Viguera, F.A. Tomas-Barberan, Analysis of anthocyanins by capillary zone electrophoresis. J. Liq. Chromatogr. Rel. Technol. **19**, 537–545 (1996)
102. J.S. Coffey, L. Castle, Analysis for caramel colour (Class III). Food Chem. **51**, 413–416 (1994)
103. L. Royle, J.M. Ames, L. Castle, H.E. Nursten, C.M. Radcliffe, Identification and quantification of Class IV caramels using capillary electrophoresis and its application to soft drinks. J. Sci. Food Agric. **76**, 579–587 (1998)
104. M. Castro-Puyana, A.L. Crego, M.L. Marina, C. Garcia-Ruiz, CE methods for the determination of non-protein amino acids in foods. Electrophoresis **28**, 4031–4045 (2007)
105. V. Garcia-Canas, A. Cifuentes, Recent advances in the application of capillary electromigration methods for food analysis. Electrophoresis **29**, 294–309 (2008)
106. M. Herrero, V. Garcia-Canas, C. Simo, A. Cifuentes, Recent advances in the application of capillary electromigration methods for food analysis and Foodomics. Electrophoresis **31**, 205–228 (2010)
107. A. Santalad, P. Terrapornchaisit, R. Burakham, S. Srijanarai, Pre-capillary derivatisation and capillary zone electrophoresis for amino acids analysis in beverages. Ann. Chim. **97**, 935–945 (2007)
108. F. Chen, S. Wang, W. Guo, M. Hu, Determination of amino acids in *Sargassum fusiforme* by high performance capillary electrophoresis. Talanta **66**, 755–761 (2005)
109. A. Carrasco-Pancorbo, A. Cifuentes, S. Cortacero-Ramirez, A. Segura-Carretero, A. Fernandez-Guiterrez, Coelectroosmotic capillary electrophoresis of phenolic acids and derivatized amino acids using N, N-dimethylacrylamide-ethylpyrrolidine methacrylate physically coated capillaries. Talanta **71**, 397–405 (2007)
110. Y.P. Lin, Y.S. Su, J.F. Jen, Capillary electrophoretic analysis of γ-aminobutyric acid and alanine in tea with in-capillary derivatization and fluorescence detection. J. Agric. Food Chem. **55**, 2103–2108 (2007)
111. S. Hunt, The Non-Protein Amino Acids, in *Chemistry and biotechnology of amino acids*, ed. by G.C. Barrett (Chapman and Hall, London, 1985), pp. 55–183
112. F. Kvasnička, Capillary electrophoresis in food authenticity. J. Sep. Sci. **28**, 813–825 (2005)
113. R.W. Peace, G.S. Gilani, Chromatographic determination of amino acids in foods. J. AOAC Int. **88**, 877–887 (2005)
114. B. Michalke, Capillary electrophoresis methods for a clear identification of seleno amino acids in complex matrices like human milk. Fresen. J. Anal. Chem. **351**, 670–677 (1995)
115. B. Michalke, P. Schramel, Application of capillary zone electrophoresis–inductively coupled plasma mass spectrometry and capillary isoelectric focusing–inductively coupled plasma mass spectrometry for selenium speciation. J. Chromatogr. A **807**, 71–80 (1998)
116. Y. Cao, X. Zhang, O. Chu, Y. Fang, J. Ye, Determination of taurine in *Lycium Barbarum* L and other foods by capillary electrophoresis with electrochemical detection. Electroanalysis **15**, 898–902 (2003)
117. C. Simo, A. Rizzi, C. Barbas, A. Cifuentes, Chiral capillary electrophoresis-mass spectrometry of amino acids in foods. Electrophoresis **26**, 1432–1441 (2005)
118. V. Prokoratova, F. Kvasnička, R. Ševčik, M. Voldřich, Capillary electrophoresis determination of carnitine in food supplements. J. Chromatogr. A **1081**, 60–64 (2005)
119. J.P. Aucamp, Y. Hara, Z. Apostolides, Simultaneous analysis of tea catechins, caffeine, gallic acid, theanine and ascorbic acid by micellar electrokinetic capillary chromatography. J. Chromatogr. A **876**, 235–242 (2000)
120. H.-M. Chang, C.-F. Tsai, C.-F. Li, Changes of amino acid composition and lysinoalanine formation in alkali-pickled duck eggs. J. Agric. Food Chem. **47**, 1495–1500 (1999)
121. X. Chen, J. Zhang, H. Zhai, X. Chen, Z. Hu, Determination of levodopa by capillary zone electrophoresis using an acidic phosphate buffer and its application in the analysis of beans. Food Chem. **92**, 381–386 (2005)
122. P. Rozan, Y.-H. Kuo, F. Lambein, Amino acids in seeds and seedlings of the genus *Lens*. Phytochemistry **58**, 281–289 (2001)

123. B. Chowdhurry, P. Rozan, Y.-H. Kuo, M. Sumino, F. Lambein, Identification and quantification of natural isoxazolinone compounds by capillary zone electrophoresis. J. Chromatogr. A **933**, 129–136 (2001)

124. V. Dolnik, Capillary electrophoresis of proteins 2003–2005. Electrophoresis **27**, 126–141 (2006)

125. P. Clement, S. Agboola, R. Bencini, A study of polymorphism in milk proteins from local and imported dairy sheep in Australia by capillary electrophoresis. Food Sci. Technol. **39**, 63–69 (2006)

126. B. Mirrales, M. Ramos, L. Amigo, Characterization of fresh cheeses by capillary electrophoresis. Milchwissenschaft **60**, 278–282 (2005)

127. B. Vallejo-Cordoba, R. Rodriguez-Ramirez, A.F. Gonzales-Cordova, Capillary electrophoresis for bovine and ostrich meat characterization. Food Chem. **120**, 304–307 (2010)

128. A.V. Jager, F.G. Tonin, M.F.M. Tavares, Comparative evaluation of extraction procedures and method validation for determination of carbohydrates in cereals and dairy products by capillary electrophoresis. J. Sep. Sci. **30**, 586–594 (2007)

129. G.M. Morales-Cid, B. Simonet, S. Cardenas, M. Valcarcel, On-capillary sample cleanup method for the electrophoretic determination of carbohydrates in juice samples. Electrophoresis **28**, 1557–1563 (2007)

130. S. Rovio, J. Yli-Kauhaluoma, H. Siren, Determination of neutral carbohydrates by CZE with direct UV detection. Electrophoresis **28**, 3129–3135 (2007)

131. P. Rodriguez-Plaza, R. Gonzales, M.V. Moreno-Arribas, M.C. Polo, G. Bravo, J.M. Martinez-Zapater, M.C. Martinez, A. Cifuentez, Combining microsatellite markers and capillary gel electrophoresis with laser-induced fluorescence to identify the grape (Vitis vinifera) variety of musts. Eur. Food Res. Technol. **223**, 625–631 (2006)

132. S. Archak, V. Lakshminarayanareddy, J. Nagaraju, High-throughput multiplex microsatellite marker assay for detection and quantification of adulteration in Basmati rice (*Oryza sativa*). Electrophoresis **28**, 2396–2405 (2007)

133. D. Melchiade, I. Foroni, G. Corrado, I. Santangelo, R. Rao, Authentication of the 'Annurca' apple in agro-food chain by amplification of microsatellite loci. Food Biotechnol. **21**, 33–43 (2007)

134. S. Spaniolas, C. Bazakos, M. Awad, P. Kalaitzis, Exploitation of the chloroplast *trn*L (UAA) intron polymorphisms for the authentication of plant oils by means of a lab-on-a-chip capillary electrophoresis system. J. Agric. Food Chem. **56**, 6886–6891 (2008)

135. M.H. Oh, Y.S. Park, S.H. Paek, H.Y. Kim, A rapid and sensitive method for detecting foodborne pathogens by capillary electrophoresis-based single-strand conformation polymorphism, Jung GY, Oh S. Food Control **19**, 1100–1104 (2008)

136. V. Garcia-Canas, R. Gonzales, A. Cifuentes, Detection of genetically modified organisms in foods by DNA amplification techniques. J. Sep. Sci. **25**, 577–583 (2002)

137. V. Garcia-Canas, R. Gonzales, A. Cifuentes, Ultrasensitive detection of genetically modified maize DNA by capillary gel electrophoresis with laser-induced fluorescence using different fluorescent intercalating dyes. J. Agric. Food Chem. **50**, 1016–1021 (2002)

138. A. Nadal, A. Coll, J.L. La Paz, T. Esteve, M. Pla, A new PCR-CGE (size and color) method for simultaneous detection of genetically modified maize events. Electrophoresis **27**, 3879–3888 (2006)

139. Y. Zhou, Y. Li, S. Pei, Determination of genetically modified soybean by multiplex PCR and CGE with LIF detection. Chromatographia **66**, 691–696 (2007)

140. A. Ehlert, F. Moreano, U. Busch, K. Engel, Development of a modular system for detection of genetically modified organisms in food based on ligation-dependent probe amplification. Eur. Food Res. Technol. **227**, 805–812 (2008)

141. N. Yan, L. Zhou, Z. Zhu, X. Chen, Determination of melamine in dairy products, fish feed, and fish by capillary zone electrophoresis with diode array detection. J. Agric. Food Chem. **57**, 807–811 (2009)

142. L. Meng, G. Shen, X. Hou, L. Wang, Determination of melamine in food by SPE and CZE with UV detection. Chromatographia **70**, 991–994 (2009)
143. Y. Wen, H. Liu, P. Han, F. Luan, X. Li, Determination of melamine in milk powder, milk and fish feed by capillary electrophoresis: a good alternative to HPLC. J. Sci. Food Agric. **90**, 2178–2182 (2010)
144. S. Baskan, F.B. Erim, NACE for the analysis of acrylamide in food. Electrophoresis **28**, 4108–4113 (2007)
145. N. Garcia-Villar, J. Saurina, S. Hernandez-Cassou, Capillary electrophoresis determination of biogenic amines by field-amplified sample stacking and in-capillary derivatization. Electrophoresis **27**, 474–483 (2006)
146. J. Ruiz-Jimenez, M.D. Luque de Castro, Pervaporation as interface between solid samples and capillary electrophoresis: Determination of biogenic amines in food. J. Chromatogr. A **1110**, 245–253 (2006)
147. S. Cortacero-Ramirez, D. Arraez-Roman, Determination of biogenic amines in beers and brewing-process samples by capillary electrophoresis coupled to laser-induced fluorescence detection, Segura-Carretero A, Fernandez-Gutierrez A. Food Chem. **100**, 383–389 (2007)
148. C. Simo, M.V. Moreno-Arribas, A. Cifuentes, Ion-trap versus time-of-flight mass spectrometry coupled to capillary electrophoresis to analyze biogenic amines in wine. J. Chromatogr. A **1195**, 150–156 (2008)
149. C.Y. Hong, Y.C. Chen, Selective enrichment of ochratoxin A using human serum albumin bound magnetic beads as the concentrating probes for capillary electrophoresis/electrospray ionization-mass spectrometric analysis. J. Chromatogr. A **1159**, 250–255 (2007)
150. M. Murillo, E. Gonzalez-Penas, S. Amezqueta, Determination of patulin in commercial apple juice by micellar electrokinetic chromatography. Food Chem. Toxicol. **46**, 57–64 (2008)
151. A. Sanchez-Arribas, E. Bermejo, A. Zapardiel, H. Tellez, J. Rodriguez-Flores, M. Zougagh, A. Rios, M. Chicharro, Screening and confirmatory methods for the analysis of macrocyclic lactone mycotoxins by CE with amperometric detection. Electrophoresis **30**, 499–506 (2009)
152. C.M. Maragos, J.I. Greer, Analysis of aflatoxin B$_1$ in corn using capillary electrophoresis with laser-induced fluorescence detection. J. Agric. Food Chem. **45**, 1139–1147 (1997)
153. C.M. Maragos, G.A. Bennett, J.L. Richard, Analysis of fumonisin B$_1$ in corn by capillary electrophoresis. Adv. Exp. Med. Biol. **392**, 105–112 (1996)
154. C.M. Maragos, M.J. Appell, V. Lippolis, A. Visconti, L. Catucci, M. Pascale, Use of cyclodextrins as modifiers of fluorescence in the detection of mycotoxins. Food Addit. Contam. **25**, 164–171 (2008)
155. S. Almeda, L. Arce, M. Valcarcel, Combined use of supported liquid membrane and solid-phase extraction to enhance selectivity and sensitivity in capillary electrophoresis for the determination of ochratoxin A in wine. Electrophoresis **29**, 1573–1581 (2008)
156. P. De la Iglesia, A. Gago-Martinez, Determination of yessotoxins and pectenotoxins in shellfish by capillary electrophoresis-electrospray ionization-mass spectrometry. Food Addit. Contam. **26**, 221–228 (2009)
157. G. Birungi, S.F. Li, Determination of cyanobacterial cyclic peptide hepatotoxins in drinking water using CE. Electrophoresis **30**, 2737–2742 (2009)
158. R. Rodriguez, Y. Pico, G. Font, J. Manes, Determination of urea-derived pesticides in fruits and vegetables by solid-phase preconcentration and capillary electrophoresis. Electrophoresis **22**, 2010–2016 (2001)
159. A. De Rossi, C. Desiderio, Application of reversed phase short end-capillary electrochromatography to herbicides residues analysis. Chromatographia **61**, 271–275 (2005)
160. M. Khrolenko, P. Dżygiel, P. Wieczorek, Combination of supported liquid membrane and solid-phase extraction for sample pretreatment of triazine herbicides in juice prior to capillary electrophoresis determination. J. Chromatogr. A **975**, 219–227 (2002)

161. H. Sun, N. Liu, L. Wang, Y. Wu, Effective separation and simultaneous detection of cyromazine and melamine in food by capillary electrophoresis. Electrophoresis **31**, 2236–2241 (2010)

162. R. Carabias-Martinez, E. Rodriguez-Gonzalo, E. Miranda-Cruz, J. Dominguez-Alvarez, J. Hernandez-Mendez, Sensitive determination of herbicides in food samples by nonaqueous CE using pressurized liquid extraction. Electrophoresis **28**, 3606–3616 (2007)

163. L. Goodwin, J.R. Stratin, B.J. Keely, D.M. Goodal, Analysis of glyphosate and glufosinate by capillary electrophoresis–mass spectrometry utilising a sheathless microelectrospray interface. J. Chromatogr. A **1004**, 107–119 (2003)

164. A.M. Rojano-Delgado, J. Ruiz-Jimenez, M.D. de Castro Luque, R. De Prado, Determination of glyphosate and its metabolites in plant material by reversed-polarity CE with indirect absorptiometric detection. Electrophoresis **31**, 1423–1430 (2010)

165. R. Rodriguez, Y. Pico, G. Font, J. Manes, Analysis of post-harvest fungicides by micellar electrokinetic chromatography. J. Chromatogr. A **924**, 387–396 (2001)

166. R. Rodriguez, Y. Pico, G. Font, J. Manes, Analysis of thiabendazole and procymidone in fruits and vegetables by capillary electrophoresis–electrospray mass spectrometry. J. Chromatogr. A **949**, 359–366 (2002)

167. R. Rodriguez, J. Manes, Y. Pico, Off-Line Solid-Phase Microextraction and capillary electrophoresis mass spectrometry to determine acidic pesticides in fruits. Anal. Chem. **75**, 452–459 (2003)

168. M. Asensio-Ramos, J. Hernandez-Borges, L.M. Ravelo-Perez, M.A. Rodriguez-Delgado, Simultaneous determination of seven pesticides in waters using multi-walled carbon nanotube SPE and NACE. Electrophoresis **29**, 4412–4421 (2008)

169. X. Wu, L. Wang, Z. Xie, J. Lu, Ch. Yan, P. Yang, Rapid separation and determination of carbamate insecticides using isocratic elution pressurized capillary electrochromatography. Electrophoresis **27**, 768–777 (2006)

170. T. Perez-Ruiz, C. Martinez-lozano, S.E. Bravo, Determination of organophosphorus pesticides in water, vegetables and grain by automated SPE and MEKC. Chromatographia **61**, 493–498 (2005)

171. Q. Chen, Y. Fung, Capillary electrophoresis with immobilized quantum dot fluorescence detection for rapid determination of organophosphorus pesticides in vegetables. Electrophoresis **31**, 3107–3114 (2010)

172. F. Ye, Z. Xie, X. Wu, X. Lin, Determination of pyrethroid pesticide residues in vegetables by pressurized capillary electrochromatography. Talanta **69**, 97–102 (2006)

173. S.M. Santos, M. Henriques, A.C. Duarte, V.I. Esteves, Development and application of a capillary electrophoresis based method for the simultaneous screening of six antibiotics in spiked milk samples. Talanta **71**, 731–737 (2007)

174. M.I. Bailon-Perez, A.M. Garcia-Campana, C. Cruces-Blanco, M. Iruela Del Olmo, Large-volume sample stacking for the analysis of seven β-lactam antibiotics in milk samples of different origins by CZE. Electrophoresis **28**, 4082–4090 (2007)

175. M.S. Fuh, S. Chu, Rapid determination of sulfonamides in milk using micellar electrokinetic chromatography with fluorescence detection. Anal. Chim. Acta. **552**, 110–115 (2005)

176. M. Ma, H.-S. Zhang, L.-Y. Ciao, L. Ciao, P. Wang, H.-R. Cui, H. Wang, Quaternary ammonium chitosan derivative dynamic coating for the separation of veterinary sulfonamide residues by CE with field-amplified sample injection. Electrophoresis **28**, 4091–4100 (2007)

177. B. Santos, A. Lista, B.M. Simonet, A. Rios, M. Valcarcel, Screening and analytical confirmation of sulfonamide residues in milk by capillary electrophoresis-mass spectrometry. Electrophoresis **26**, 1567–1575 (2005)

178. C. Horstkotter, E. Jimenez-Lozano, D. Barron, J. Barbosa, G. Blaschke, Determination of residues of enrofloxacin and its metabolite ciprofloxacin in chicken muscle by capillary electrophoresis using laser-induced fluorescence detection. Electrophoresis **23**, 3078–3083 (2002)

179. H. Lu, X. Wu, Z. Xie, L. Guo, C. Yan, G. Chen, Separation and determination of seven fluoroquinolones by pressurized capillary electrochromatography. J. Sep. Sci. **28**, 2210–2217 (2005)
180. Y. Fan, X. Gan, S. Li, W. Qin, A rapid CE-potential gradient detection method for determination of quinolones. Electrophoresis **28**, 4101–4107 (2007)
181. F.J. Lara, A.M. Garcia-Campana, F. Ales-Barrero, J.M. Bosque-Sendra, L.E. Garcia-Ayuso, Multiresidue method for the determination of quinolone antibiotics in bovine Raw milk by capillary electrophoresis–tandem mass spectrometry. Anal. Chem. **78**, 7665–7673 (2006)
182. J.M. Serrano, M. Silva, Trace analysis of aminoglycoside antibiotics in bovine milk by MEKC with LIF detection. Electrophoresis **27**, 4703–4710 (2006)
183. S. Wang, P. Yang, Y. Cheng, Analysis of tetracycline residues in bovine milk by CE-MS with field-amplified sample stacking. Electrophoresis **28**, 4173–4179 (2007)
184. P.U. Wickramanayake, T.C. Tran, J.G. Hughes, M. Macka, N. Simpson, P.J. Marriott, Simultaneous separation of nitrofuran antibiotics and their metabolites by using micellar electrokinetic capillary chromatography. Electrophoresis **27**, 4069–4077 (2006)
185. A. Juan-Garcia, G. Font, Y. Pico, Simultaneous determination of different classes of antibiotics in fish and livestock by CE-MS. Electrophoresis **28**, 4180–4191 (2007)

Chapter 18
Application of Electromigration Techniques in Environmental Analysis

Edward Bald, Paweł Kubalczyk, Sylwia Studzińska,
Ewelina Dziubakiewicz and Bogusław Buszewski

Abstract Inherently trace-level concentration of pollutants in the environment, together with the complexity of sample matrices, place a strong demand on the detection capabilities of electromigration methods. Significant progress is continually being made, widening the applicability of these techniques, mostly capillary zone electrophoresis, micellar electrokinetic chromatography, and capillary electrochromatography, to the analysis of real-world environmental samples, including the concentration sensitivity and robustness of the developed analytical procedures. This chapter covers the recent major developments in the domain of capillary electrophoresis analysis of environmental samples for pesticides, polycyclic aromatic hydrocarbons, phenols, amines, carboxylic acids, explosives, pharmaceuticals, and ionic liquids. Emphasis is made on pre-capillary and on-capillary chromatography and electrophoresis-based concentration of analytes and detection improvement.

18.1 Introduction

Recently, electromigration capillary techniques, both in their classical and microchip form, have been commonly applied in environmental analysis. Low concentration sensitivity of the methods based on these techniques has seriously limited their environmental applicability since they were developed. Detectors of high detection limits were used. However, within the last few years, a remarkable advance has been

E. Bald (✉) · P. Kubalczyk
Faculty of Chemistry, University of Łódź, Łódź, Poland
e-mail: ebald@uni.lodz.pl

S. Studzińska · E. Dziubakiewicz · B. Buszewski
Faculty of Chemistry, Nicolaus Copernicus University, Toruń, Poland

B. Buszewski et al. (eds.), *Electromigration Techniques*, Springer Series
in Chemical Physics 105, DOI: 10.1007/978-3-642-35043-6_18,
© Springer-Verlag Berlin Heidelberg 2013

made in this field. Procedures based on electromigration principles have become complementary to chromatographic methods in the analysis of real environmental samples for the presence of numerous pollutants. These are, for instance, pharmaceuticals, organic acids, polycyclic aromatic hydrocarbons, phenols, pesticides, amines, nitrosamines, explosives, chemical weapon residues, surfactants, or dyes. The advance was made due to (1) detection improvement and (2) new methods of analytes concentration before or in a capillary. New methods of environmental samples analysis based on electromigration techniques have become the topics of numerous review papers [1–16]. Approximately 3,000 papers concerning applications, including those on environmental sample analysis, have been published within the last 20 years. Hence, the authors found it reasonable to quote only selected papers, those discussing detection procedures and determination of main groups of compounds emitted to the environment, published within the last 5 years.

18.2 Detectors

A huge environmental complexity of samples and low analytes concentration (often in the presence of structurally similar compounds) poses a challenge to detectors. Therefore, detection methods improvement is a priority to electrophoretic research. Optical detection methods are commonly applied in commercially available CE apparatus as well as in a majority of application procedures. Research in the domain of absorption detection in the range of UV–Vis is focused on the improvement of precision, detection limit, and miniaturization. Attempts to lower detection limits are made by using extended light-path capillaries within a detector, derivatization, or applying light emitting diodes (LED). Beams of light emitted by LED do not, however, exhibit characteristics as advantageous as those emitted by laser sources.

Using laser-induced fluorescence (LIF) is usually connected with derivatization carried out prevalently outside an analytical system, as in the case of herbicides determination in river water by micellar electrokinetic chromatography (MEKC) [17]. Derivatization in a capillary was applied to determine organophosphorous pesticides in river water [18].

The application of electrochemical detection is justified by relatively low cost, high sensitivity, and susceptibility to miniaturization. Chen et al. [7] made a review of microchip capillary procedures applied in a environmental monitoring. In most cases, amperometric detection is applied; it is particularly useful for electro-active analytes identification (e.g. chlorinated phenols) [19]. Conductivity detectors are implemented to determine amines and phenolic acids [20, 21].

Connecting CE with a mass detector that provides much information gives an opportunity to improve competitiveness of electromigration methods in environmental analysis [16]. A development of an interface that will provide compatibility of an electrolyte used in CE with a kind of ionization applied in a mass detector poses a challenge to apparatus constructors. The success mainly depends on the mode of an electrophoretic separation. Completely validated procedures of capillary zone electrophoresis—mass spectroscopy are already available. For MEKC and capillary electro-chromatography (CEC), validation is still in progress [1]. Many teams made attempts to separate environmentally important compounds by means of MEKC and CEC–MS. In most cases, however, these were only model samples. Recently, determination procedures for organic acids, nitrosamines, explosives, or chemical weapon products degradation by CZE–MS have been formulated [22–25].

18.3 Analytes Concentration in a Sample

In order to raise concentration sensitivity of electromigration-based techniques, besides amelioration of detectors, researchers make efforts to concentrate analytes in samples. Within the last few years, numerous new concentration methods based on chromatographic and electrophoretic mechanisms as well as their combinations have been practically implemented.

A review of the chromatographic methods for concentration in environmental electrophoretic analysis was published by Puig et al. [26]. The methods include solid phase extraction or micro-extraction (SPE and SPME, respectively), liquid–liquid micro-extraction and stirring bar sorptive extraction. In the simplest, manual version outside a CE system, the method was successfully applied to concentrate alkyl phenols in sewage by micro-extraction with the use of methyl chloroacetate followed by a separation according to MEKC [27]. Automatic micro-extraction in a CE analytical system is performed with the use of interfaces connecting the concentrating section with the separating section [28, 29].

Solid phase extraction can be performed as a manual operation before a sample is injected to a CE system. A sorbent can also be placed into a capillary (concentration in CE measuring system). In the manual concentration, which is apparently effective, a concentrated sample in a volume ranging from a few microliters to a few millilitres is obtained. At such a large sample volume, microanalytical CE potential is not exploited. Nevertheless, the use of CE can be justified by its high potential (in comparison to HPLC) for the separation of problematic mixtures of analytes such as diastereoisomers. Concentration by SPE in the CE system is carried out by a sorbent injection into a capillary. Then, the whole eluate, together with a sample matrix passes through the capillary causing a significant growth in the analysis sensitivity, and a risk of the capillary blocking and electroosmotic flow interruption or reversal with all the negative consequences of these for the analysis quality. In order to avoid the above-mentioned situations, Zhang and Wu [30] proposed an SPE-CE combination for the determination of chlorophenols in river water. There, analytes are retained on the sorbent and a sample matrix, together with a washing solution is removed by a side hole in the capillary prior to applying voltage.

Stirring Bar Sorptive Extraction (SBSE) with liquid desorption was used to determine polycyclic aromatic hydrocarbons in environmental samples by MEKC [31, 32].

Micro-analytical potential of electromigration techniques can be fully exploited if analytes concentration is carried out in accordance with electrophoretic principles. The concentration takes place in a separation capillary and can usually be used for anions, cations, and uncharged analytes. The concentration can occur in many kinds of stacking and sweeping ways which were widely described in a number of reviews [33, 35] as well as in other chapters of this book.

18.4 Microchips

As in conventional CE, research on microchip electrophoresis application in the environmental analysis is focused on analytes concentration and detection improvement. Additionally, in microchip CE, studies are aimed at the improvement of limited separation efficiency caused by short distances of separation channels. Since chip CE systems are compact, which is their main advantage, they are particularly appreciated by environmental analysts. The systems enable analyzing samples at the place they are collected. Various detectors find their use in the chip analysis, but the amperometric detector is the most often implemented due to its high sensitivity and compatibility with microchip devices. The absorptive detector was used in the determination of phosphates after derivatization in the form of vanadium-molybdenum-phosphate acid at the analytical wavelength equal to 370 nm [36]. A unique electrochemical detector with a composite electrode made of carbon nanotubes and polymethyl methacrylate was applied to analyze water for phenolic contaminations content [37]. The use of chip electrophoresis for environmental pollutants determination has been discussed in a number of papers published within the last few years [7, 38].

18.5 Environmental Contaminants Determination

Day by day, thousands of chemical compounds of both natural and anthropogenic origin are emitted to various environmental elements. These compounds exhibit many properties which can be environmentally hazardous and degrading. Compounds of anthropogenic origin, as opposed to those of natural origin, are often of high chemical stability. It adversely influences their degradation process and prolongs their environmental presence with all the negative results such as bioaccumulation and biomagnification.

18.6 Pesticides

Pesticides are compounds of the anthropogenic origin most commonly applied to control pests. Agriculture is the main area in which insecticides, herbicides, and fungicides are applied. Other kinds of pesticides are a point of interest in branches other than agriculture. Pesticides and some of their degradation products are characterized by three harmful properties: toxicity, lipophilicity, and chemical stability. The latter one causes their comparatively prolonged presence in environment due to long degradation time. They can also permeate to the food chain. The importance of the above-mentioned problem results in the fact that modern separation techniques, including electromigration ones (characterized by high separation efficiency) are implemented in order to monitor these pollutants. Recently, a broad range of papers have been published [9, 13, 39–43] on applying CE for the analysis of environmental samples. Analyses for the presence of pesticides are carried out with the use of concentration and detection systems typical of the trace analysis.

Organo-phosphoric pesticides were assayed in water after derivatization in a capillary and LIF detection [44] at the LOD ca. 3 µg/L for aminomethylphosphonate acid. Enantioselective electromigration analytical techniques in the context of organo-phosphoric pesticides content and chemical fate in environment have been discussed in recently published reviews [42, 43].

MEKC-MS was successfully applied in the determination of N-methylcarbamate in water [45]. After the analytes concentration in SPE columns and the MS detector friendly pseudo-stationary phase application, the LOD at 10 µg/L was obtained. A number of papers have been written on pesticides or their metabolites determination in soil [46–50]. In order to determine amino acid-based herbicides [46], samples of soil were extracted, derivatized to obtain fluorescent derivatives, and electrophoretic separation with LIF detection was performed. Following such a procedure enabled the achievement of the LOD in the range from 0.02 to 0.1 µg/L. The metribuzin separation from the soil matrix components was achieved by MEKC [47] and after the previous analyte concentration in the SPE column. The method can be successfully applied with the use of a UV detector for samples containing only micrograms of analytes in 1 kg of soil.

A sample injection to a capillary by FESI, after SPE initial concentration, made it possible to determine triazopyrimidine sulfonanilides by means of the capillary zone electrophoresis separation technique. The method enabled the determination of these pesticides at the concentration higher than 30 µg/kg of soil [48].

Carbamate insecticides, after concentration in the SPE column, were separated by MEKC using electrochemical detection with a platinum electrode. For five kinds of pesticides in water and soil, the LOD in the range from 0.5 to 0.01 µmol/L was achieved [49].

MEKC was also applied for researching atrazine sorption in soil [50]. Atrazine, a popular herbicide applied to control annual grass and latifolious weeds, permeates to groundwater because of its stability in soil and the ease of elution.

18.7 Polycyclic Aromatic Hydrocarbons

Polycyclic aromatic hydrocarbons (PAHs) are present in the environment as a result of natural and human activity. Their resolute hydrophobicity and no charge drew scientific attention to electromigration hybrid techniques, namely EKC and CEC.

PAHs were assayed in water and sewage residues after separation according to MEKC-UV. Before being injected into a capillary, samples were extracted by the SBSE technique [31, 32]. Implementation of the procedure enabled achievement of the LOD in the range between 2 and 10 µg/L. In other research work, the MEKC application with analytes concentration in a capillary by FESI [51] enabled the PAHs determination at the LOD between 0.5 and 5 µg/L.

Mono-methylated positional benzo(a)pyrene isomers were separated and determined by the combined CEC-MS technique with the use of the APPI interface (atmospheric pressure photo-ionization) [52]. The method LOD was established at the level of 400 µg/L.

18.8 Phenols

The phenols' presence in natural water is not problematic; their low concentration is a result of plant material decay. Problems become serious when defectively purified industrial wastes supply natural reservoirs. Phenol and substituted phenols can remain in water for a very long period of time, dependently on temperature and pH. Even at low concentrations, they negatively influence water taste and smell. They are toxic to most aquatic organisms and are placed on the official lists of the main environmental pollutants. Electromigration analytical techniques, together with the improvement in concentration sensitivity, play an increasingly important role in determination and monitoring of phenols in environmental samples [30, 53–62].

Non-aqueous capillary electrophoresis (NACE) was applied [53–55] to determine 26 phenols [53] from the official environmental pollutants list approved by EPA (Environmental Protection Agency) and the corresponding EU directive. Samples with low analyte concentrations were concentrated in SPE columns with styrene–divinylbenzene packing. The LOD obtained met the sewage phenol detection criteria.

In another research [54], NACE-UV was used to assay halogenated phenols with the assumed or confirmed ability to cause hormonal disorders. Water samples were initially concentrated in SPE columns and subjected to LVSS.

Phenol compounds, including flavonoids, were also determined [55] by NACE-ESI-TOF-MS after concentration in SPE columns.

The idea of chip electrophoresis was also fulfilled in environmental sample analysis to test the presence of phenols [56–60]. Toxic nitrophenols were determined [56] in groundwater on amperometric detection chips. The background buffer used for electrophoresis included α-cyclodextrin. The electropherogram of the water

Fig. 18.1 Groundwater sample electropherogram before (**a**) and after (**b**) phenols addition to the final concentration: 3NP (3-nitrophenol), 30 μmol/L; 2NP (2-nitrophenol), 4NP (4-nitrophenol), DNC (2-methyl-4,6-dinitrophenol), DNP (2,4-dinitrophenol), 40 μmol/L. Reprinted with permission from Fischer et al. [56] Copyright 2006, Wiley–VCH

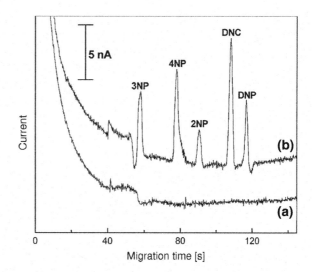

sample is depicted in Fig. 18.1. Under optimum conditions, the LOD was ca. 5 μmol/L. In another paper [57] on phenols chip CE, pulsation amperometric detection with the gold electrode was applied. Under optimum conditions, polynitro- and polychlorophenols were determined in water at the LOD in the range between 1 and 2.5 μmol/L.

Phenols in urban water mains and groundwater were determined in a three-channel chip [58]. Two channels were used to concentrate analytes by FASS and FASI; in the third, electrophoretic separation with amperometric detection occurred.

18.9 Amines

Amines are important intermediates for manufacturing dyes, pesticides, rubber products, man-made fibers, or pharmaceuticals. Their production wastes are released into the environment polluting water and air. They are hazardous to human and animal health. Some amines exhibit carcinogenic properties [63, 64]. Chromatographic and electromigration techniques are applied to monitor their environmental concentrations. In a review published recently [65], methods based on these techniques were described with some experimental details.

Nitroaniline positional isomers in dye waste were assayed by CE with ampero-metric detection at the LOD ca. 10 nmol/L [66]. Eight amines, aniline and naph-thylamine derivatives, were separated by CE after concentration by liquid–liquid extraction with the use of micro-porous membranes [67]. The diode UV detector application enabled the LOD determination in the range between 0.16 and 7.5 ng/ml. The method was tested on water from city water mains, rivers, and sewage.

Volatile amines in indoors and outdoors air were assayed by CE with indirect UV or MS detection. In the CE-UV procedure the imidazole buffer was applied, whereas in the CE-MS procedure the imidazole buffer was replaced by the ammonium buffer. The LOD was determined at the 1–2 µg level per 40 l of air [68].

Cis- and trans-1,2-diaminodicyclohexane isomers and other organic amines were separated by CE with conductometric detection. For isomers separation, the background buffer contained a crown-ether [69].

18.10 Carboxylic Acids

Within the last few years, a series of new procedures have been developed to implement CE for aliphatic and aromatic carboxylic acids determination [21, 70–77].

Benzoic and cinnamic acids derivatives were separated and determined [21] with the use of a contactless conductometric detector. Depending on analytes concentration in a sample, two procedures were proposed: analytes stacking in a capillary at the LOD 0.12–0.14 µmol/L and without stacking at the LOD 2.3–3.3 µmol/L.

Numerous papers discussed the organic acids determination in atmospheric aerosols [70–73]. After their methanol extraction and evaporation to dryness, low-molecular organic acids in aerosols were electrophoretically determined [70]. Disturbing and highly mobile anions such as chlorides, nitrates, and sulfates, commonly present in aerosols, at concentration values hundreds of times higher than analytes values, were removed by reverse pre-electrophoresis. Analytes loss which cannot be avoided in the operation was mathematically corrected on the basis of the linear dependence between analytes loss and their electrophoretic mobility. C_5–C_6 dicarboxylic acids such as adipic and glutar acids, being the main methylene cyclohexane ozonolysis products, were detected by CE-ESI–MS/MS [71].

In another paper [72], a simple CE-UV method for the detection of isomeric benzoic acid derivatives in atmospheric aerosols and car exhaust fumes was described. Optimum conditions were achieved by FASI into a dynamically modified capillary.

Perfluorinated carboxylic acids were also determined with the use of CE [74]. The indirect UV detection was possible after introducing 2,4-dinitrobenzoic acid to the background buffer. For perfluorinated C_6–C_{12} acids, the LOD was obtained in the range between 0.6 and 24 mg/L.

18.11 Explosives

Increasingly common terrorist attacks create the demand for methods of quick military, industrial, or home-made explosives determination. Capillary electrophoresis, particularly its chip version, creates the possibility to develop compact devices that allow the performance of fast analyses even at the attack site. Such fast analyses will allow the authorities to make adequate decisions concerning civil security. Concerning the above, as well as due to the need to analyze huge amounts of soil and water samples, many laboratories study the CE applications for explosives detection and determination. As a result, many papers presenting trends in the environmental samples analysis for the presence of explosives with the use of the conventional and microchip CE are published. Some of them are original [24, 78–86], or reviews [87–91].

MEKC and CEC were employed to separate a mixture of 14 explosive substances from the EPA list [78]. The mixture contained nitroaromatic and nitroamine compounds. Nitroaromatic compounds were detected by a UV detector as well as LIF with indirect detection. Nitroamines do not exhibit a sufficient ability to quench the background fluorescence and must be detected within the ultraviolet range. Independently from a kind of detection and a separation technique, sensitivity expressed in the LOD is between 1 and 10 µg/L. The same hybrid techniques, MEKC and CEC, were applied for nitroaromatic and nitroamine explosives determination in seawater [79]. In the MEKC version, taking advantage of salts presence in seawater, a technique of the analytes concentration in capillary by high-salt stacking was used. It significantly improved the separability and sensitivity of the process. In the CEC procedure, prolonged, electrokinetic sample injection was applied, which enabled the analytes detection at the 100 µg/L level. In another paper [85], the determination of nitroaromatic compounds and their degradation products in seawater by MEKC with direct untreated sample injection into a capillary was described. A large seawater zone, electrokinetically injected (injection time: 5–100 s) and a high surfactant concentration allowed the analytes determination at the 70–800 ppb concentration level.

River water and sewage samples [86] were analyzed for organic nitro compounds presence by CE with amperometric detection. Optimum results, i.e., the LOD in the 3–47 mg/L range, were obtained with the use of an electrode modified with a mesoporous, nanostructural carbon material. The amperometric detector was employed also in the microchip CE during the separation of a nitro esters mixture [84]. According to the authors, the method is quick and sensitive. It enables detection on a picogram level with the LOD for nitroglycerine equal to 0.3 ppm.

Phosphonate acids, products of chemical weapon decomposition, were assayed in samples taken from soil, concrete, or granite by means of CE with conductometric detection [82, 83] or CE-MS/MS [84]. In both the methods, while searching for optimum analytical conditions, various ways of a sample injection into a capillary were tested.

18.12 Pharmaceuticals

Pharmaceuticals traces are detected in groundwater in all the countries in which they are commonly consumed. Pharmaceuticals are excreted in the form of metabolites or as unaffected substances. They are released to sewage, and then, through sewage treatment plants, to the environment. Veterinary drugs are released into the soil, and then to water as a result of fertilization with animal excrements. The drugs are also directly added to water in fish breeding ponds. The amounts of drugs produced and distributed are comparable with the amounts of pesticides. Within the last few years, some reviews on pharmaceutical determination methods, also electromigration ones, have been published [4–6, 92]. Non-steroidal anti-inflammatory drugs belong to those most often detected and common in the environment [4, 5]. It results from the fact that they are often given to people and animals in order to relieve pain, decrease raised body temperature, or eliminate inflammation. Since the early 1990s many teams have been doing research to determine these drugs' content with electromigration techniques such as CZE, MEKC, MEEKC, ITP, tITP [5, 93]. All the efforts were focused on improving concentration sensitivity to the level corresponding with the environmental analysis. Almost all the analytes concentration methods based on chromatographic and electrophoretic interactions as well as their combinations inside and outside measuring systems were applied.

In the last published paper dealing with the problem [93], a combination of electrokinetic sample injection and transient isotachophoresis called electrokinetic supercharging (EKS) was employed. Such stacking made the method 2,000 times more sensitive in comparison to hydrodynamic sample injection. This kind of stacking enabled the development of a procedure with the use of a commercially available CE set with the UV-DAD detector at the LOD in the range between 0.9 and 2 μg/L to check the presence of five anti-inflammatory drugs (i.e., naproxen, fenoprofen, ketoprofen, diclofenac, and piroxicam) in river water.

CE coupled with mass spectrometry (TOF/MS) was utilised for determination of antidepressants in groundwater and water from a treatment plant [94]. The samples were prepared by SPE outside a measuring system. The method allowed the antidepressants to be determined (Fig. 18.2) in real environmental samples at the level of a few hundreds ng/L. In another paper, the NACE-TOF/MS [95] was used to assay 20 antidepressants.

Hypo-lipidemic drugs in river water were determined by CE with ESI/MS [96]. Electrophoresis was performed with the use of an EOF reversing reagent. The applied procedure allowed the concentration sensitivity to be 1,000 times higher in comparison to field amplified sample injection. Sulphonamide drugs in groundwater were determined by concentration in a capillary by large volume sample stacking and UV detection [97]. With additional SPE concentration, outside the measuring system, the LOD obtained for nine sulphonamides was in the range between 2.5 and 23 μg/L. One year later [98], the same research team improved the method replacing the UV detection with MS (ESI–MS/MS). The achieved results enabled the sulphonamides determination in environmental samples and

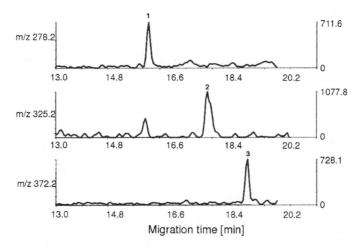

Fig. 18.2 Electropherograms of water samples from water treatment plant obtained by CE-TOF/MS, after concentration by extraction to the solid phase. Signals: 1–venlafaxine (240 ng/L); 2–citalopram (190 ng/L); 3–trazodone (140 ng/L). Reprinted with permission from Himmelsbach et al. [94] Copyright 2006, Wiley–VCH

food products of animal origin at concentrations lower than those established in the EU regulations.

Combining micro-extraction in the measuring system with MS detection (SPME-CE-ESI/MS) allowed tetracycline antibiotics to be assayed in soil [99]. A problematic task of combining SPME with a separation column and detector was solved by means of a special interface with an additional channel for adsorptive fibre placing. The procedure is characterized by the LOD at approximately 3 µg/kg of soil.

18.13 Ionic Liquids

Ionic Liquids are substances in the liquid state, made of ions. They differ from typical liquid salts having low melting points usually up to 100 °C [100–102]. These are salts usually built of large and asymmetric cations (e.g., alkylimidazolium, alkylpiridine ones) and smaller anions of organic and inorganic character [100–103]. The gradual industrial implementation of ionic liquids can be observed. However, it causes a serious risk of their spread and accumulation in the environment. Thus, it is necessary to develop selective and cheap analytical tools that would enable the determination of ionic liquids [103]. The first of the techniques used for the imidazolium ionic liquids determination was capillary zone electrophoresis. Qin et al. [104] optimized the separation conditions of seven imidazole derivatives (the choice of a proper buffer, its pH and concentration). A short analysis time was obtained (8 min) and the LOD between 0.42 and 1.36 ppm [105]. The method presents good linearity and repeatability. Its applicability was checked in the

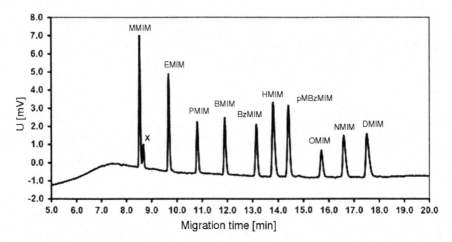

Fig. 18.3 Electropherogram obtained for imidazole cations in ionic liquids mixture. Analytical conditions: 200 mmol/L citrate buffer (pH 4); temp. 20 °C; voltage 12 kV; ionic liquids concentration 0.1 mg/mL; notations: MMIM—1-methyl-3-methylimidazolium cation, EMIM—1-ethyl-3-methylimidazolium cation, PMIM—1-propyl-3-methylimidazolium cation, BMIM—1-butyl-3-methylimidazolium cation, BzMIM—1-benzyl-3-methylimidazolium cation, HMIM—1-hexyl-3-methylimidazolium cation, pMBzMIM—1-(p-methylbenzyl)-3-methylimidazolium cation, OMIM—1-octyl-3-methylimidazolium cation, NMIM—1-nonyl-3-methylimidazolium cation, DMIM—1-decyl-3-methylimidazoloum cation. Reprinted with permission from Markuszewski et al. [105] Copyright 2004 Wiley–VCH

analysis of commercially available ionic liquids impurities. Therefore, it was possible to determine the quantity of 1-methylimidazolium in 1-ethyl-3-methylimidazolium chloride. The method developed was also employed to control the 1-butyl-3-methylimidazolium chloride synthesis process [105].

Electromigration techniques were also applied for controlling the 1-butyl-3-methylimidazolium tetrafluoroborane photo-degradation process. In its solution, the examined ionic liquid as well as its degradation products were determined. Thus, capillary electrophoresis was employed. The method of electrophoretic imidazole salts cations separation optimised by Markuszewski et al. [105] was found to be simple, quick, and repeatable. The developed procedure, however, is unreliable in the case of ionic liquids of similar molecular weights separation [105]. Figure 18.3 presents an example of an examined cation mixture separation electropherogram.

So far, ionic liquids analysis was based mainly on the cations analysis. Recently, the results of both ionic liquids cations and anions determination by ITP with conductometric detection [106] have been published. Developed by Kosobudzki et al. [106], the procedure enabled the separation of five imidazolium cations. The procedure was repeated for chloride and tetrafluoroborate anions which were qualitatively and quantitatively determined. The method is characterized by adequate analytical parameters such as linearity, repeatability, precision, or LODs. Since the method is not costly, it may be an alternative to ionic liquids determination by CZE or HPLC [106]. Moreover, water samples can be analyzed readily, with no previous preparation, which is a remarkable advantage, the fact being impossible in HPLC [106].

18.14 Summary

Chemical compounds present at low concentrations, often in trace amounts, are the subject of environmental analysis. When we become aware that volumes of samples available to analysts are rarely limited, we will soon come to the conclusion that capillary electrophoresis is not the ideal technique to implement in environmental analysis. Conventional concentration methods such as liquid–liquid extraction or solid phase extraction provide samples with concentrated analytes in the volume of a few microliters up to a few millilitres, prepared to perform the final analysis. In such a case, it is difficult to take advantage of micro-analytical possibilities of conventional capillary or chip electrophoresis. Due to the above, all the efforts made by research groups are focused on the development of analytes concentration techniques in measuring systems based on chromatographic and electrophoretic principles. Still, much attention is paid to the improvement of chip electrophoresis which could enable the construction of compact analyzers used in areas where ecological disasters or terrorist attacks take place.

References

1. S. Rubio, D. Perez-Bendito, Recent advances in environmental analysis. Anal. Chem. **81**, 4601–4622 (2009)
2. E. Dabek-Zlotorzynska, V. Celo, M.M. Yassine, Recent advances in CE and CEC of pollutants. Electrophoresis **29**, 310–323 (2008)
3. E. Dabek-Zlotorzynska, V. Celo, Recent advances in capillary electrophoresis and capillary electrochromatography of pollutants. Electrophoresis **27**, 304–322 (2006)
4. A. Gentili, Determination of non-steroidal anti-inflammatory drugs in environmental samples by chromatographic and electrophoretic techniques. Anal. Bioanal. Chem. **387**, 1185–1202 (2007)
5. A. Macia, F. Borrull, M. Calull, C. Aguilar, Capillary electrophoresis for the analysis of non-steroidal anti-inflammatory drugs. Trends Anal. Chem. **26**, 133–153 (2007)
6. W.W. Buchberger, Novel analytical procedures for screening of drug residues in water, waste water, sediment and sludge. Anal. Chim. Acta **593**, 129–139 (2007)
7. G. Chen, Y. Lin, J. Wang, Monitoring environmental pollutants by microchip capillary electrophoresis with electrochemical detection. Talanta **68**, 497–503 (2006)
8. J. Wang, Electrochemical detection for capillary electrophoresis microchips: A review. Electroanalysis **17**, 1133–1140 (2005)
9. J. Hernandez-Borges, M.A. Rodrigues-Delgado, F.J. Garcia-Montelongo, A. Cifuentes, Chiral analysis of pollutants and their metabolites by capillary electromigration methods. Electrophoresis **26**, 3799–3813 (2005)

10. S. Eeltink, W.T. Kok, Recent applications in capillary electrochromatography. Electrophoresis **27**, 84–96 (2006)
11. M.C. Breadmore, Recent advances in enhancing the sensitivity of electrophoresis and electrochromatography in capillaries and microchips. Electrophoresis **28**, 254–281 (2007)
12. K. Stulik, V. Pacakova, J. Suchankova, P. Coufal, Monolithic organic polymeric columns for capillary liquid chromatography and electrochromatography. J. Chromatogr. B **841**, 79–87 (2006)
13. O.H. Szolar, Environmental and pharmaceutical analysis of dithiocarbamates. Anal. Chim. Acta **582**, 191–200 (2007)
14. C.W. Klampfl, Recent advances in the application of capillary electrophoresis with mass spectrometric detection. Electrophoresis **27**, 3–34 (2006)
15. J.J. Xu, A.J. Wang, H.Y. Chen, Electrochemical detection modes for microchip capillary electrophoresis. Trends Anal. Chem. **26**, 125–132 (2007)
16. C.W. Klampfl, W. Buchberger, Coupling of capillary electroseparation techniques with mass spectrometric detection. Anal. Bioanal. Chem. **388**, 533–536 (2007)
17. J. Jiang, C.A. Lucy, Determination of glyphosate using off-line ion exchange preconcentration and capillary electrophoresis-laser induced fluorescence detection. Talanta **72**, 113–118 (2007)
18. L. Zhou, Z. Luo, S. Wang, Y. Hui, Z. Hu, X. Chen, In-capillary derivatization and laser-induced fluorescence detection for the analysis of organophosphorus pesticides by micellar electrokinetic chromatography. J. Chromatogr. A **1149**, 377–384 (2007)
19. G.W. Muna, V. Quaiserova-Mocko, G.M. Swain, Chlorinated phenol analysis using off-line solid-phase extraction and capillary electrophoresis coupled with amperometric detection and a boron-doped diamond microelectrode. Anal. Chem. **77**, 6542–6548 (2005)
20. X.Y. Gong, P.C. Hauser, Determination of different classes of amines with capillary zone electrophoresis and contactless conductivity detection. Electrophoresis **27**, 468–473 (2006)
21. P. Kuban, D. Sterbova, V. Kuban, Separation of phenolic acids by capillary electrophoresis with indirect contactless conductometric detection. Electrophoresis **27**, 1368–1375 (2006)
22. C. Muller, Y. Linuma, O. Boge, H. Herrmann, Applications of CE-ESI-MS/MS analysis to structural elucidation of methylenecyclohexane ozonolysis products in the particle phase. Electrophoresis **28**, 1364–1370 (2007)
23. X. Liu, Y.Y. Zhao, K. Chan, S.E. Hrudey, X.F. Li, J. Li, Analysis of nitrosamines by capillary electrospray-high-field asymmetric waveform ion mobility spectrometry-MS with programmed compensation voltage. Electrophoresis **28**, 1327–1334 (2007)
24. M. Lagarrigue, A. Bossee, A. Begos, A. Varennent, P. Gareil, B. Bruno, CE-MS-MS for the identification of chemical warfare agent degradation products. LCGC **25**, 292–301 (2007)
25. M. Lagarrigue, A. Bossee, A. Begos, N. Delaunay, A. Varennent, P. Gareil, B. Bellier, Field-amplified sample stacking for the detection of chemical warfare agent degradation products in low-conductivity matrices by capillary electrophoresis-mass spectrometry. J. Chromatogr. A **1178**, 239–247 (2008)
26. P. Puig, F. Borrull, M. Calull, C. Aguilar, Sorbent preconcentration procedures coupled to capillary electrophoresis for environmental and biological applications. Anal. Chim. Acta **616**, 1–18 (2008)
27. T. Wang, J. Tang, W. Wan, S. Zhao, Methyl chloroacetate as an extraction solvent for coupling liquid-liquid semi micro extraction with micellar electrokinetic chromatography through on-capillary decomposition for the separation of neutral compounds with concentration enhancement. J. Chromatogr. A **1147**, 05–110 (2007)
28. S. Almeda, L. Nozal, L. Arce, M. Valcarcel, Direct determination of chlorophenols present in liquid samples by using a supported liquid membrane coupled in-line with capillary electrophoresis equipment. Anal. Chim. Acta **587**, 97–103 (2007)
29. P. Puig, F.W.A. Tempels, G.W. Somsen, G.J. de Jong, F. Borrull, C. Aguilar, M. Calull, Use of large-volume sample stacking in on-line solid-phase extraction-capillary electrophoresis for improved sensitivity. Electrophoresis **29**, 1339–1346 (2008)

30. L.H. Zhang, X.Z. Wu, Capillary electrophoresis with in-capillary solid-phase extraction sample cleanup. Anal. Chem. **79**, 2562–2569 (2007)
31. P.M.A. do Rosario, J.M.F. Nogueira, Combining stir bar sorptive extraction and MEKC for the determination of polynuclear aromatic hydrocarbons in environmental and biological matrices. Electrophoresis **27**, 4694–4702 (2006)
32. A.R.M. Silva, F.C. Portugal, J.M. Noqueiro, Advances in stir bar sorptive extraction for the determination of acidic pharmaceuticals in environmental water matrices Comparison between polyurethane and polydimethylsiloxane polymeric phases. J. Chromatogr. A **1209**, 10–16 (2008)
33. M.C. Breadmore, J.R.E. Thabano, M. Davod, A.A. Kazarian, J.P. Quirino, R.M. Guijt, Recent advances in enhancing the sensitivity of electrophoresis and electrochromatography in capillaries and microchips (2006–2008). Electrophoresis **30**, 230–248 (2009)
34. Z. Mala, A. Slampova, P. Gebauer, P. Bocek, Contemporary sample stacking in CE. Electrophoresis **30**, 215–229 (2009)
35. S.L. Simpson Jr, J.P. Quirino, S. Terabe, On-line sample preconcentration in capillary electrophoresis fundamentals and applications. J. Chromatogr. A **1184**, 504–541 (2008)
36. C.M. McGraw, S.E. Stitzel, J. Cleary, C. Slater, D. Diamond, Autonomous microfluidic system for phosphate detection. Talanta **71**, 1180–1185 (2007)
37. X. Yao, H. Wu, J. Wang, S. Qu, G. Chen, Carbon nanotube/poly(methyl methacrylate) (CNT/PMMA) composite electrode fabricated by in situ polymerization for microchip capillary electrophoresis. Chem. Eur. J. **13**, 846–853 (2007)
38. H.F. Li, J.M. Lin, Applications of microfluidic systems in environmental analysis. Anal. Bioanal. Chem. **393**, 555–567 (2009)
39. L.M. Ravelo-Perez, J. Hernandez-Borges, M.A. Rodrigues-Delgado, Pesticides analysis by liquid chromatography and capillary electrophoresis. J. Sep. Sci. **29**, 2557–2577 (2006)
40. V.R. Robledo, W.F. Smyth, The application of CE-MS in the trace analysis of environmental pollutants and food contaminants. Electrophoresis **30**, 1647–1660 (2009)
41. A. Kumar, A.K. Malik, Y. Pico, Sample preparation methods for the determination of pesticides in foods using CE-UV/MS. Electrophoresis **31**, 2115–2125 (2010)
42. L. Li, S. Zhau, L. Jin, C. Zhang, W. Liu, Enantiomeric separation of organophosphorus pesticides by high-performance liquid chromatography, gas chromatography and capillary electrophoresis and their applications to environmental fate and toxicity assays. J. Chromatogr. B **878**, 1264–1276 (2010)
43. M.G. Nillos, J. Gan, D. Schlenk, Chirality of organophosphorus pesticides: analysis and toxicity. J. Chromatogr. B **878**, 1277–1284 (2010)
44. L. Zhou, Z. Luo, S. Wang, Y. Hui, Z. Hu, X. Chen, In-capillary derivatization and laser-induced fluorescence detection for the analysis of organophosphorus pesticides by micellar electrokinetic chromatography. J. Chromatogr. A **1149**, 377–384 (2007)
45. G.V. Biesen, C.S. Bottaro, Ammonium perfluorooctanoate as a volatile surfactant for the analysis of N-methylcarbamates by MEKC-ESI-MS. Electrophoresis **27**, 4456–4468 (2006)
46. E. Orejuela, M. Silva, Rapid and sensitive determination of phosphorus-containing amino acid herbicides in soil samples by capillary zone electrophoresis with diode laser-induced fluorescence detection. Electrophoresis **26**, 4478–4485 (2005)
47. J.F. Huertas-Perez, M.D.O. Iruela, A.M. Garcia-Campana, A. Gonzalez-Casado, A. Sanchez-Navarro, Determination of the herbicide metribuzin and its major conversion products in soil by micellar electrokinetic chromatography. J. Chromatogr. A **1102**, 280–286 (2006)
48. J. Hernandez-Borges, F.J. Garcia-Montelongo, A. Cifuentes, M.A. Rodriguez-Delgado, Analysis of triazolopyrimidine herbicides in soils using field-enhanced sample injection-coelectroosmotic capillary electrophoresis combined with solid-phase extraction. J. Chromatogr. A **1100**, 236–242 (2005)
49. A. Santalad, L. Zhou, F. Shang, D. Fitzpatrick, R. Burakham, S. Srijaranai, J.D. Glennon, J.H.T. Luong, Micellar electrokinetic chromatography with amperometric detection and off-

line solid-phase extraction for analysis of carbamate insecticides. J. Chromatogr. A **1217**, 5288–5297 (2010)

50. D.L. Lima, G.L. Erny, V.I. Esteves, Application of MEKC to the monitoring of atrazine sorption behaviour on soils. J. Sep. Sci. **32**, 4241–4246 (2009)

51. G.Q. Song, Z.L. Peng, J.M. Lin, Comparison of two capillary electrophoresis online stacking modes by analysis of polycyclic aromatic hydrocarbons in airborne particulates. J. Sep. Sci. **29**, 2065–2071 (2006)

52. J. Zheng, S.A. Shamsi, Capillary electrochromatography coupled to atmospheric pressure photoionization mass spectrometry for methylated benzo[a]pyrene isomers. Anal. Chem. **78**, 6921–6927 (2006)

53. S. Morales, R. Cela, Highly selective and efficient determination of US Environmental Protection Agency priority phenols employing solid-phase extraction and non-aqueous capillary electrophoresis. J. Chromatogr. A **896**, 95–104 (2000)

54. E. Blanco, M.C. Casais, M.C. Mejuto, R. Cela, Comparative study of aqueous and non-aqueous capillary electrophoresis in the separation of halogenated phenolic and bisphenolic compounds in water samples. J. Chromatogr. A **1068**, 189–199 (2005)

55. A.M. Gomez-Caravaca, A. Carrasco-Pancorbo, A. Segura-Carretero, A. Fernandez-Gutierrez, NACE-ESI-TOF MS to reveal phenolic compounds from olive oil: introducing enriched olive oil directly inside capillary. Electrophoresis **30**, 3099–3109 (2009)

56. J. Fischer, J. Barek, J. Wang, Separation and detection of nitrophenols at capillary electrophoresis microchips with amperometric detection. Electroanalysis **18**, 195–199 (2006)

57. Y. Ding, C.D. Garcia, Pulsed amperometric detection with poly(dimethylsiloxane)-fabricated capillary electrophoresis microchips for the determination of EPA priority pollutants. Analyst **131**, 208–214 (2006)

58. M.J.A. Shiddiky, H. Park, Y.-B. Shim, Direct analysis of trace phenolics with a microchip: in-channel sample preconcentration, separation, and electrochemical detection. Anal. Chem. **78**, 6809–6817 (2006)

59. X. Yao, H. Wu, J. Wang, S. Qu, G. Chen, Carbon nanotube/poly(methyl methacrylate) (CNT/PMMA) composite electrode fabricated by in situ polymerization for microchip capillary electrophoresis. Chem. Eur. J. **13**, 846–853 (2007)

60. S.I. Wakida, K. Fujimoto, H. Nagai, T. Miyado, Y. Shibutani, S. Takeda, On-chip micellar electrokinetic chromatographic separation of phenolic chemicals in waters. J. Chromatogr. A **1109**, 179–182 (2006)

61. L.J. Yan, Q.H. Zhang, Y.Q. Feng, W.B. Zhang, T. Li, L.H. Zhang, Y.K. Zhang, Octyl-functionalized hybrid silica monolithic column for reversed-phase capillary electrochromatography. J. Chromatogr. A **1121**, 92–98 (2006)

62. M.F. Mora, C.D. Garcia, Electrophoretic separation of environmentally important phenolic compounds using montomorillonite-coated fused-silica capillaries. Electrophoresis **28**, 1197–1203 (2007)

63. T.S. Scott, *Carcinogenic and toxic hazards of aromatic amines* (Elsevier, Amsterdam, 1962)

64. A. Pielesz, I. Baranowska, A. Rybak, A. Wlochowicz, Detection and determination of aromatic amines as products of reductive splitting from selected azo dyes. Ecotoxicol Env. Saf. **53**, 42–47 (2002)

65. A. Fekete, A.K. Malik, A. Kumar, P. Schmitt-Kopplin, Amines in the environment. Crit. Rev. Anal. Chem. **40**, 102–121 (2010)

66. X. Guo, J. Lv, W. Zhang, Q. Wang, P. He, Y. Fang, Separation and determination of nitroaniline isomers by capillary zone electrophoresis with amperometric detection. Talanta **69**, 121–125 (2006)

67. Q. Zhou, G. Jiang, J. Liu, Y. Cai, Combination of microporous membrane liquid–liquid extraction and capillary electrophoresis for the analysis of aromatic amines in water samples. Anal. Chim. Acta **509**, 55–62 (2004)

68. A. Fekete, M. Frommberger, G. Ping, M.R. Lahaniatis, J. Lintelman, J. Fekete, J. Gebefugi, A.K. Malik, A. Kettrup, P. Schmitt-Kopplin, Development of a capillary electrophoretic method for the analysis of low-molecular-weight amines from metal working fluid aerosols and ambient air. Electrophoresis **27**, 1237–1247 (2006)

69. X.Y. Gong, P.C. Hauser, Determination of different classes of amines with capillary zone electrophoresis and contactless conductivity detection. Electrophoresis **27**, 468–473 (2006)

70. G.A. Blanco-Heras, M.I. Turnes-Carou, P. Lopez-Mahia, S. Muniategui-Lorenzo, D. Prada-Rodriguez, E. Fernandez-Fernandez, Determination of organic anions in atmospheric aerosol samples by capillary electrophoresis after reversed pre-electrophoresis. Electrophoresis **29**, 1347–1354 (2008)

71. C. Muller, Y. Iinuma, O. Boge, H. Herrmann, Applications of CE-ESI-MS/MS analysis to structural elucidation of methylenecyclohexane ozonolysis products in the particle phase. Electrophoresis **28**, 1364–1370 (2007)

72. E. Dabek-Zlotorzynska, M. Piechowski, Application of CE with novel dynamic coatings and field-amplified sample injection to the sensitive determination of isomeric benzoic acids in atmospheric aerosols and vehicular emission. Electrophoresis **28**, 3526–3534 (2007)

73. S.D. Noblitt, L.R. Mazzoleni, S.V. Hering, J.L.J. Collett, C.S. Henry, Separation of common organic and inorganic anions in atmospheric aerosols using a piperazine buffer and capillary electrophoresis. J. Chromatogr. A **1154**, 400–406 (2007)

74. L. Wojcik, K. Korczak, B. Szostek, M. Trojanowicz, Separation and determination of perfluorinated carboxylic acids using capillary zone electrophoresis with indirect photometric detection. J. Chromatogr. A **1128**, 290–297 (2006)

75. P.-L. Laamanen, E. Blanco, R. Cela, R. Matilainen, Improving sensitivity in simultaneous determination of copper carboxylates by nonaqueous capillary electrophoresis. J. Chromatogr. A **1110**, 261–267 (2006)

76. J. Hu, C. Xie, R. Tian, Z. He, H. Zou, Hybrid silica monolithic column for capillary electrochromatography with enhanced cathodic electroosmotic flow. Electrophoresis **27**, 4266–4272 (2006)

77. D.L.D. Lima, A.C. Duarte, V.I. Esteves, Optimization of phenolic compounds analysis by capillary electrophoresis. Talanta **72**, 1404–1409 (2007)

78. C.G. Bailey, S.R. Wallenborg, Indirect laser-induced fluorescence detection of explosive compounds using capillary electrochromatography and micellar electrokinetic chromatography. Electrophoresis **21**, 3081–3087 (2000)

79. B.C. Giordano, C.L. Copper, G.E. Collins, Micellar electrokinetic chromatography and capillary electrochromatography of nitroaromatic explosives in seawater. Electrophoresis **27**, 778–786 (2006)

80. X. Yao, J. Wang, L. Zhang, P. Yang, G. Chen, A three-dimensionally adjustable amperometric detector for microchip electrophoretic measurement of nitroaromatic pollutants. Talanta **69**, 1285–1291 (2006)

81. B.C. Giordano, A. Terray, G.E. Collins, Microchip-based CEC of nitroaromatic and nitramine explosives using silica-based sol-gel stationary phases from methyl- and ethyl-trimethoxysilane precursors. Electrophoresis **27**, 4295–4302 (2006)

82. A. Seiman, M. Jaanus, M. Vaher, M. Kaljurand, A portable capillary electropherograph equipped with a cross-sampler and a contactless-conductivity detector for the detection of the degradation products of chemical warfare agents in soil extracts. Electrophoresis **30**, 507–514 (2009)

83. N. Makarotseva, A. Seiman, M. Vaher, M. Kaljurand, Analysis of the degradation products of chemical warfare agents using a portable capillary electrophoresis instrument with various sample injection devices. Procedia Chem **2**, 20–25 (2010)

84. E. Picein, N. Dossi, A. Cagan, E. Carrilho, J. Wang, Rapid and sensitive measurements of nitrate ester explosives using microchip electrophoresis with electrochemical detection. Analyst **134**, 528–532 (2009)

85. B.C. Giordano, D.S. Burgi, G.E. Collins, Direct injection of seawater for the analysis of nitroaromatic explosives and their degradation products by micellar electrokinetic chromatography. J. Chromatogr. A **1217**, 4487–4493 (2010)

86. D. Nie, P. Li, D. Zhang, T. Zhau, Y. Liang, G. Shi, Simultaneous determination of nitroaromatic compounds in water using capillary electrophoresis with amperometric detection on an electrode modified with a mesoporous nano-structured carbon material. Electrophoresis **31**, 2981–2988 (2010)

87. M. Pumera, Analysis of explosives via microchip electrophoresis and conventional capillary electrophoresis: a review. Electrophoresis **27**, 244–256 (2006)

88. D.S. Moore, Recent advances in trace explosives detection instrumentation. Sens. Imag. **8**, 9–38 (2007)

89. M. Pumera, Trends in analysis of explosives by microchip electrophoresis and conventional CE. Electrophoresis **29**, 269–273 (2008)

90. R.M. Burks, D.S. Hage, Current trends in the detection of peroxide-based explosives. Anal. Bioanal. Chem. **395**, 301–313 (2009)

91. K.G. Lahoda, O.L. Collin, J.A. Mathis, H.E. LeClair, S.H. Wise, B.R. McCord, A survey of background levels of explosives and related compounds in the environment. J. Forens. Sci. **53**, 802–806 (2008)

92. C. Garcia-Ruiz, M.L. Marina, Recent advances in the analysis of antibiotics by capillary electrophoresis. Electrophoresis **27**, 266–282 (2006)

93. I. Botello, F. Borrull, C. Aguilar, M. Calull, Electrokinetic supercharging focusing in capillary zone electrophoresis of weakly ionizable analytes in environmental and biological samples. Electrophoresis **31**, 2964–2973 (2010)

94. M. Himmelsbach, W. Buchberger, C.W. Klampfl, Determination of antidepressants in surface and waste water samples by capillary electrophoresis with electrospray ionization mass spectrometric detection after preconcentration using off-line solid-phase extraction. Electrophoresis **27**, 1220–1226 (2006)

95. Y. Sasajima, L.W. Lim, T. Takeuchi, K. Suenami, K. Sato, Y. Takekoshi, Simultaneous determination of antidepressants by non-aqueous capillary electrophoresis-time of flight mass spectrometry. J. Chromatogr. A **1217**, 7598–7604 (2010)

96. M. Dawod, M.C. Breadmore, R.M. Guijt, P.R. Haddad, Electrokinetic supercharging-electrospray ionisation-mass spectrometry for separation and on-line preconcentration of hypolipidaemic drugs in water samples. Electrophoresis **31**, 1184–1193 (2010)

97. J.J. Soto-Chinchilla, A.M. Garcia-Campana, L. Gamiz-Gracia, C. Cruces-Blanco, Application of capillary zone electrophoresis with large-volume sample stacking to the sensitive determination of sulfonamides in meat and ground water. Electrophoresis **27**, 4060–4068 (2006)

98. J.J. Soto-Chinchilla, A.M. Garcia-Campana, L. Gamiz-Gracia, Analytical methods for multiresidue determination of sulfonamides and trimethoprim in meat and ground water samples by CE-MS and CE-MS/MS. Electrophoresis **28**, 4164–4172 (2007)

99. B. Santos, B.M. Simonet, A. Rios, M. Valcarcel, On-line coupling of solid-phase micro extraction to commercial CE-MS equipment. Electrophoresis **28**, 1312–1318 (2007)

100. J. Liu, J.A. Jönsson, G. Jiang, Application of ionic liquids in analytical chemistry. Trends Anal. Chem. **24**, 20–27 (2005)

101. K.N. Marsh, J.A. Boxall, R. Lichtenthaler, Room temperature ionic liquids and their mixtures-a review. Fluid Phase Equilibr. **219**, 93–98 (2004)

102. C.F. Poole, Chromatographic and spectroscopic methods for the determination of solvent properties of room temperature ionic liquids. J. Chromatogr. A **1037**, 49–82 (2004)

103. B. Buszewski, S. Studzińska, A review of ionic liquids in chromatographic and electromigration techniques. Chromatographia **68**, 1–10 (2008)

104. W. Qin, H. Wei, S.F.Y. Li, Separation of ionic liquid cations and related imidazole derivatives by alpha-cyclodextrin modified capillary zone electrophoresis. Analyst **127**, 490–493 (2002)

105. M. Markuszewski, P. Stepnowski, M.P. Marszałł, Capillary electrophoresis separations of cationic constituents of imidazolium ionic liquids. Electrophoresis **25**, 3450–3454 (2004)
106. P. Kosobucki, B. Buszewski, Isotachophoretic separation of selected midazolinium ionic liquids. Talanta **74**, 1670–1674 (2008)

Index

A

Ampholyte, 120, 121, 123, 124, 129, 130, 146, 147, 151
Analyte concentration
    EFGF, 225
    FASI, 219, 220, 341, 342
    FASS, 217, 219–221, 225, 226, 341
    FESI, 225, 230, 339, 340
    IEF, 141, 143–145, 225
    LVSS, 219–221, 340
    tITP, 221, 222, 224, 344
    TGF, 225
An electrical double layer, 288
Anthocyanins, 301
Antioxidants, 301, 302, 307, 308
Automation, 27, 33, 62, 63, 113, 115

B

Background electrolyte (BGE), 208, 209, 224, 227
Bioinformatics, 135, 271, 274, 279
Biomarker, 157, 272–275
Buffer, 5–8, 11, 14, 19–22, 28, 31, 38, 44, 46, 49, 51–54, 56, 60, 62–64, 73–75, 83, 94, 99, 120, 125, 141–144, 150, 161, 166, 172, 193, 218–220, 228, 230, 240, 245, 277, 283, 289, 300, 307, 314, 317, 340, 342, 345

C

Calibration
    external standard method, 69, 71
    internal standard method, 69, 71, 72, 97

multipoint calibration, 72
    one-point calibration, 71
Carboxylic acids, 246, 335, 342
Carotenoids, 304
Casting, 257, 258
Catechins, 301, 302, 304
Cell engineering, 253
Chaotropic substances, 142
Chip, 3, 28, 37, 40, 42–44, 52, 53, 55–57, 62, 63, 113, 163, 178, 254, 256, 266
Chiral crown ethers, 241–243
Chiral selectors, 84, 212, 241–245, 249
Chiral surfactants, 237, 243, 244
Chymotrypsin, 246
Columns
    capillary columns, 30, 162, 168, 169, 171, 174–177, 181, 186
    monolithic columns, 178–181
    polymeric columns, 178, 179, 181, 185, 186
    silica columns, 182, 185
Critical micellar concentretion, 82–84
Cyclodextrins, 82, 86, 87, 212, 237, 241, 242, 304
Cytotoxicity, 255, 264, 266

D

Detection limits, 335
Detector
    electrochemical detector
        amperometric, 34, 50, 56, 57, 63, 104, 302, 336, 338, 340, 341, 343
        conductometric, 50, 52–54, 95, 114, 342, 346

B. Buszewski et al. (eds.), *Electromigration Techniques*, Springer Series in Chemical Physics 105, DOI: 10.1007/978-3-642-35043-6,

D (*cont.*)
    potentiometric, 27, 51–53
    fluorescence detector, 36, 40, 41, 304
    LIF detector, 42, 64, 307
    MS detector, 27, 58, 318, 342, 109, 153,
        155, 256
    NMR detector, 60, 111
    spectrophotometric detector, 100
Derivatization, 43, 46, 47, 62, 63, 125, 308,
    314, 319, 336, 338, 339
Dyeing, 145, 153

E
Efficiency, 3, 5, 13, 14, 18, 40, 61, 62, 77, 81,
        82, 122, 162, 174, 176, 177, 179,
        192–194, 198, 217, 220, 227, 260,
        279, 289, 338, 339
Electric double layer, 8, 21, 160, 192
Electric field gradient focusing (EFGF), 225
Electrochemical detector, 216
Electrochemical methods, 255
Electrodisperssion, 69, 75
Electrokinetically injected, 221
Electrokinetic sample injection, 225, 228
Electrolyte
    background electrolyte, 5, 78, 126, 216,
        218–226, 228, 275, 277
    leading electrolyte, 94, 95, 98, 100, 101,
        107, 221
    terminating electrolyte, 94, 95, 104, 113
Electroosmosis, 70, 99, 159, 160, 162, 166,
        191, 219
Electroosmotic flow (EOF), 2, 7–12, 15, 16,
        19, 21, 34, 120, 159–166, 169, 176,
        180, 192–194, 196, 198–200, 204,
        219, 220, 223, 224, 260
Electrophoresis, 1, 2, 5–7, 13–15, 22, 28, 41,
        52, 62, 69–71, 74, 78, 83, 90, 99,
        101, 108, 113, 126, 134, 137, 144,
        145, 152–154, 157, 167, 192, 198,
        203, 215, 217–222, 239–241,
        243–246, 249, 254, 275, 283
Electrophoretic mobility
    absolute mobility, 11
    effective mobility, 11, 95, 98, 99
    observed (apparent) mobility, 10, 11, 13,
        78, 240, 337
Electrophoretic velocity, 7, 216
Enantiomers, 87, 167, 210, 212, 238–241,
        243–246, 249, 304
EOF modifier, 18
Explosives, 336, 337, 342, 343
Extraction

    solid phase extraction, 216, 247, 300, 321,
        337

F
Field amplified sample injection (FASI), 219
Field amplified sample stacking (FASS), 217,
        219, 225, 226
Field enhanced sample injection (FESI), 225,
        230
Flavines, 307
Flavonoids, 300–302, 304, 340
Food analysis, 287, 299, 308, 323
Fractionation, 1, 137, 138, 144, 145
Fritless column, 177

H
Heparin, 145, 246

I
Injection
    electrokinetic, 33, 34, 169, 173, 219, 221,
        225, 228, 229, 343, 344
    hydrodynamic, 33, 34, 63, 173, 218–220,
        226, 227, 303, 344
Ionic liquids, 345–346
Ionic strength, 8, 20, 120, 136, 142, 148, 164,
        172, 228, 245
Isoelectric point, 119, 122, 126, 134–136
Isotachophoresis, 6, 28, 93–95, 97, 99, 100,
        102, 103, 106–109, 111–115, 221,
        222, 344

J
Joule heat, 1, 16, 19, 20, 28, 29, 31, 49, 62, 88,
        160, 167, 171, 193, 200, 206

L
Lab-on-a-chip, 253, 254, 256, 257
Lab-on-a-robot, 64, 66
Laminar flow, 9, 28, 123, 162, 260, 261
Large volume sample stacking (LVSS), 219
Laser-induced fluorescence (LIF), 216, 336
LIF detection, 283
Limit of detection (LOD), 34, 42, 101, 153,
        215, 217, 224, 277, 278, 319
Linearity, 16, 35, 39, 52, 71, 74, 301, 314, 345,
        346
Liquid–liquid extraction (LLE), 216, 302, 321,
        323

M

Mass spectrometry, 277
Metabolomics, 271–273, 278, 279
Micelle, 3, 77–84, 87–90, 172, 209, 225–229, 275
Microchannel, 257–261, 263
Microchip, 220, 230, 254, 255, 257, 258, 262, 263, 265, 336, 338, 343
Microorganisms, 137, 145, 239, 287, 288, 289, 295, 318, 321
Microsystem, 254–260, 262, 264–267
Miniaturization, 295
Migration time, 12, 14, 16, 17, 32, 69, 70, 74, 79, 81, 84, 87, 90, 97, 123, 126, 130, 196, 197, 225, 279
Mobilization
  chemical mobilization, 122
  hydraulic mobilization, 122, 123
Monolithic bed, 162
Monolithic silica beds, 181

O

Ohm's law, 16, 52, 137

P

Peak shape, 17, 69, 74, 75, 163
Pesticides, 209, 228, 238, 242, 249, 299, 300, 319, 336, 339, 340, 344
Phenols, 340
pH*, 203, 205
pH gradient, 121–122, 130, 135, 141, 146–149
Photolithography, 257, 258
Polycyclic aromatic hydrocarbons (PAHs), 209, 336, 338, 340
Polymeric monolithic columns, 180, 181
Polymeric monoliths, 178
Polyphenols, 141, 300–302
Protein solubilization, 140–142, 147, 149, 153

Q

Qualitative analysis, 36, 69, 70, 95–97
Quantitative analysis, 35, 69, 71, 74, 83, 96, 109, 307

R

Rehydration, 149–152
Relative step height (RSH), 96, 97, 149
Resolution, 1, 3, 14, 16–18, 20, 21, 28, 77, 82, 90, 124–126, 128, 130, 134, 141, 149, 155, 205, 206, 216, 219, 275, 278
Retention coefficient, 80–82, 86, 87, 227

S

Selectivity, 14, 17, 35, 51, 81–83, 87, 88, 90, 167, 192, 198, 199, 203, 204, 240, 243, 301, 321
Sensitivity, 31, 34, 37, 44, 49, 50, 54, 56, 57, 59, 60, 90, 102, 111, 121, 124, 153, 216, 219, 224, 254, 274, 277, 278, 283, 301, 307, 308, 314, 321, 335–338, 340, 344
Silanol groups, 7, 8, 18, 19, 21, 161, 179
Silica monoliths, 181
Silica open tubular columns, 185
Siphon effect, 32, 63, 206, 207
Solid phase extraction (SPE), 216, 300
Solvent
  amphiprotic, 204
  aprotic, 87
Spacers, 104–106
Stereoisomers, 237
Stereoisomerism, 237
Surfactant, 78, 82–90, 142, 143, 172, 209, 225, 241, 243, 302, 343

T

Temperature gradient focusing (TGF), 225
Terminating, 221
Transient isotachophoresis (tITP), 222
Transient pseudo-isotachophoresis, 222

Z

Zeta potential, 161

CPSIA information can be obtained at www.ICGtesting.com
Printed in the USA
BVOW05*1645231016

465803BV00002B/7/P